THE HAWKMOTHS
OF THE WESTERN PALAEARCTIC

THE HAWKMOTHS
OF THE
WESTERN
PALAEARCTIC

A. R. Pittaway F.R.E.S.

in association with
The Natural History Museum, London

Harley Books (B. H. & A. Harley Ltd.),
Martins, Great Horkesley,
Colchester, Essex CO6 4AH, England

Text set in Linotron 202 Plantin by
Rowland Phototypesetting Ltd.,
Bury St Edmunds, Suffolk

Text printed by St Edmundsbury Press Ltd.,
Bury St Edmunds, Suffolk

Colour originated by Adroit Photo Litho Ltd.,
Birmingham, and printed by Jolly & Barber
Ltd., Rugby, Warwickshire

Bound by WBC Bookbinders Ltd.,
Bridgend, Wales

The Hawkmoths of the Western Palaearctic

British Library Cataloguing-in-Publication Data available

ISBN 0 946589 21 6

Contents

Foreword

Rarely is the natural history of a group of moths as effectively presented as it is here for the hawkmoths of the western Palaearctic. Three major components comprise the work.

The first contains an extended introduction (including discussions on geography, population dynamics, historical origin of the species, biomes in the area, hostplant/moth interactions, and competition), which sets the scene for the second section. In this second part a taxon by taxon treatment of the 57 species and 29 subspecies of hawkmoths details the particulars of the taxa. For genera and higher taxa, the structure of adults, larvae and pupae is characterized; and broad distribution is summarized. At the species' level, distribution, morphological and colour characterization of all stages, flight time of adults, life history emphasizing adults and larvae, appearance and behaviour of immature stages, pupation sites, full citation of plant hosts, parasitoids, personal information on rearing or breeding species, and geographic distribution that reveals the vagrant nature of many species are combined to portray extant knowledge of each species and subspecies. The third component is the illustrative material. Colour illustrations of the immature stages and adults (both living and set specimens) and of their habitats, line illustrations of pupae as well as of pertinent structures, and distribution maps effectively expand and complement the text. Bibliographic citation of all names adds greatly to the usefulness of this work in which an impressive series of more than eighty accounts has been written on all or on geographic subsets of this sphingid fauna.

A species concept based on population biology, coupled with a concomitant grasp of infraspecific variation that may be expressed in any life stage, the definition of subspecies as geographic races, the author's extensive field work in the critical southern and south-eastern parts of the western Palaearctic over extended periods of time, and his experience of rearing representatives of populations of many species taken together lay the groundwork for several taxonomic decisions at the specific or subspecific level. These conclusions refute or corroborate earlier taxonomic decisions resulting from study of limited samples. This information is extremely significant in resolving taxonomic problems in the formerly confused and confusing *Hyles euphorbiae* complex.

The Hawkmoths of the Western Palaearctic is an exemplar for treating relatively well-known taxa in a fully scientific manner and providing much basic observation so that both professional and avocational lepidopterists can make new observations that will further the knowledge of hawkmoths.

Systematic Entomology Laboratory,   Ronald W. Hodges
U.S. Department of Agriculture,
Washington, D.C.
August, 1992

This book is dedicated to the memory of
L. Hugh Newman
(3 February 1909–23 January 1993)
whose enthusiasm for hawkmoths fired my own

Preface

I first became aware of hawkmoths at the tender age of three, when I discovered a magnificent horned creature crawling across the road outside my parents' home in Gillingham, Kent. Being concerned for its welfare, I duly placed it on a tree beside the road. Only later in life did I recognize it for what it was, namely, a caterpillar of the Lime Hawkmoth (*Mimas tiliae*) on its way to find a pupation site. From such beginnings arose a passionate love of these noble insects, which has driven me to seek them out across Europe, North Africa, the Middle East and beyond. Although the adult moths are undoubtedly creatures of great beauty, my main interest in the Sphingidae has always remained their early stages.

L. Hugh Newman, whose book *Hawk-moths of Great Britain and Europe* forced me in 1965 to plunder 50 shillings (£2.50) from my meagre store of pocket-money, stated that 'Counting the hairs on a newly hatched caterpillar under a magnifying glass may be necessary in order to determine its exact place in the complicated relationship of insects, but to me such details are not important. What matters infinitely more is an understanding of the behaviour of [these] moths in the various stages of their life cycle'. I share such sentiments, but also realise that without the detailed taxonomic work carried out over the last 200 years by many dedicated professional scientists and amateurs, much of the accumulated data on sphingid ecology and biology could be misinterpreted. My primary intention, therefore, is not to produce a taxonomic treatise but rather a work on the ecology and biology of the Sphingidae of the western Palaearctic. However, at the same time, the taxonomy of the species covered has been revised in line with modern practice and recent discoveries.

Because of their ecological importance, the larval stages of most species have been figured in colour and drawn from life rather than described. Most adults are easily recognized from the illustrations, and descriptions have been omitted except where of importance. Genitalia and pupae have been described and figured only if of special significance for identification.

This book could never have been written without the invaluable advice and assistance received from many people in very many ways. I would like to express my thanks to the staff (past and present) of the Entomology Department of the Natural History Museum, London, for allowing me free access to the superb National Collection. Certain members of staff deserve special mention for frequently going out of their way to help in the early days, in particular L. A. Mound (formerly Keeper of Entomology), the late A. H. Hayes, M. R. Honey, I. W. B. Nye, and A. Watson. To I. J. Kitching and W. G. Tremewan, whose expert knowledge and painstaking editorial assistance have been of incalculable value in preparing the text throughout all its stages, I acknowledge deep indebtedness. I have also received constructive comments from Willem Hogenes of Instituut voor Taxonomische Zoölogie, Amsterdam, and A. Rodger Waterston (formerly Keeper of Entomology, Royal Scottish Museum), who kindly read the initial drafts of the manuscript.

For their great assistance in tracing sources in the literature, my special thanks go to Mrs Brenda Leonard, former librarian of the Royal Entomological Society of London, for putting up with me for so long as I rummaged through the endless rows of books in the Society's superb library; also to Mrs Jacqueline Ruffle, the present librarian; to Miss Pam Gilbert, formerly Librarian of the Entomology Library, Natural History Museum, London; and to Mrs Julie M. V. Harvey and Mrs Kathie Martin also of that Library.

Additional information on various aspects of the work has been supplied by J.-M. Cadiou (Belgium); S. V. Beschkow (Bulgaria); V. Pinkava (Czech Republic); P. Skou and B. Skule

(Denmark); U. Eitschberger, M. Geck, S. Heinig, T. Kaltenbach, C.M. Naumann, M.A. Pelzer, H. Schröder and E. Thomson (Germany); D. Benyamini (Israel); S. Sugi (Japan); the late A. Valetta (Malta); W. Hogenes, E.A. Loeliger and J.C. Meerman (Netherlands); F. König and A. Popescu-Gorj (Romania); V.V. Zolotuhin (Russia); and the late R. Agenjo, the late M.R. Gómez Bustillo, J.L. Viejo Montesinos and J.L. Yela Garcia (Spain), to all of whom I express my warmest thanks. I am especially indebted to E.P. Wiltshire, not only for supplying information, but also for directing me to his extensive Middle Eastern collection in the Natural History Museum, London. Many thanks are also due to A. Wood and 'Systems' of CAB International for their help.

R.W. Crosskey, M.G. Fitton and T. Huddleston (Natural History Museum, London); T.H. Ford (Sheffield); M.R. Shaw (Royal Scottish Museum, Edinburgh); and B. Herting and H.-P. Tschorsnig (Staatliches Museum für Naturkunde, Stuttgart), all provided specialist information and guidance on the vexed subject of sphingid parasitoids.

The larval plates are the work of Allan Walker (Pls I–III) and myself (Pl. IV). The adult plates (V–XIII) have been rearranged from photographs taken at the Natural History Museum, London, by Frank Greenaway, using set specimens from the National Collection, that of E.P. Wiltshire, the Zoologische Staatssammlung, München, the private collection of Jean-Marie Cadiou, Brussels, and my own collection. With a number of exceptions, acknowledged below, the colour habitat and life-history photographs and monochrome text figures are my own work, as are most of the line drawings; however text figures 7–10 and 27 have been kindly drawn by Richard Lewington. For permission to use their photographs, I would like to express my gratitude to K. Bibbing (Text fig. 11), H. Harbich of Germany (Text figs 25 a,b), the late G.E. Hyde (Text fig. 24), F. Karrer of Switzerland (Plate D, fig. 3), J.C. Meerman of The Netherlands (Plate D, fig. 2) and C.M. Naumann of Germany (Plate G, fig. 5). The genitalia drawings of *Kentrochrysalis elegans* (Text fig. 32) are from originals by A. Popescu-Gorj of the Romanian Natural History Museum, Bucharest; the map in Text fig. 13 is reproduced from *World Vegetation* by kind permission of Cambridge University Press. Text fig. 15a is reproduced from *The Moths and Butterflies of Great Britain and Ireland* by permission of the publishers. Text fig. 15b was most kindly made available by Paul Waring, on behalf of the National Recording Network for the Rarer British Macro-moths.

The distribution maps, broadly showing the whole range of a species, including subspecies, are based on published and unpublished records by workers in the field, on data labels from specimens preserved in the Natural History Museum, London, and on my own records. I apologize to any recorder whose published work has inadvertently not been included in the reference.

The indexes were prepared by Annette Harley. I would like to thank both her and Basil Harley, for their encouragement, tolerance, constructive suggestions, editorial help and indefatigable attention to detail over the years, without which this book would not be what it has finally become. Any errors which remain are mine alone.

Moulsford, Oxfordshire   A. R. PITTAWAY
February, 1993

I. INTRODUCTION

This work is a detailed study of the 86 species and subspecies of Sphingidae that occur regularly within the boundaries of the western Palaearctic region, with particular emphasis on their ecology, biology and biogeography. The introductory chapters contain a general survey of the family; details of the morphology and biology of early stages (including hostplants, predators and parasites); adult morphology; adult biology; details of distribution, how it originated and how it is maintained; and notes on the classification of the group. This is followed by a check list and a systematic account of each species and subspecies (including the lectotype designations of *Smerinthus ocellatus atlanticus* and *Deilephila rivularis*), with maps of their distribution.

Although the hawkmoths were first formally classified by Carl von Linné (Linnaeus) in 1758 under the name *Sphinx*, it was not until 1903 that a complete revision of the 722 species of this family then recognized was published by Rothschild & Jordan. By 1911 some 850 species were known. As with most insects, the sphingids reach their greatest diversity in the tropics. The number of species and genera diminishes with increasing latitude and none is resident in the Arctic and Antarctic regions. With the exception of these regions, the family is found throughout the world, even on small oceanic islands. Today there are approximately 1200 known species, of which 100 are resident in the Palaearctic and approximately 54 in the western portion of this biogeographical area.

The western boundary of the western Palaearctic covered by this work is 30°W, and includes the Azores. The southern boundary runs along the Tropic of Cancer. The northern limit is 80°N and the eastern limit 80°E as far south as the Tian Shan range. From there the boundary is less distinct, running south-westwards through the Pamir, Hindu Kush and along the south-eastern edge of the Iran-Afghanistan plateau and thence across the Gulf of Oman to Muscat. The fauna and flora within this natural region forms a recognizable, distinct community governed by distinct climatic parameters.

The definition in this work of what constitutes the western Palaearctic thus differs somewhat from that in most other works, which have tended to select features such as mountain chains or rivers as its boundaries. Europe is a geopolitical entity and of little significance to sphingid ecology. The western Palaearctic, on the other hand, is a distinct biogeographical region with, for the most part, clearly defined boundaries. To the west it is cut off from North America by the vast expanse of the Atlantic Ocean. To the south the hot and dry Sahara desert forms a very effective barrier to all but the most determined Afrotropical species. To the north lie the frozen Arctic wastes, and to the east the vast inhospitable swamps, forests and cold deserts of central Siberia. (The Urals are regarded by many as the natural eastern boundary of the western Palaearctic, but these low mountains do not, in fact, form a very effective barrier). Only to the south-east does the flora and fauna merge with that of the Oriental region along a mountainous boundary which stretches from Afghanistan and Kashmir down through western Pakistan to Baluchistan.

The distribution maps are of two main types: one depicts the greater part of the western Palaearctic (Text fig. 1, p. 12); the other, at more than twice the scale, the important area of eastern Iran, Afghanistan, Kashmir and the Tian Shan (Text fig. 2, p. 13). There is also a special map (p. 142) for *Hyles tithymali himyarensis* which is restricted to the Arabian Peninsula. For each species, a single map is provided showing the main area of distribution, with both resident (solid black) and migrant (---) ranges indicated. However, because of the scale of the maps accompanying the text, precise definition of areas of distribution is, of course, impossible:

some areas indicated, such as mountainous, wetland or urban regions, may be unsuitable for the species to which the map relates. In addition, the distribution includes areas where a species formerly existed but is, or may be, extinct (such as that for *Hemaris tityus* on Map 24, p. 113); the maps therefore show the optimum range of the species. Some regions where there is doubt are indicated by a question mark. To facilitate interpretation of the distribution range given in the text, national and some major provincial boundaries are shown on each map corresponding to those in Text figs 1 and 2, where the individual countries are identified by name. The map numbers in the text correspond with the check list numbers on pp. 75–77.

This work covers those sphingids which are resident within the above-mentioned boundaries

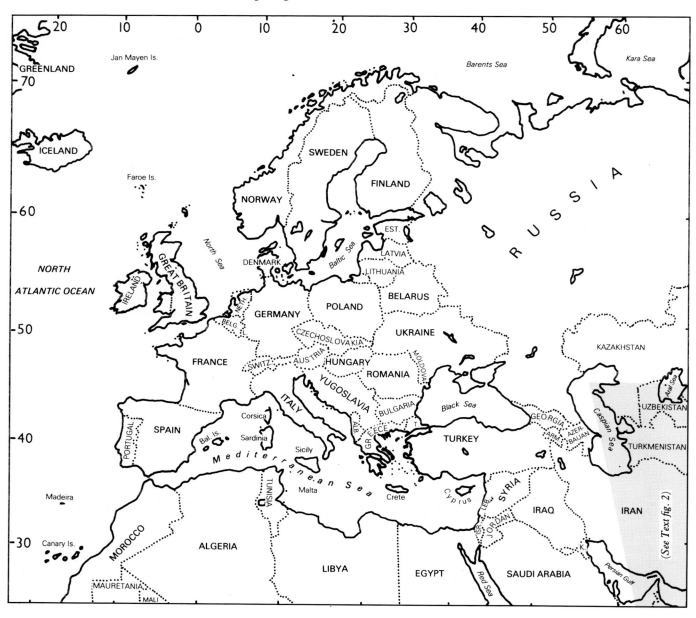

Key to Abbreviations

ALB. = ALBANIA
ARM. = ARMENIA
Bal. Is. = Balearic Islands
BELG. = BELGIUM
EST. = ESTONIA
ISR. = ISRAEL

K. = KUWAIT
L. = LUXEMBOURG
LEB. = LEBANON
SWITZ. = SWITZERLAND
T. = TURKEY (in Europe)

Notes (i) The constituent parts of the former YUGOSLAVIA are not shown separately (i.e. Bosnia and Herzegovina, Croatia, Macedonia, Slovenia, and Serbia and Montenegro).

(ii) MOLDOVA is indicated between ROMANIA and UKRAINE but its boundaries are not shown on the map.

(iii) CZECHOSLOVAKIA has now divided into the CZECH REPUBLIC and SLOVAKIA.

Text Figure 1 Map of the western Palaearctic region, excluding part of the south-eastern sector

of the western Palaearctic. It also covers those species which are regular migrants into this area from the Afrotropical region. Where relict populations of species or genera which are obviously of western Palaearctic origin are present outside the region, such as on the Cape Verde Islands and in south-western Saudi Arabia and Yemen, these also have been included. It does not cover what are clearly Oriental species which penetrate into the warmer areas of Kashmir and Afghanistan; these are listed under vagrants. The mixed fauna of Pakistan is also excluded except for clearly recognizable Palaearctic species whose distribution extends into the cool western highlands of that country.

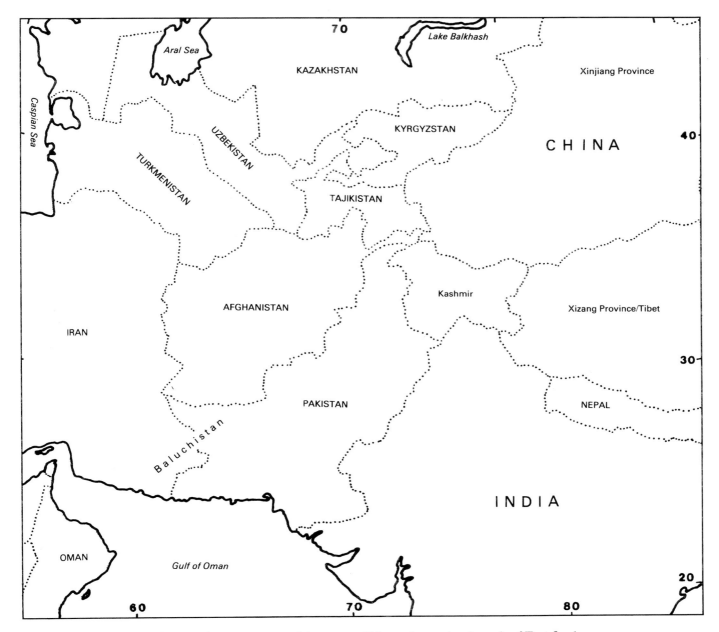

Text Figure 2 Map of the south-eastern sector of the western Palaearctic at twice the scale of Text fig. 1

II. HISTORICAL ACCOUNT OF THE STUDY OF THE SPHINGIDAE

With the flowering of the Renaissance in Europe came a desire to understand more of 'God's Kingdom'. Prior to this period, nature was regarded by Europeans almost entirely as something to be overcome or utilized; survival was the order of the day. In the 13th-century work *De Natura Animalium*, by Albertus Magnus, butterflies were described simply as 'flying worms', with no eye to their beauty. The development of printing enabled information to be more widely disseminated, which in turn promoted a greater interest in natural history. It is fascinating to trace chronologically the publication of works in which mention is made of hawkmoth species within the region covered by this book – a period which spans four centuries.

Early works (1589–1850)

The first English work in which hawkmoths were recognizably described was *Insectorum sive Minimorum Animalium Theatrum* by Thomas Mouffet. Although completed in 1589, it was not published until 1634 – thirty years after Mouffet's death – at the expense of Sir Theodore de Mayerne, one of the royal physicians. A number of butterflies, moths and their caterpillars were given names of which one, *Porcellus*, was used by Linnaeus (1758) for *Deilephila porcellus* when describing both adult and larva. The work proved so popular that an English edition was published in 1658 as part of a larger work incorporating Edward Topsell's *The History of Four-footed Beasts and Serpents*.

J. Hoefnagel, in parts 1–4 of *Archetypa Studiaque Patris* (1592), illustrated adults and larvae of *Agrius convolvuli*, *Acherontia atropos*, *Smerinthus ocellatus*, *Laothoe populi*, *Macroglossum stellatarum* and *Hyles euphorbiae*; there was no text. The illustrations in the book were uncoloured, but a copy at the Natural History Museum, London, has been coloured by a contemporary hand.

Although more famous for her spectacular work *De Metamorphosis Insectorum Surinamensium* (1705), Maria Sibylla Merian had also produced a similar and earlier work on European Lepidoptera, first published in German as *Der Raupen wunderbare Verwandelung* (1679). A second volume, which appeared four years after the first in 1683, contained fine copperplate illustrations of eight European hawkmoths with hostplants. The text makes clear her powers of observation and interest in insect metamorphosis. Maria Merian was so fascinated by the parasites which developed from the caterpillars she collected that she also illustrated and described many of them. The plate of *Smerinthus ocellatus* shows quite clearly the cocoons and adults of *Microplitis ocellatae*; the tachinid *Masicera sphingivora*, or a closely related species, is depicted on the plate of *Hyles euphorbiae*. Dutch, Latin and French editions of this work were also published between 1713 and 1730.

Johann Jacob Swammerdam, in *Historia Insectorum generalis* (1669), later translated into English as *The Book of Nature; or, the History of Insects* (1758), described the adult, larva and pupa of *Agrius convolvuli*, which he called 'Pernix'. In the English edition, this species, which he states was originally described by John Bauhinius in 1693 in his treatise on 'hurtful winged animals', is easily recognized from the uncoloured line engraving.

John Ray, in *Historia Insectorum* published posthumously in 1710, described the biology and morphology of the larvae and adults of several British species, of which only *Sphinx ligustri* and *Deilephila porcellus* are recognizable from the text; there were no plates.

Eleazar Albin, in *A Natural History of English Insects* (1720), described the ecology and morphology and listed the hostplants of six sphingid species found in England, depicting in the fine colour plates their complete life cycles and hostplants. One plate – that of *Deilephila elpenor* – even illustrated the parasite *Amblyjoppa proteus*; another yellow-and-black ichneumonid is depicted on the plate illustrating *Sphinx ligustri*. He also wrote of larvae of *Mimas tiliae* being so numerous as to constitute a pest of lime trees (*Tilia*), and stated correctly that the larva of *Deilephila elpenor* could swim. It appears probable from the similarity of the species of hawkmoths selected and the nature of their depictions that Benjamin Wilkes, James Dutfield and Moses Harris all borrowed freely from this beautiful work in preparing their own plates.

Jacob L'Admiral, in *Naauwkeurige Waarneemingen van Veele Gestaltverwisselende Gekorvene Diertjes* (1740), illustrated and described the ecology and biology of two European hawkmoths, *Smerinthus ocellatus* and *Laothoe populi*, and also listed their hostplants. It was republished in 1744, after the artist's death, as *Naauwkeurige Waarneemingen Omtrent de Veranderingen van Veele Insekten*.

Benjamin Wilkes, in *Twelve new Designs of English Butterflies* (1742), depicted seven hawkmoths found in Britain, arranged geometrically with other moths and butterflies in his twelve plates. The captions to each provided information on the time of emergence, hostplants and occurrence of the caterpillar, of which four were depicted on the title-page. The English names for each species are almost the same as those in use today. Wilkes' more ambitious *The English Moths and Butterflies* [1749] contained descriptions of the ecology, morphology and hostplants of eight British species of hawkmoths, with good colour plates of their complete life-cycles on the hostplants. Like Albin, he depicted the ichneumonid *Amblyjoppa proteus* on the plate of *D. elpenor*. The work ran to three editions of which the last, incorporating Linnaean nomenclature, was published in 1824.

August Johann Rösel von Rosenhof, in Vols I, III and IV of the four-volume work, *Insecten Belustigung* ([1746]–61), provided the biology, hostplants and descriptions of fifteen central European species. Good colour plates depicted both larvae and adults. He died before the publication of the fourth volume which was completed by Christian Friedrich Carl Kleeman.

During 1748–49, James Dutfield started on the publication of *A new and complete Natural History of English Moths and Butterflies* but only six parts are known and probably no more were issued. Several species of hawkmoths were illustrated and described. Similarities between these plates and those of Eleazar Albin suggest that Dutfield plagiarized Albin's work.

Carl DeGeer, in the first two volumes of *Mémoires pour servir à l'Histoire des Insectes* (1752, 1771), described the biology and hostplants of seven species found in Sweden, with uncoloured plates of larvae and adults of *Sphinx pinastri*, *Deilephila elpenor*, *Hyles gallii*, *Laothoe populi* and *Sphinx ligustri* in Volume I, and of *Mimas tiliae* and *Smerinthus ocellatus* in Volume II.

Until this time, species illustrated and described had been given vernacular names, many of which survive to this day. René Antoine Ferchault de Réaumur had been the first to use the word 'Sphinx' in 1736, in reference to the larva of *Sphinx ligustri*, a name which Linnaeus, on page 248 of the first edition of *Fauna Suecica* (1746), retained for this species. That edition of *Fauna Suecica* did not contain binominal nomenclature but, with the publication of the tenth edition of *Systema Naturae* in 1758, Linnaeus applied his Latin binominal system to the animal kingdom which contains the Lepidoptera. He selected Réaumur's name *Sphinx* as the generic name for all the hawkmoths but gave each a trivial or specific name. Sixteen European sphingid species were included in this category, as were several of the Zygaenidae and Sesiidae – an association which was maintained, at least in part, for about 100 years until Butler (1876) removed all remaining species of these families from the Sphingidae. It is interesting to note that Linnaeus had first adopted a short binominal system in 1745 in the index to a travel book, and used it again in 1749 as a shorthand when referring to Swedish plants. In 1753 he applied the binominal system to the plant kingdom in his *Species Plantarum*. The concept of a uniform system of classification was not new, but Linnaeus was the first person to apply it universally (Usinger, 1964).

Nicolaus Poda, in *Insecta Musei Graecensis* (1761), adopted the Linnaean system to name and describe the adults of eight central European sphingid species, as well as providing information on their distribution and hostplants. He included an uncoloured engraving of *Hemaris fuciformis* but, like many early authors, he confused this species with *H. tityus*.

The father and son team of Christian and Jan Christiaan Sepp, in the first four volumes of *De Nederlandsche Insekten* (1762–1811), described the biology and ecology and gave hostplants for twelve species found in the Netherlands. Superb colour plates depicted the life-cycles of most of these. Binominal as well as Dutch names were given for all species but, like many of their contemporaries, the authors treated *Hemaris tityus* and *H. fuciformis* as one species. After the death of Jan Christiaan in 1811, the work was published by further generations of the Sepp family until the death of Jan Christiaan's grandson, Cornelis Sepp in 1868.

Johann Anton Scopoli included no plates of hawkmoths in his *Entomologia Carniolica* (1763), but gave the distribution, adult morphology and hostplants of eleven central European species 'methodo Linnaeana', although also confusing *H. fuciformis* with *H. tityus*.

In 1766, Moses Harris published *The Aurelian: Or, Natural History of English Insects; Namely Moths and Butterflies*, a magnificent work, with 44 beautiful colour plates which included the ecology, descriptions, life cycles and hostplants of nine hawkmoths he had found in England. Following Albin and Wilkes, Harris also depicted the parasite *Amblyjoppa proteus* on the plate of *D. elpenor*. In 1775, he published the first pocket guide to Lepidoptera entitled *The English Lepidoptera: Or, The Aurelian's Pocket Companion*. In tabular form, he listed numerous species of butterflies and moths by their common names, including eleven hawkmoths. Where known, hostplants, dates of occurrence, habitats, descriptions and Linnaean names were given. Unfortunately, no plates were included.

Johann Christian Fabricius, in *Systema Entomologiae* (1775), described fifteen European hawkmoth species. He split up Linnaeus' Sphinges into three distinct subgroups, namely *Sphinx*, *Sesia* (in which he included the humming-bird and bee hawkmoths) and *Zygaena*.

J. J. Ernst, in Volume III of his seven-volume work, *Papillons d'Europe peints d'après Nature* (1779–1793), described the biology and listed the hostplants of twenty European species.

Etienne Louis Geoffroy, in the second volume of *Histoire abrégée des Insectes* (1799), dealt with aspects of ecology and the hostplants and morphology of nine species found in France, but with a poor colour illustration of only one, *Macroglossum stellatarum*.

Pierre André Latreille ([1802]), in Sonnini's edition of Buffon's *Histoire Naturelle*, proposed the name Sphingides for the whole group based on Linnaeus' original group Sphinges. It is from this term that the modern family name Sphingidae is derived.

In 1837, Volume 1 of *Illustrations of Exotic Entomology* was published. This work by Dru Drury had originally appeared in three volumes which were issued between 1770 and 1772 entitled *Illustrations of Natural History*. Although it dealt mainly with tropical moths and butterflies, the British status of *Hyles euphorbiae* and *Sphinx pinastri* was discussed. It recorded that the former was rare in Devon, but that larvae could sometimes be common on *Euphorbia paralias* at Braunton Burrows and Barnstaple, and had been plentiful in 1814. *Sphinx pinastri* was recorded as being but a rare vagrant to England and Scotland.

By the end of the first half of the nineteenth century, the binominal system of nomenclature with recognizable genera and species had become firmly established and the first regional faunal works began to appear. One of the best of these was *Fauna Lepidopterologica Volgo-Uralensis* (1844) by Eduard Eversmann, in which the distribution of twenty species of hawkmoths found in the European part of Russia was recorded.

From 1850 to the present

Since 1850, many other works have appeared on some or all of the Sphingidae of the whole of the western Palaearctic region or of its parts -- Europe, individual countries, or even smaller areas. These are too numerous to list here in their entirety but the following are those which have been most frequently consulted in the preparation of this book. It is recognized that the literature cited is far from exhaustive but it is hoped that it will provide a useful source of reference. Authors, with date of publication only, are given here, divided according to their regional coverage; the full titles of the works will be found in the references towards the end of this volume. It should be noted that many references also give details of distribution beyond the region under which they are cited.

GENERAL: Rothschild W. & Jordan, K. (1903); Hodges, R. W. (1971); Carcasson, R. H. (1968); Fletcher, D. S. & Nye, I. W. B. (1982); D'Abrera, B. (1986).

WESTERN PALAEARCTIC: Hofmann, E. (1893, 1894); Staudinger, O. & Rebel, H. (1901); Kuznetsova, N. Ya. (1906); Jordan, K. (1912); Denso, P. (1913b); Gehlen, B. (1932b); Pittaway, A. R. (1983b).

EUROPE & NORTH AFRICA: Roüast, G. (1883); Hofmann, E. (1893, 1894); Tutt, J. W. (1902, 1904); Kirby, W. F. (1903); Spuler, A. (1908); Newman, L. H. (1965); Rougeot, P.-C. & Viette, P. (1978); Sokoloff, P. (1984); Freina, J. J. de & Witt, T. J. (1987).

AFGHANISTAN: Ebert, G. (1969, 1974); Daniel, F. (1963, 1971).

ALBANIA: Rebel, H. & Zerny, H. (1934); Eichler, F. & Friese, G. (1965).

ALGERIA: Rothschild, W. (1917); Speidel, W. & Hassler, M. (1989).

ARMENIA: Derzhavets, Yu. A. (1984).

AUSTRIA: Hoffmann, F. & Klos, R. (1914).

AZERBAIJAN: Derzhavets, Yu. A. (1984).

AZORES: Meyer, M. (1991).

BALEARIC ISLANDS: Rebel, H. (1926, 1934).

BELARUS: Merzheevskaya, I., Litvinova, A. N. & Molchanova, R. V. (1976); Derzhavets, Yu. A. (1984).

BELGIUM: Herbulot, C. (1971).

BULGARIA: Ganev, J. (1984).

CANARY ISLANDS: van der Heyden, T. (1990, 1991)

CAPE VERDE ISLANDS: Bauer, E. & Traub, B. (1980).

CHINA: Chu, H. F. & Wang, L. Y. (1980b).

CORSICA: Mann, J. (1855); Bretherton, R. F. & Worms C. G. M. de (1964); Rungs, C. E. (1977, 1988).

CZECHOSLOVAKIA [CZECH REPUBLIC and SLOVAKIA]: Komárek, O. & Tykač, J. (1952); Hrubý, K. (1964).

DENMARK: Nordström, F. (1955); Hoffmeyer, S. (1960); Nordström, F., Opheim, M. & Sotavalta, O. (1961).

EGYPT: Badr, M. A., Oshaibah, A. A., Nawabi, A. El- & Gamal, M. M. Al- (1985).

ESTONIA: Petersen, W. (1924); Thomson, E. (1967).

FINLAND: Seppänen, E. J. (1954); Nordström, F. (1955); Nordström, F., Opheim, M. & Sotavalta, O. (1961).

FRANCE: Rondou, J.-P. (1903); Frionnet, M. C. (1910); Herbulot, C. (1971); Feltwell, J. & Ducros, P. (1989).

GEORGIA: Derzhavets, Yu. A. (1984).

GERMANY: Heinemann, H. von (1859); Bergmann, A. (1953); Forster, W. & Wohlfahrt, T. A. (1960).

GREAT BRITAIN: Buckler, W. (1887); Barrett, C. G. (1893); Lucas, W. J. (1895); Tutt, J. W. (1902, 1904); South, R. (1907, 1961); Newman, L. H. (1965); Heath, J. & Skelton, M. J. (1973); Gilchrist W. L. R. E. (1979); Skinner, B. (1984); Easterbrook, M. (1985).

GREECE: Koutsaftikis, A. (1970, 1973, 1974).

HUNGARY: Abafi-Aigner, L., Pável, J. & Uhryk, F. (1896); Kovács, L. (1953).

ICELAND: Wolff. N. L. (1971).

IRAN: Bienert, T. (1870); Watkins, H. T. G. & Buxton, P. A. (1923); Brandt, W. (1938); Sutton, S. L. (1963); Barou, J. ([1967]); Daniel, F. (1963, 1971); Ebert, G. (1976); Kalali, Gh.-H. (1976).

IRAQ: Boulenger, G. A. *et. al.* ([1923]); Watkins, H. T. G. & Buxton, P. A. (1923); Wiltshire, E. P. (1957).

IRELAND: Heath, J. & Skelton, M. J. (1973); Gilchrist, W. L. R. E., (1979); Skinner, B. (1984); Lavery, T. A. (1991).

ISRAEL: Eisenstein, I. (1984).

ITALY: Herbulot, C. (1971).

KASHMIR: Bell, T. R. D. & Scott, F. B. (1937).

KAZAKHSTAN: Derzhevets, Yu. A. (1984).

KYRGYSTAN: Derzhevets, Yu. A. (1984).

LEBANON: Zerny, H. (1933); Ellison, R. E. & Wiltshire, E. P. (1939).

LITHUANIA: Kazlauskas, R. (1984).

MADEIRA: Baker, G. T. (1891); Cockerell, T. D. A. (1923); Gardner, A. E. & Classey, E. W. (1960); Worms, C. G. M. de (1964).

MALTA: Valetta, A. (1973).

MOROCCO: Rothschild, W. (1917); Zerny, H. (1936); Rungs, C. E. (1981).

NETHERLANDS: Lempke, B. J. (1937); Meerman, J. C. (1987).

NORWAY: Nordström, F. (1955); Nordström, F., Opheim, M. & Sotavalta, O. (1961).

OMAN: Wiltshire, E. P. (1975a,b).

PAKISTAN: Bell, T. R. D. & Scott, F. B. (1937).

PORTUGAL: Gómez Bustillo, M. R. & Fernández-Rubio, F. (1976).

ROMANIA: Fleck, E. (1901).

RUSSIA: Alphéraky, S. (1882); Grum-Grshimailo, G. E. (1890); Kumakov, A. P. (1977); Derzhevets, Yu. A. (1984).

SAUDI ARABIA: Pittaway, A. R. (1979b, 1981, 1987); Wiltshire, E. P. (1980, 1986, 1990).

SICILY: Mina-Palumbo, F. & Failla-Tedaldi, L. (1889); Mariani, M. (1939).

SPAIN: Gómez Bustillo, M. R. & Fernández-Rubio, F. (1974, 1976); Lacasa, A., Garrido, A., Rivero, J. M. *et al.* (1979); Pérez Lopez, F. J. (1989).

SWEDEN: Nordström, F. (1955); Nordström, F., Opheim, M. & Sotavalta, O. (1961).

SWITZERLAND: Heinemann, H. von (1859); Vorbrodt, K. & Müller-Rutz, J. (1911); Herbulot, C. (1971).

SYRIA: Hariri, G. El- (1971).

TAJIKISTAN: Grum-Grshimailo, G. E. (1890); Shchetkin, Yu. L. (1949), Derzhevets, Yu. A. (1984).

TUNISIA: Rothschild, W. (1917).

TURKEY: Staudinger, O. ([1979]–1881); Daniel, F. (1932, 1939); Rebel, H. (1933); Freina, J. J. de (1979).

TURKMENISTAN: Myartzeva, S. N. & Tokeaew, R. T. (1972); Tashliev, A. O. (1973); Derzhavets, Yu. A. (1984).

UKRAINE: Derzhavets, Yu. A. (1984).

UZBEKISTAN: Grum-Grshimailo, G. E. (1890); Derzhavets, Yu. A. (1984).

YUGOSLAVIA [BOSNIA & HERZEGOVINA, CROATIA, MACEDONIA, SLOVENIA, SERBIA and MONTENEGRO]: Daniel, F., Forster, W. & Osthelder, L. (1951); Andus, L. (1986).

III. LIFE HISTORY

The Sphingidae, like all Lepidoptera (Scoble, 1992), are endopterygote insects, i.e. they undergo a complete metamorphosis, passing through four distinct stages – the ovum (egg), larva (caterpillar), pupa (chrysalis) and imago (adult).

The ovum

Sphingid ova are generally green or yellow in colour, smooth and shiny with no obvious surface sculpturing, although under high magnification a slight reticulate pattern may be evident. They are either spherical or oval, slightly flattened dorso-ventrally (Pl. C, fig. 1).

The micropyle, through which the male sperm fertilizes the egg, is lateral. The egg is not always proportionate to the size of the moth which lays it, that of *Agrius convolvuli*, for example, being about the same size as that of *Macroglossum stellatarum*, i.e. about 1mm in diameter. The largest eggs are those of species in which the adult does not feed, such as *Marumba quercus* and *Laothoe populi*. These vary in diameter from 2–3mm. Freshly laid, most are some shade of green or yellow, although the waterproof sticky substance which attaches them to the hostplant may discolour them. Very few are patterned, an exception being those of *Rethera komarovi*, which are green with an encircling white band (Pittaway, 1979a). The normally thick, translucent egg-shell or chorion changes to some shade of grey or yellow prior to hatching, such coloration being due to the colours of the developing larva. When the larva is about to emerge it becomes very active, twisting and turning within the egg-shell; the dark mandibles of green larvae can be seen biting at the interior at this stage.

The speed of egg development is very variable between species and is not always dependent upon temperature. The eggs of *Hyles livornica* have been known to hatch in three days; during cold weather those of *Sphinx ligustri* may take 21 days. No Palaearctic sphingids undergo diapause in this stage.

Few western Palaearctic sphingids intentionally lay their eggs on anything but their larval hostplants. However, it has been observed that *Hyles vespertilio* will occasionally oviposit on stones at the base of *Epilobium* plants, and *Rethera komarovi* and *Macroglossum stellatarum* may lay on dead stems protruding through the main hostplant, *Galium* (pers. obs.). Guided by visual, olfactory, tactile and other stimuli, the female moth normally chooses the spot with great care. However, to quote Reavey & Gaston (1991), 'Where the female moth lays eggs depends on how likely she is to be eaten by predators in the process, how far away she is from plants with nectar for her food, how urgently she wants to lay and where the egg is most likely to survive'. Between one and three ova are usually laid at any one time (up to ten in the *Hyles euphorbiae* complex), several females sometimes selecting the same shoot. Apart from those species which feed on *Galium*, where eggs may be placed on the delicate flower-heads, ova are normally deposited on the underside of a leaf where they are better protected from parasitoids, extremes of temperature and desiccation.

The larva

The larval stage is primarily concerned with the efficient acquisition of food to promote growth. As the outer cuticle has a limited ability to stretch, the larva can grow only by moulting, which it does three, four or five times. Each moult is called an ecdysis and the intervals between these are known as instars (often expressed as L_1, L_2, L_3, etc.).

The body (Text fig. 3) is cylindrical, or anteriorly tapering and consists of thirteen segments in addition to a capsular head which itself consists of six fused sclerotized plates. Caterpillars are polypodous larvae, that is, they have a series of false legs (prolegs) on the abdomen in addition to the three pairs of jointed thoracic limbs, the true legs. The former terminate in a single semicircular crown of hooks (crotchets) and are situated on abdominal segments 3–6 and on segment 10. The latter terminate in sharp claws and are found on thoracic segments 1, 2 and 3. The head is either oval, rounded or triangular, although most first-instar larvae have spherical heads. The highest region of the head is called the vertex; behind the vertex is the occiput, with a small triangular sinus situated dorsally on the hinder margin and called the occipital sinus; the front of the head is called the face, the side the cheek, and the underside of the head behind the mouthparts the gula or throat. A triangular clypeus occupies the centre of the face. The mouthparts and immediate surrounding area contain most of the gustatory and sensory organs. Six stemmata (often incorrectly referred to as ocelli) are situated at the base of each cheek and gula, just above the base of the antennae. They are simple, round and convex, and are capable of limited vision. The antennae, one at the base of each cheek, are each composed of three segments, all of which can be retracted telescopically. These are used by the caterpillar to feel its way and for taste. A pair of strong, truncated, hollow mandibles are set in sockets near the base of the antennae. These are used not only for eating but also, in some species, for self-defence. Directly in front of the mandibles lies a transverse, grooved plate composed of the labrum and ligula, which is used to guide plant material between the mandibles, for tasting, and to stop food particles falling out from between the mandibles. The maxillolabial complex below and behind the mandibles is responsible not only for the main sense of taste but also bears pairs of maxillary and labial palpi, and the all important spinneret.

The eighth abdominal segment bears a caudal horn, but in the final instar of some species this may be absent or replaced by a tubercle or 'button'. There is a spiracle situated laterally on either side of thoracic segment 1 and abdominal segments 1–8. The cuticle bears sparse, fine, secondary hairs and, in the Sphinginae, the larvae may also have raised transverse ridges edged with small tubercles in some or all instars. Larvae of all the western Palaearctic Sphingidae are generally similar in shape although they may exhibit an amazing variety of patterns and colours, not only between species but also intraspecifically. In a significant proportion of species, the larva rests with the anterior segments raised clear of the substrate and the head tucked underneath. This feature gives the group its alternative common name of 'sphinx moths'.

Unlike birds, young caterpillars eat their way out of the eggshell. After a short rest most consume the remaining chorion and then proceed to a suitable resting-site. No western Palaearctic species is dependent on its eggshell for its first meal. Even the larva of *Marumba quercus*, which in its first instar does not feed on leaves at all, often ignores the vacated chorion and wanders off.

In some species, the young larva is inactive and sits sphinx-like along the midrib on the underside of a leaf, hidden from predators. In *Hemaris fuciformis*, small holes are chewed through the leaf on either side of the midrib (Meerman, 1987). In *M. quercus* and *Sphinx ligustri*, feeding is always from the leaf margin inwards. Many of the Macroglossini select the flowers of their hostplants and sit among them, at least while still small. As they grow larger some become strictly nocturnal (*Hyles vespertilio* passes the day on the ground), or, like *Deilephila elpenor* and *D. porcellus*, hide in the tangled mass of hostplant when not feeding. Many of these intermittent feeders change colour from green to a cryptic brown so as to blend in with the dead leaves, or even stones, amongst which they rest. Others, such as the *Euphorbia*-feeding *Hyles* species – *H. euphorbiae*, *H. tithymali*, *H. dahlii* and *H. nicaea* – feed quite openly and seem to rely on brilliant aposematic coloration to warn vertebrate predators that they are poisonous (see Plate D, fig. 2). Some larvae bear eyespots (*Deilephila elpenor*, *D. porcellus*, *Daphnis nerii* and *Rethera komarovi*) which may be located so as to make them appear snake-like, thus deterring attack.

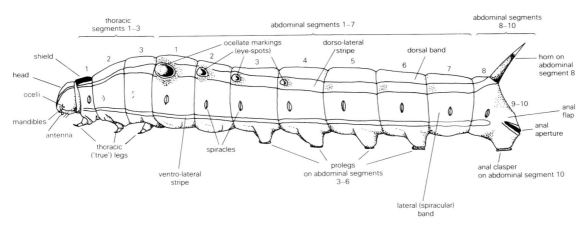

Text Figure 3 Generalized sphingid larva indicating typical body patterning and morphology

Interestingly, mature larvae of many of the smaller Macroglossinae which remain feeding on their hostplants (*Hemaris croatica*, *Sphingonaepiopsis gorgoniades* and *Macroglossum stellatarum*) are cryptically patterned in some shade of green with longitudinal white lines and resemble the younger stages of those which later change colour (*Rethera komarovi*, *Proserpinus proserpina* and *Hyles vespertilio*) (Pl. D, fig. 5).

Fully-grown larvae of the Sphinginae practice a different strategy and tend to rely totally on cryptic coloration and patterns to blend in with their hostplant. Most hang suspended under a leaf (Pl. C, fig. 4) and move only to feed or relocate. So as not to have their position given away by denuded twigs, larvae of many species (e.g. *Laothoe populi*) change position frequently.

The growth rate of all caterpillars is very variable and highly dependent on temperature as well as the quality and quantity of food. Little work has been carried out on the nutritional quality of the hostplants of the Palaearctic Sphingidae; *Marumba quercus* has a higher growth- and survival-rate on *Quercus cerris* (Turkey oak) than on *Quercus robur* (English oak) (pers. obs.). Detailed research on the North American *Papilio* (*Pterourus*) *glaucus* (L.) (tiger swallow-tail) (Papilionidae) (Scriber, Lintereur & Evans, 1982) suggests that this would be a fertile field for further study (see p. 50).

Lack of food is rarely a problem for larvae unless too many females select the same small hostplant on which to oviposit, in which case the resultant starving larvae generally wander off in search of new plants before they are mature. Many succeed, some fall victim to predators and a few, if sufficiently developed, will pupate prematurely. However, *Hyles livornica*, being a true desert species, has developed some remarkable survival techniques to overcome this problem. Although ova are laid on preferred hostplants (*Rumex*, *Asphodelus* and *Zygophyllum*), the resulting larvae can complete their development on a large number of other neighbouring plants should drought or overgrazing eliminate their normal hosts. In fact, under such conditions many larvae actively shorten their instars and pupate when still very undersized (pers. obs.). The resulting dwarf adults are, however, still perfectly formed and fertile.

Unlike those of many other lepidopterous families, no Palaearctic sphingid larvae undergo diapause, although those of *Acherontia atropos* will become torpid during short, cold spells in the North African winter.

Like all insects, the Sphingidae are poikilothermic (i.e. cold-blooded) and are incapable of maintaining for long an internal body-temperature much higher than their surroundings. As the biochemical processes necessary for growth are faster and more efficient between 20°C and 30°C, both the adults and larvae of the Palaearctic Sphingidae need either to generate or to acquire heat for many activities. At a controlled temperature of 22°C, the larvae of *Hyles hippophaes* mature in 11–18 days; those of *Marumba quercus* in 30 days. In central Europe, under field conditions, these species take, on average, 35 and 42 days respectively (pers. obs.). Usually, newly emerged larvae try to avoid intense sunlight but, during cold weather or in northern latitudes, they will actively sunbathe. This activity becomes very noticeable in more mature caterpillars such as those of *Macroglossum stellatarum*, *Hyles euphorbiae*, *H. hippophaes* and *H. gallii* and allows them to complete their development more rapidly (see p. 45). Large caterpillars can raise their body temperature to 10°C or more above that of their surroundings by this method (Reavey & Gaston, 1991). This ability to condense their life cycles with increasing temperatures and sunshine-rates allows many species to undergo multiple generations in southern Europe and North Africa. It is also possible that *H. euphorbiae*, or even *H. croatica*, could not survive so far north in Europe if they did not sunbathe; the amount of available sunlight at critical times of the year may be the main factor limiting the distribution of these two species. In Lebanese populations of *H. euphorbiae*, this process has been taken one stage further in that larvae occurring at high altitudes have more black coloration than those from lower, hotter zones and hence can absorb more heat. The same applies to a form of the larva of *H. nicaea castissima* found in the Atlas Mountains (Pl. D, fig. 4) (pers. obs.). This black coloration may also afford protection against high levels of ultra-violet radiation found at high altitude.

Once mature, the caterpillar ceases feeding and rests for up to 24 hours. Its colour darkens – green species generally turn a shade of brown or purple – and it voids any remaining gut-contents. Many tree-feeding species will then descend under cover of darkness and make off rapidly in search of a suitable pupation site.

The pupa

The pupa or chrysalis (Text fig. 4) also consists of the head and thirteen segments, although many of these are indistinct and the divisions apparently arbitrary. The head is anterior to the thorax, which comprises the prothorax (thoracic segment 1), the mesothorax (segment 2), and the metathorax (segment 3); the abdomen comprises abdominal segments 1–10. The pupal body is usually smooth, short and fusiform, with the head often distinctly narrower than the thorax; the fifth, sixth and seventh abdominal segments are movable.

The front of the head is called the frons, the lower part the clypeus. This is not the same structure as in the larva, although the name is used for a similar region. The base of the proboscis adjoins the clypeus, and this structure can be cariniform (keel-shaped), enlarged basally, or a free, jughandle-like lobe. However, in most species it lies flush with the surrounding segments and extends to the tip of, and narrowly separates, the two wings. The top of the head adjoining segment 2 is termed the vertex; this separates the two large eyes and gives rise to a pair of antennae which run down either side between the wings and the midlegs. The mesothorax merges laterally with the forewings, these covering the ventral surface of the first four abdominal segments (1–4). The abdomen terminates in a prominent triangular cremaster. A spiracle is found either side on the prothorax, and there is a spiracle on each side of abdominal segments 2–8.

Sphingid pupae have characteristic differences (Text figs 5a–c). Cuticle colours range from a pale cream in *Macroglossum stellatarum*, through various browns to a deep black in *Laothoe populi*. The pupa may be very glossy (e.g. *Rethera komarovi*), have a shiny, rugose finish (e.g. *Marumba quercus*), or be matt black or brown (e.g. *Mimas tiliae*).

At the onset of the pupal stage of metamorphosis, activity is reduced and feeding and excretion cease. Pupation may occur in a subterranean chamber (*A. atropos*), under stones (*Hyles vespertilio*), or within a loose cocoon on the surface of the soil (*Hemaris*, *Hyles* and *Deilephila* spp.), or even within leaves on the hostplant (summer broods of *Theretra alecto*). Larvae of *Mimas tiliae* and *Laothoe populi* wander no further than the base of 'their tree' and pupate just beneath surface debris. Occasionally, pupae of *M. tiliae* may even be found high in

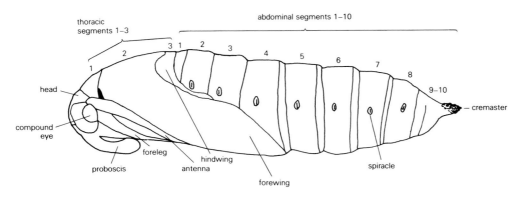

Text Figure 4 Morphology of a typical sphingid pupa (*Sphinx l. ligustri* (Linnaeus), indicating segments

species of *Ulmus* (elm) trees under loose bark. This is probably a method of reducing mortality, for if a caterpillar spends too long a time and travels too far searching for a safe place to pupate it risks being eaten or losing body weight (Reavey & Gaston, 1991).

During this stage there is a very active reorganisation and transformation of the larval organs; larval tissues are broken down and, in some cases, immediately rebuilt into those of the adult. However, in those pupae undergoing diapause (the overwintering phase for most species) the tissues pass the winter as clusters of nuclear cells in a sea of liquid. (An exception is the pupa of *Proserpinus proserpina*, which overwinters in an almost fully-formed state.) Only with the onset of warmer weather, coupled with other appropriate stimuli, are they transformed into the adult insect.

Adult emergence

As soon as the transformation is complete, the adult is ready to emerge. The chrysalis splits dorsally and laterally and the adult crawls out, propelled by waves of abdominal contractions. If underground, these same movements enable the insect to reach the surface. In *Deilephila elpenor*, abdominal hooks on the pupa allow it to wriggle out of its cocoon and free of leaf debris, before the moth emerges. However, these hooks appear not to have evolved for this purpose but as an aid to escape submergence in water (Brock, 1990).

Once free, the adult moth climbs to some suitable support from which to hang and then expands its wings with haemolymph, a process which takes approximately 30 minutes, after which they dry and harden. Most species emerge in the late morning or early afternoon so that their wings have sufficient time to dry before nightfall.

Text Figure 5 Three live sphingid pupae, showing different characteristics of form and texture

(a) *Sphinx p. pinastri* Linnaeus: glossy, elongate with small free proboscis parallel to body

(b) *Mimas tiliae* (Linnaeus): rough and rugose, not glossy; tapering anteriorly

(c) *Laothoe p. populi* (Linnaeus): matt, rough, stout and anteriorly blunt

IV. ADULT BIOLOGY

Prior to their first flight most hawkmoths expel the remaining meconium – waste matter accumulated during the pupal stage. Apart from the diurnal *Hemaris* species and *Macroglossum stellatarum*, and also *Hyles gallii* and *H. livornica* which occasionally fly in bright sunshine, all western Palaearctic sphingids are crepuscular or nocturnal. None is active the entire day or night and many show definite sexual and feeding behaviour patterns (see below). Some do not feed, but most are avid visitors to flowers. Many are powerful and skilful fliers (Heinig, 1982) and undertake prodigious migrations.

Little work has been done in the field on adult lifespan but in captivity most European species live from 10–30 days. The exception is *Macroglossum stellatarum*, which hibernates as an adult from about October to April.

Before flight, adults must increase the temperature of their flight-muscles. This is achieved by simultaneously contracting both the upbeat and downbeat wing-muscles, which causes the wings to vibrate. The amount of heat generated can be positively correlated with wing area and body size and is independent of ambient temperatures. However, the latter have a marked influence on how long a moth needs to warm up. For example, the North American hawkmoth *Manduca sexta* (L.) takes twelve minutes to warm up at 15°C but only 1 minute at 30°C (Sbordoni & Forestiero, 1985). At 5°C, this species will not fly at all, although the European *Smerinthus ocellatus* will (pers. obs.). However, the Sphingidae are not as cold-blooded as many believe. Temperatures within the thorax during flight can exceed 40°C; heat-loss from the thoracic muscles is reduced by a dense covering of insulating, fur-like scales.

Another method of acquiring body heat is adopted by adults of the diurnal genus *Hemaris* and *Macroglossum stellatarum*. Individuals may elect to warm up as normal or, alternatively, bask in the sun. The latter is more energy efficient and it is not uncommon to observe several *M. stellatarum* sunbathing on a house wall after hibernation (pers. obs.). In many respects these diurnal species behave like butterflies in response to ambient temperatures and weather conditions.

In larvae, feeding is straightforward and concerned with the efficient acquisition and utilization of food, resulting in growth (see p. 19). In the adult, growth no longer occurs but food is still required to supply energy for activity, to allow ova within the abdomen to mature, and to provide moisture.

Flower nectar is the main source of energy and moisture, but provides little in the way of protein and vitamins. The latter are supplied by the breakdown of body tissues accumulated during the larval stage. Superficially, most adult moths visit flowers indiscriminately, showing little preference for certain colours or species. Careful observations have shown, however, that a moth will not only learn the location of suitable nectar-bearing flowers within its territory, but will also visit each patch at the same time each day or night (pers. obs.). Additionally, many species exhibit distinct periods of feeding activity. In *Sphinx pinastri*, this is between dusk and midnight, when it avidly visits pale, deep-throated, tubular flowers such as *Lonicera*. *Hemaris tityus* prefers blue or purple, shallow-throated flowers such as *Ajuga*, *Salvia* and *Knautia*, which it frequents between 10.00–15.00 h.

Acherontia atropos and *A. styx* are unique in that they rarely visit flowers, preferring to enter beehives in search of honey. Their proboscises are modified into short, dagger-like tubes capable of piercing honeycombs.

However, feeding and egg maturation take up valuable time and can expose the female moth to increased predation with the subsequent loss of all reproductive potential. The Smerinthini have solved this problem by emerging with large fat reserves and bearing fully-developed eggs. They have no need to feed and can initiate egg-laying immediately after pairing. Many have atrophied proboscises.

Unlike butterflies, moths are mainly nocturnal and cannot rely on visual stimuli to find a mate. Virgin females of all Palaearctic Sphingidae (even the diurnal species) produce powerful airborne sex pheromones capable of 'calling' males of their own species from some distance. The male approaches upwind, alights and immediately copulates, although in the Macro-glossini and Sphingini he may first douse the female in a pungent male pheromone from body hair-pencils (Birch, Poppy & Baker, 1990). *Laothoe populi* remains paired for up to twenty hours; *Hyles euphorbiae* for rarely more than two hours. In the diurnal Macroglossinae, the pairing process has developed one stage further – the sexes may engage in pursuit courtship-flights before copulation and may even dispense with the calling phase altogether (pers. obs.).

V. ADULT MORPHOLOGY

The adult moth or imago, the final stage of the lepidopterous life cycle, has an easily distinguishable head, thorax and abdomen.

With very few exceptions, species of the Sphingidae can be distinguished from other Lepidoptera by their general appearance. They are medium- to large-sized moths with relatively long and slender wings, a stout thorax and a long conical abdomen: in bulk, these moths are amongst the largest in the world.

The head

The head bears a pair of large compound eyes, above each of which arises a long, three-segmented antenna composed of the first segment (scape), the second segment (pedicel) and the third segment (flagellum), which often terminates in a small reflexed tip. The antennae are well provided with sensory receptors (sensilla) along the unscaled ventral surface. In the male, the ventral surface usually also has pronounced fasciculate setae (Text fig. 6), but in the female these setae are simple.

Text Figure 6 Head region of *Smerinthus k. kindermanni* Lederer ♂ showing the pronounced fasciculate setae on the flagellar segments of the antennae

The compound eyes are separated posteriorly by the occiput, dorsally by the vertex (which may bear ocelli) and anteriorly by the frons. At the base of the frons (clypeus), either side of the proboscis base, lie a pair of pilifers and atrophied maxillary palps. The coiled and tubular proboscis (haustellum) is derived from modified maxillary galeae. When coiled, this is protected on either side by an enlarged, heavily scaled, three-segmented labial palp. In those species with a functional proboscis, the tip is provided with chemoreceptors (papillae).

In comparison with other moths, one internal feature of sphingid anatomy is remarkable – namely the head musculature (Fleming, 1968). Most sphingids have powerful cranial muscles, the origins of which appear to be independent of ectodermal structures. The musculature varies between species but as many as eighteen muscles may be found inside the head in addition to the intrinsic proboscis and antennal muscles. Most are paired and can be divided into six groups, the details of which are beyond the scope of this book (see Eaton, 1988).

The thorax

The thorax is composed of three more or less fused segments, each bearing a pair of legs. The prothorax (which bears a pair of membranous patagia which protect the 'neck') is greatly reduced and lies in front of the highly developed mesothorax which, in addition to the paired forewings, supports two tegulae covering the wing bases. Lastly, the smaller metathorax bears the smaller hindwings. Each thoracic segment is composed of a series of rigid plates joined together by flexible membranes; dorsally the plates are known as tergites, ventrally as sternites and laterally as pleurites. The whole thorax is densely covered with insulating hair-like scales.

Typically, the three pairs of legs consist of five sections (Text fig. 7a): the coxa, trochanter, femur, tibia and tarsus. The foretibia on the inner surface bears a movable lamellate spur or lobe (epiphysis) which is used to clean the eyes and antennae. The tarsi are usually five-segmented with a pretarsus, or terminal segment, of paired apical claws (Text fig. 7b). In most species, a pad-like pulvillus is present between the claws which is richly endowed with chemoreceptors. Each claw is also flanked on its outer side by a paronychium. Large spurs are frequently present on the mid- and hindtibiae.

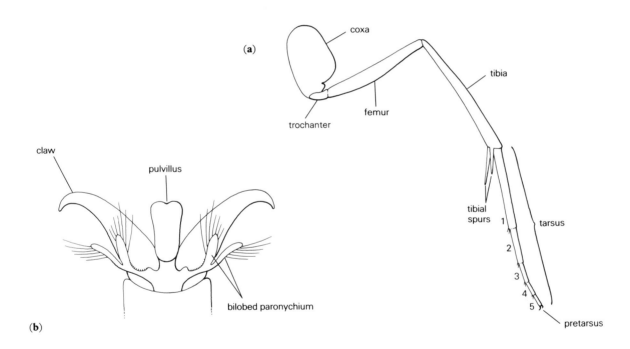

Text Figure 7 Midleg of generalized adult sphingid showing (a) segments and tibial spurs; (b) detail of pretarsus

The wings

The wings of the Sphingidae are heavily scaled and generally triangular in shape: the forewing is long and narrow; the hindwing is small. As in all Lepidoptera, the wings consist of a double membrane supported by a network of tubular veins. At both family and generic level the wing venation has common features (Text fig. 8). For taxonomic and descriptive purposes, wings are divided into clearly defined areas (Text fig. 9) and the veins designated by letter and number according to one or other system of nomenclature. That of Comstock & Needham (1898), with minor modifications, has been adopted in this work. Among Macrolepidoptera, all Sphingidae have the following unique combination of features: vein R_1 of the hindwing crosses to vein Sc from the middle of the cell, the base of vein M_1 is missing, and vein 1A (regarded as the posterior cubital vein CuP in some modern works) is absent. The forewing lacks the first anal vein 1A. Following Rothschild & Jordan (1903) and Bell & Scott (1937), the three parts of the angled distal cross-vein bordering the discoidal cell separating M_1 and M_2, M_2 and M_3, and M_3 and Cu_1 are referred to as D_2, D_3 and D_4 respectively. The frenulum (a brush of bristles in the female, a single strong bristle in the male), which links the hindwing to the forewing, is usually well developed, although atrophied in a number of the Smerinthini.

The abdomen

The abdomen is cylindrical, fusiform and consists of ten segments (as in the larva); in the male, the last two are fused together and modified to form the external genitalia. Each segment bears a dorsal plate, the tergum, and a ventral plate, the sternum, joined laterally by a membranous pleural area bearing the spiracles.

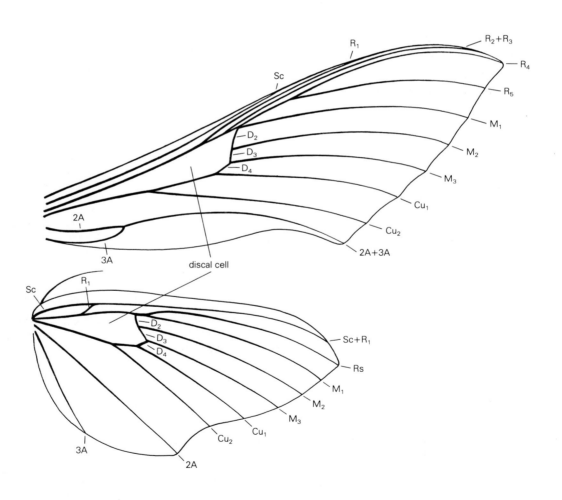

Text Figure 8 Sphingid wing venation (*Hippotion celerio* (Linnaeus)). In this diagram, for ease of labelling, forewing veins Sc, R_1, and $R_2 + R_3$ are shown as closely parallel but separate along the apical third of the costa. In fact, these veins unite along the costa towards the apex, as shown in Text figs 26a,b (p. 78).

Text Figure 9 Sphingid wing patterning
(a) *Laothoe p. populi* (Linnaeus)
(b) *Hyles g. gallii* (Rottemburg)

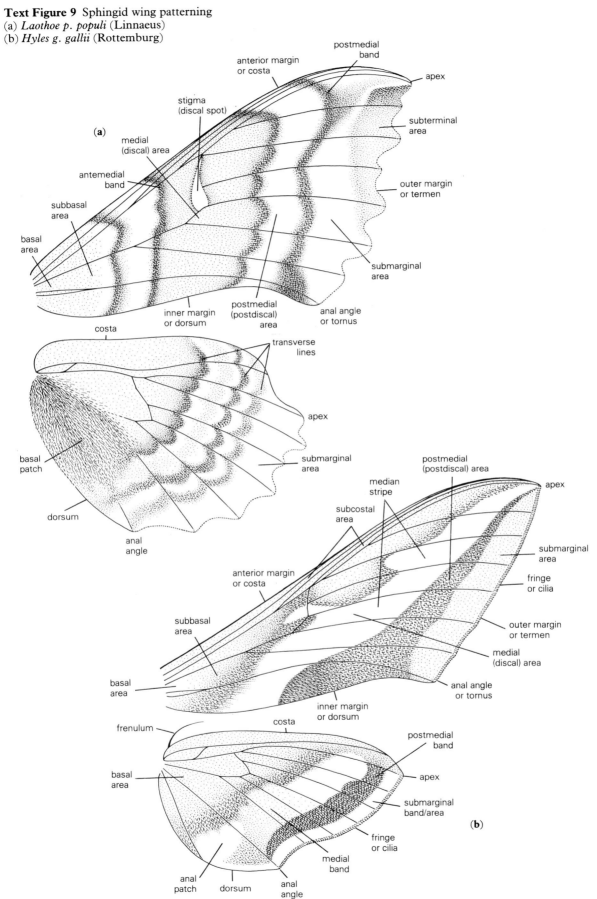

The genitalia

The genital organs are of diagnostic importance in the taxonomic study of the Sphingidae as they differ both between genera and species. In the male (Text figs 10a,b), there is a roughly triangular dorsal tegumen derived from the ninth tergum (T9). It abuts the tenth tergum (T10), which is a single or double hooked process (the uncus). Ventrally, the ninth sternum (S9) has been transformed into an arch-like vinculum, which may extend anteriorly as a median process (or saccus). Laterally, the ninth sternum bears plate-like claspers (valvae), which vary in form. The tenth sternum (S10) is usually jaw-shaped – hence the name 'gnathos' – and with the uncus it forms a sclerotized ring around the anal tube. The penis (aedeagus) lies in the centre of this structure between the valvae.

In the female (Text fig. 10c), the genitalia are much simpler. The anus and oviposition pore are flanked by a pair of sensitive, hairy pads (papillae anales) situated on the ninth segment. The copulatory opening (ostium bursae) lies between the seventh and eighth segments and leads into the abdomen along a duct (ductus bursae) to the corpus bursae. It is here that the spermatophore is deposited during mating.

Morphological variation

Interspecific variation is used for taxonomic purposes and this is discussed in the chapter on classification (see page 60). There are also, however, small differences between individuals of the same species to the extent that no two hawkmoths are exactly alike. Two main factors produce individual variation:

(a) genetic constitution (internal) – genotype ⎫
(b) the environment (external) ⎬ phenotype
⎭

Unlike genetic variability, that produced by environmental factors cannot be passed on to the next generation, although selection pressures can alter the genetic bias of a population by favouring certain genotypes. Several populations may demonstrate a range of individual variability and still interbreed to give viable offspring; these constitute a species (or subspecies).

A species in which three or more colour forms, or morphs, share the same range is termed polymorphic. For example, adult *Mimas tiliae* are polymorphic and many of these morphs have been given names.

Sexual dimorphism is the commonest variation within a population and is due to the different complement of sex chromosomes – in Lepidoptera, males are XX, females XY or XO. The main differences are usually in the colour, shape and size of the wings and antennae. For example, females of *Laothoe populi* are always larger than the males, lighter in colour (often rosy), and bear narrow, filiform antennae, whereas the antennae of the male have fasciculate cilia ventrally.

The same genotype can produce, under differing environmental conditions, different phenotypes. Heat applied to the developing pupae of *Hyles euphorbiae* results in beautiful pink-flushed moths. Those of *Mimas tiliae*, if chilled considerably, yield the dark montane form, f. *montana* Daniel. Similarly, the dry parched conditions of the eastern Mediterranean produce smaller, paler individuals of *Theretra alecto* than is normal in India. However, when reared in more humid environments and given succulent hostplants, Greek specimens are indistinguishable from Asian material. Conversely, specimens of *Hyles tithymali deserticola* reared under optimum conditions retain their unique patterning and coloration; they do not take on the appearance of *H. t. mauretanica* (pers. obs.). In central Europe all second generation specimens of *Hemaris fuciformis* have a toothed border in the forewings (f. *milesiformis* Trimen), which is almost always absent in first generation individuals.

Infrequently, individuals occur in which both male and female characters are present; these are termed gynandromorphs. Some have these characters scattered over the body in a mosaic;

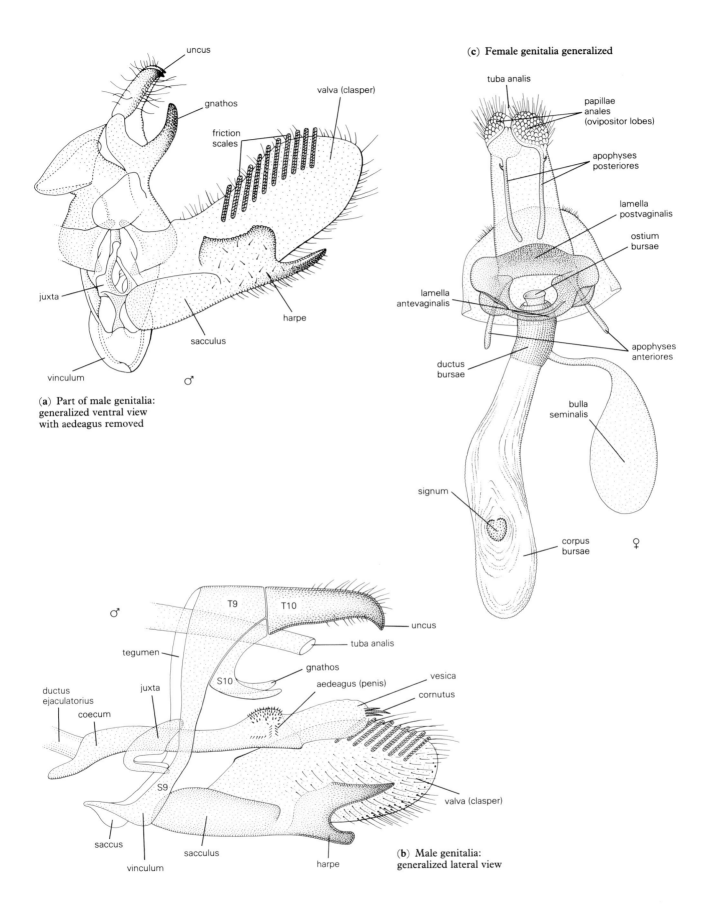

Text Figure 10 Morphology of sphingid genitalia: (a,b) male; (c) female

Text Figure 11 Gynandromorph of *Laothoe p. populi* (Linnaeus) from the British Isles: (a) upperside; (b) underside. The most conspicuous difference is in the antennae

others are divided bilaterally, with one side male and the other wholly female. Bilateral gynandromorphs are found not infrequently in *Laothoe populi* (Text fig. 11).

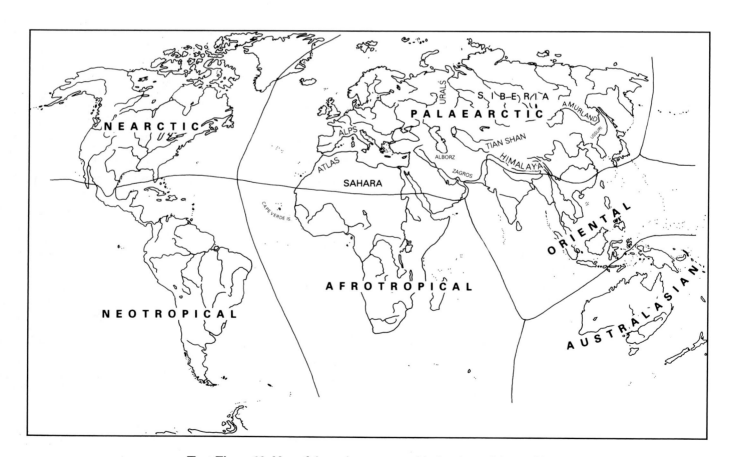

Text Figure 12 Map of the main zoogeographical regions of the world

VI. ECOLOGY OF WESTERN PALAEARCTIC SPHINGIDAE

Geographical factors affecting distribution

The world can be broadly divided into six convenient zoogeographic regions (Text fig. 12, p. 32) (Sclater, 1858) although this division is very generalized, with recognizable sub-units and areas of endemism. Boundaries between the regions are sometimes not very clearly defined and alternative delineations exist.

Palaearctic – Europe, North Africa, Asia Minor, the Himalayas and northern Asia

Nearctic – Canada, the U.S.A. and temperate Mexico

Neotropical – Central and South America

Afrotropical – Sub-Saharan Africa

Oriental – India to Indonesia

Australasian – Australia, New Zealand and New Guinea

The world is divided in this way on the basis that certain regions have evolved their own characteristic fauna which, in combination with unique floral, climatic and ecological parameters, are found nowhere else. Within all these regions there are a number of main world vegetation types (Text fig. 13) which are governed by geological and climatic factors.

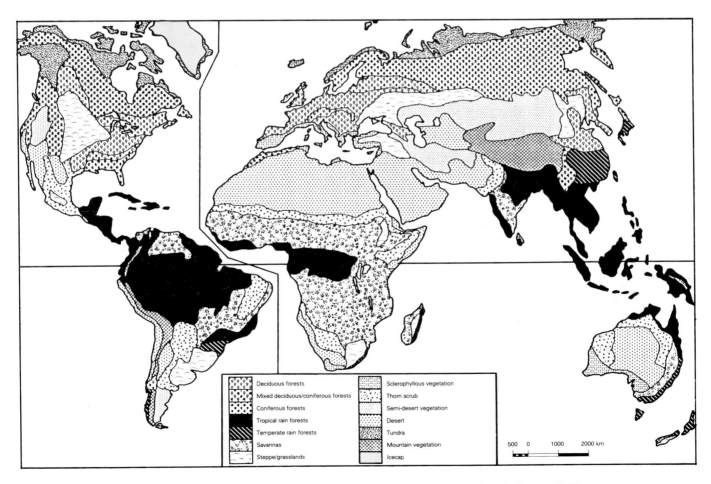

Text Figure 13 Map of the principal world vegetation types (after Riley & Young, 1968)

Due to their close relationship, the Palaearctic and Nearctic regions are often grouped together as one region – the Holarctic. This region has four distinct seasons – spring, summer, autumn and winter, the last of which is relatively long and cold, with severe frosts. These climatic characteristics are not confined to the Holarctic region as they can also be found in the southern temperate areas of the Australasian and Neotropical regions; however, the fauna and flora of these regions differ markedly from that of the Holarctic.

Zoogeographical regions often have definable boundaries due to physical barriers, such as mountains, deserts or water. The Palaearctic is now separated from the Nearctic by the Atlantic Ocean on one side and the Pacific Ocean and Bering Straits on the other. However, where no such barriers exist, each region gradually merges with the next, pockets of one extending some way into the other due to environmental conditions. Such transitional or eremic zones may themselves have certain definable characteristics and are often classified as distinct regions. The desert between the Palaearctic and Afrotropical regions is one such zone.

Climatic factors affecting distribution

Apart from geographical barriers, climate is the other major influence that determines these regions and, consequently, hawkmoth distribution. The main climatic factors are maximum and minimum average temperatures; hours and strength of sunshine; rainfall (amount and periodicity); humidity; length of seasons; severity of frosts; evaporation rates; and wind. Individual species have their own genetic tolerance to climate and it is a combination of this and the indirect effect of biomes that restricts the potential range of a species. Competition, geographical barriers, climatic barriers and historical factors will determine if that range can be fully utilized. Temperature is probably the main factor affecting distribution, as can be seen by correlating January and July isotherms (Text figs 14a,b) with the distribution of individual species; however, laboratory experiments are needed to establish if such deductions are correct. *Agrius convolvuli* appears unable to establish itself in areas where the January mean temperature falls below +4°C, although it is quite capable of breeding over most of Europe during the summer months. *Sphinx ligustri* appears to have the reverse problem in that it cannot survive in areas where the July and January temperature means exceed +30°C and +4°C, respectively. *Marumba quercus* appears to be limited both by temperature and hostplant distribution. Its

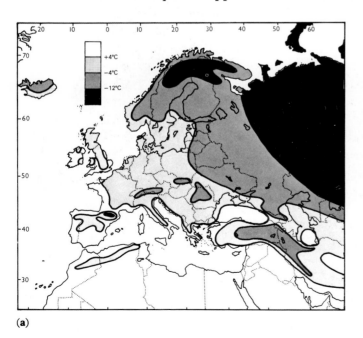

(a)

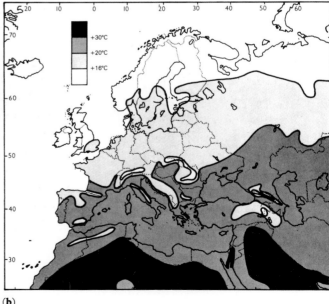

(b)

Text Figure 14 Isotherms for the western Palaearctic region: (a) January; (b) July southern distribution is confined to mountain chains where oak (*Quercus*) grows. Its northern limit correlates very well with the +20°C July isotherm; however, its eastern limit in the Ukraine and Russia appears to be limited by the −12°C January isotherm.

Unlike some moth species which are found worldwide, no hawkmoth is cosmopolitan although some genera such as *Sphinx* are very widely distributed (Eurasia, North America and South America). Some have large, continuous ranges, e.g. *Laothoe amurensis* which occurs right across the Palaearctic region from Poland to Japan. If such a range is interrupted in places, as in the case of *Mimas tiliae*, which is found in Europe, Asia Minor and then eastern Asia with no population in between, it is termed disjunct (see pp. 61ff.). Some sphingids are Holarctic, occurring right across the temperate regions of both Eurasia and North America, e.g. *Hyles gallii* (see p. 147). A species which is confined to a specific area and is found nowhere else is termed endemic: *Hyles vespertilio* is endemic in the western Palaearctic; *H. dahlii* in Corsica, Sardinia and the Balearic Islands. Where a small population of one species has been left isolated when the parent population retreated in the face of changing climatic conditions, it is known as a relict population. *H. euphorbiae* has one such in the mountains of south-west Arabia (Pittaway, 1987; Meerman, 1988b) and *Sphingonaepiopsis gorgoniades* another in Yugoslavia (Mentzer, 1974).

Because a species has a certain range today does not mean that this has always been so nor that it will remain so. However, apart from relict populations, subspecies and generic relationships, there is little direct evidence to support this view. Nevertheless, the dependence of specific sphingid species on certain hostplants, ecosystems and biomes is recognized (see p. 43) and the study of plant remains (particularly pollen) can provide a great deal of indirect evidence.

During part of the Tertiary epoch (65–2 million years BP), Europe alternated between a typical Indo-Malay climate (wet, warm and humid) and warm, temperate conditions, characterized by the presence of *Magnolia* species (Polunin & Walters, 1985). There then followed a period of progressive cooling with distinct changes in vegetation. Many plant species characteristic of present-day conditions appeared, as well as others which today are confined to the mountains of western China, eastern Tibet and Japan, e.g. *Picea*, *Carpinus*, *Pterocarya*, *Liriodendron* and *Tsuga*. Some became extinct due to further cooling, or retreated to refuge areas in the Tian Shan, Turkey and Iran, e.g. *Pterocarya*. Further changes produced a more contemporary European flora of *Quercus*, *Pinus*, *Picea* and *Abies* intermingled with *Carya*, *Juglans*, *Pterocarya* and *Magnolia*. By the end of the Tertiary epoch most eastern Asiatic species had vanished from Europe (*c.* 1 million years BP) and few plants were present which do not exist in the western Palaearctic today. During the Quaternary epoch there were at least four major and several minor ice-ages interrupted by warm interglacial periods. In the warmest periods, forests of *Quercus*, *Ulmus*, *Fraxinus* and *Corylus* formed. Wetter interludes favoured the spread of *Alnus*, whereas drier conditions favoured *Pinus*. Conversely, the glacial phases produced a decidedly tundra-like community of dwarf *Salix* and *Betula* and numerous herbaceous plants. During these periods the former forests clung precariously to the Mediterranean shores and the mountains of Central Asia, Iran and North Africa. The penultimate major glaciation commenced 150,000 years BP and produced a very distinctive tundra biome in Europe dominated by *Dryas octopetala*. This was followed by a short interglacial and another ice-age, the Weichselian, which commenced 110,000 years BP. This finally ended approximately 10,000 BP, to be followed by a series of well-defined warming stages, characterized by various floral regimes (Polunin & Walters, 1985).

Pre-Boreal	10,000–9500 BP	Sub-Boreal	5000–2500 BP
Boreal	9500–7500 BP	Sub-Atlantic	2500 BP–today
Atlantic	7500–5000 BP		

With such drastic changes occurring in the biome regimes of Europe, it is safe to assume that sphingid ranges were also profoundly affected. Even to this day, two essentially east Asian butterfly taxa, *Vanessa indica* (Herbst) and *Apatura metis* Freyer, still have relict populations in the western Palaearctic (Higgins & Riley, 1980). Did a number of Chinese hawkmoths once occur in Europe, only to be exterminated by successive ice-ages? Are the Central Asian mountains and deserts preventing them from recolonizing Europe? After all, many of their hostplants are present as are suitable climatic conditions: the Oriental *Marumba gaschkewitschii* (Bremer & Grey), *M. maacki* (Bremer) and *M. jankowskii* (Oberthür) would probably feel quite at home in Europe today, if they could get there.

An expansion in range can occur when a geographical/environmental barrier is overcome, when there is a minute change in climate or when a hawkmoth's genotype alters due to advantageous mutations. Within the last ten years, *Hyles hippophaes* has colonized the Aegean from Romania in response to widespread planting of the amenity shrub *Elaeagnus angustifolia* (Pittaway, 1982a). The popularity of species of *Tilia* (limes), *Ulmus* (elms) and *Populus* (poplars) as ornamental trees in towns and cities has, no doubt, benefited both *Mimas tiliae* and *Laothoe populi*. Both are very common in city suburbs and rarer in rural areas. Within recent times *Acherontia styx* has spread right across Saudi Arabia. Originally confined to the Arabian Gulf oases, with the development of Saudi Arabia and the spread of gardens and farms, it adopted various garden shrubs as larval hostplants, appearing first in 1979 in Riyadh and then in 1982 in Jeddah (Wiltshire, 1986). How long before it crosses the Red Sea and enters Africa? Even the normally sedentary *Proserpinus proserpina* appears to be expanding its range in Europe in response to some, as yet unknown, change in its biology or ecology. During the last decade, the species has colonized eastern Belgium and individuals have been captured in Britain (Pratt, 1985) and Corsica (Guyot, 1990).

Other factors affecting distribution

Four main factors may determine a contraction in range: a change of climate; human activity; a change of genetic constitution; and competition. Unfortunately, in recent times, the second factor has proved a major force.

In the 1960s, when I was cycling around country lanes in south-east Austria during August, several fully grown larvae of *Hemaris tityus* and *Deilephila porcellus* could be seen and picked up on the road every day as they went in search of pupation sites. At that time the roadside vegetation was cut twice each summer. Additionally, overgrown boundary strips between the fields yielded countless larvae of *Macroglossum stellatarum*, *D. elpenor* and *D. porcellus* at night, and numerous adults at garden flowers and lights. By the 1980s the roadside verges were being cut every four weeks and the boundary strips had vanished, and so had most of the hawkmoths – both larvae and adults. In this area no increase in the use of insecticides was noted and the damage appeared to be purely mechanical (pers. obs.). Similarly, flood-control measures in the Swiss and French alpine valleys have reduced the amount of *Hippophae rhamnoides* and with it the numbers of *Hyles hippophaes*. In most cases it is the physical destruction of habitats which is more important than the much discussed use of pesticides. This is especially so at the periphery of a hawkmoth's range. Over the last 40 years *Hemaris tityus* has suffered a dramatic drop in numbers in Britain (Text figs 15a,b) (Gilchrist, 1979; Waring, 1992) and is all but extinct in Holland (Meerman, 1987). This species favours as breeding sites those lowland meadows near rivers which are now regarded as prime agricultural or building land. The same fate may have befallen relict populations of *H. croatica* in central Europe (see species account for more detail).

Although much has been said about the expansion and contraction of a species' range in response to changing environmental conditions (see p. 34), some species have managed to survive intermittent inhospitable regimes and recolonize much of their former range thereafter. But where did the refugees survive?

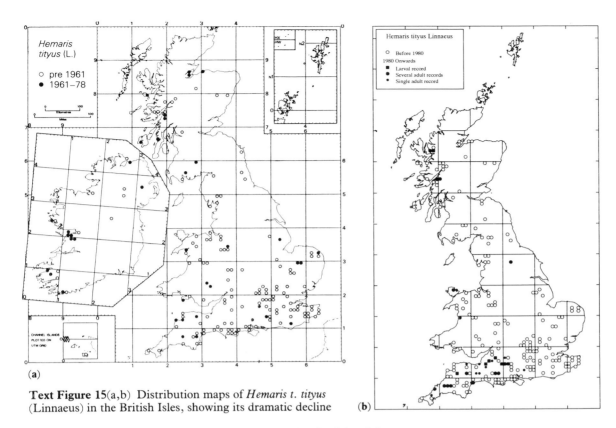

Text Figure 15(a,b) Distribution maps of *Hemaris t. tityus* (Linnaeus) in the British Isles, showing its dramatic decline

The origins of the western Palaearctic Sphingidae

During the last ice age that biome richest in hawkmoth species, the deciduous temperate forest, clung precariously to the Mediterranean and Black Sea shores, and the Caucasus and Tian Shan, and so, presumably, did many Sphingidae. These are known as 'refuge' areas (refugia). Each had its own characteristic fauna which developed in isolation. With a gradual increase in temperature many species spread out from these land-locked islands to colonize their territories again. This type of expansion is clearly evident in Scandinavia where no hawkmoths survived the glaciations. Since then, the Siberian refugium species *Laothoe amurensis* and the Mediterranean refugium species *L. populi* have recolonized that area. This has been due to a breakdown in barriers between the refugia. Of course, during the periods of isolation speciation may or may not have occurred if one species had been trapped in two or more refugia, i.e. it was polycentric (see p. 38). The area encompassing the Tian Shan and Pamir Mountains gave rise to several new species and subspecies, such as *Laothoe philerema*, *Sphingonaepiopsis kuldjaensis*, *Hyles chamyla* and *Acosmeryx naga hissarica*. The first three are closely related to *L. populi*, *S. gorgoniades* and *H. hippophaes* respectively. *L. philerema* now appears unable to spread into neighbouring regions occupied by *L. populi* or *L. amurensis* and the last two species have been unable to penetrate the Tian Shan.

In the western Palaearctic a number of Pleistocene refugia are recognized – the northern Mediterranean littoral (Italy, Greece); the Ponto-Mediterranean refugium (N. Turkey); the southern Mediterranean littoral (North Africa); the Syrian refugium (N. Syria and N. Iraq); the Iranian refugium (Iran); and the Tian Shan/Pamir refugium (Kyrgyzstan/Tajikistan/China). Most of the region's hawkmoths survived the ice-ages in one or more of these areas. Some, such as *Daphnis nerii* and *Acherontia atropos*, became extinct and recolonized them from their tropical ranges. Two, *Smerinthus caecus* and *Laothoe amurensis*, came from the very important Siberian refugium. Several species evolved distinct subspecies during their period of isolation, such as *Hemaris tityus aksana* in North Africa and *Deilephila porcellus suellus* in Iran. Whereas the former has remained isolated by retreating into the isolated Atlas Mountains, the latter has spread westwards to meet *D. p. porcellus*, which itself was spreading eastward. The two subspecies are now intergrading in the Levant and eastern Anatolia. However, it should be remembered that, on occasions, the terms 'refuge population' and 'relict population' become synonymous. But what are the origins of most of the present western Palaearctic sphingid fauna?

Compared with the number of species, relatively many genera of hawkmoths occur in the Palaearctic due to the fact that numerous Oriental taxa extend their range northwards in the eastern part, unhindered by natural barriers. Farther west the North African, Central Asian and Middle Eastern deserts are very effective barriers, leading to only four migrant Afrotropical species reaching Europe in any numbers – *Agrius convolvuli*, *Acherontia atropos*, *Daphnis nerii* and *Hippotion celerio*. Only two Oriental species have penetrated from east to west, namely *Laothoe amurensis* and *Theretra alecto*. However, this has not always been the case, for few species appear to have evolved in the western Palaearctic; most have arrived as offshoots from cold-tolerant tropical species, or when a species has expanded its range from other areas of similar climatic conditions.

When considering the origins of a region's hawkmoth population one must take into consideration not only the evolutionary factors involved but also the fact that the Sphingidae are an old group of Lepidoptera, having been in existence for about 60 million years. A very small fossil record has led to much speculation as to the origins of the present western Palaearctic Sphingidae. Most conclusions are based on their current biogeographical distributions. Circumstantial evidence supplied by palaeobotany, especially pollen records, can also indicate possible distributions but such methods are prone to erroneous conclusions. However, reasoned guesses can be made as to the origin of some species.

Acherontia, particularly the common ancestor of *A. atropos* and *A. styx*, arose in the Old World tropics and spread right across them. These two species probably separated when shallow seas and deserts cut the ancestral range in two across the Middle East. It is only in recent times (geologically speaking) that a habitable strip has developed across north-east Africa and the northern Middle East, allowing *A. atropos* to extend its range into southern Europe, Mesopotamia and Iran, where it now overlaps with *A. styx*. At the same time, the latter has expanded its range right across Saudi Arabia in the opposite direction (Wiltshire, 1986). A similar sequence of events may also apply to *Daphnis nerii* and its Oriental counterpart *D. placida* Walker, 1856, except that *D. nerii* has been able to push its distribution much farther east to Hawaii (Beardsley, 1979). *Agrius convolvuli* also appears to have arisen in the Old World tropics. It is probable that this species colonized the tropical Americas where it gave rise to the very closely related *A. cingulatus*. Interestingly, *A. cingulatus* has recently crossed over the Atlantic Ocean in the other direction to colonize the Cape Verde Islands (Bauer & Traub, 1980), probably from Brazil.

The cold-tolerant genus *Sphinx* arose in North America. During repeated interglacial periods some species managed to cross from the Nearctic to the eastern Palaearctic region and radiate adaptively in Asia, giving rise to the two adaptive Asian species, *S. ligustri* and *S. pinastri*, which then spread across Asia to western Europe.

Laothoe is an offshoot of the Oriental branch of the Smerinthini which first spread northwards into the eastern Palaearctic region. The same applies to *Marumba*, *Mimas* and *Smerinthus*. The last has managed to invade North America – an expansion of range which the other three genera have not achieved.

Hyles appears to be an old genus of indeterminate origin, occurring all around the world. This is one taxon which has adopted the western Palaearctic region as its main home, where it has evolved into numerous species, and is still radiating.

Theretra, a relative of *Deilephila*, is of Oriental origin and has spread to Europe and Africa in recent times via the Middle East.

Population dynamics

The distribution of hawkmoth species over the western Palaearctic is neither temporally nor spatially evenly distributed. Each species occupies a particular ecological niche comprising an array of environmental factors with which it interacts and which are utilized by no other species in that combination (see p. 51). These environmental factors are an amalgam of temperature, rainfall, seasons, availability of habitat, presence of hostplants, and numbers of predators, amongst others. Combined with the genetic parameters of a species, these have a marked effect on its abundance and distribution.

Although the climate of the western Palaearctic has four distinct seasons, these do vary from year to year and a given species may experience marked fluctuations in both emergence times and abundance. Many species even stagger emergence to prevent short spells of inclement weather from destroying whole populations. A sphingid species will emerge when:

  (a) it is warm enough for adult activity;
  (b) there are sufficient nectar flowers for the adults;
  (c) there is sufficient good-quality food for larval development;
  (d) there is enough time to complete development;
  (e) competition is at a minimum;
  (f) predation and parasitism are at a minimum.

Thus, in northern Europe most resident species such as *Mimas tiliae* (Text fig. 16) emerge in May or June. Farther south many of these fly as early as April and, with sufficient warmth, are able to complete several generations in a season. In southern Austria, adults of *Deilephila porcellus* are on the wing in April and August; in Britain, in June only.

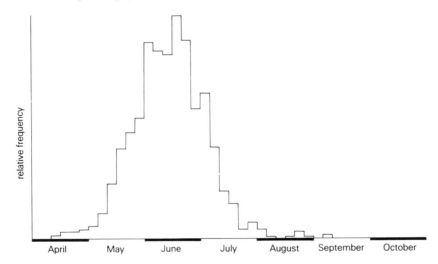

Text Figure 16 Typical seasonal emergence in the Netherlands of *Mimas tiliae* (Linnaeus)
(after Meerman, 1987)

Although the population dynamics of a given sphingid can depend on such factors as predation and parasitism rates, migration, and success in mating and oviposition, by far the most important factor is the weather.

Spatial variation in habitat quality is probably the main reason why regional populations seldom die out. Within a given area there tend to be certain optimum localities where conditions are favourable every year but colonies produce varying numbers of adults from year to year. Other areas may contain poor localities which, in bad seasons, will fail to produce any adults and are recolonized from adjacent colonies only in subsequent good years.

A warm spring may encourage a species to emerge early. On 6th May 1986, females of *Hemaris tityus* were seen ovipositing on *Knautia arvensis* in the Vienna Woods of north-east

Austria after a warm April; a cool spring in 1987 delayed this activity by 20 days (pers. obs.). Such delays may produce only a single generation in a normally double-brooded species in critical areas whereas hot weather may encourage the development of an extra brood. Poor weather during an entire season may even prevent larvae from completing their development. During 1986, in London, southern England, most of the eggs of *Laothoe populi* were deposited on *Populus* spp. (poplar) during early July, but a cool summer coupled with early leaf-fall resulted in significant numbers of immature larvae meeting an untimely end (pers. obs.). Conversely, the warm summer of 1989 allowed this species to complete two generations successfully. Whilst a species can usually recover during later, more favourable seasons, a succession of poor summers or even a slight but gradual deterioration of climate can locally exterminate a species occurring at the limit of its range. This appears to have happened to *Hyles euphorbiae* in the Netherlands after 1950. During the same period *L. populi*, a species which is not so dependent on warm, sunny summers, increased in numbers, probably due to the widespread planting of *Populus* species as amenity trees (Text fig. 17a). A similar study of *Mimas tiliae* shows a steady decline during the present century (Text fig. 17b) (Meerman, 1987).

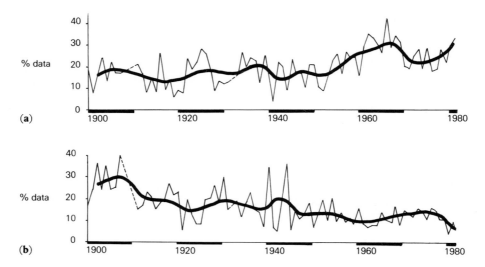

Text Figure 17 80-year population fluctuations in the Netherlands for (a) *Laothoe p. populi* (Linnaeus); (b) *Mimas tiliae* (Linnaeus). The heavy line represents 5-year averages (after Meerman, 1987)

The number of generations a hawkmoth species can produce is directly related to temperature. Usually, the warmer the climate the faster the development. *Smerinthus ocellatus* can produce up to three generations in North Africa but only one in Britain. There is, however, another influencing factor – rainfall. Rarely a problem in northern Europe, a lack of rain in sunnier regions may shrivel hostplants and terminate all breeding activity. In North Africa, *Hyles livornica* first emerges in the Sahara during October and breeds until May. Thereafter this species either aestivates or migrates north to take advantage of the European summer.

In some sphingid species, parasitoids are equally as important as the weather in determining seasonal numbers. One colony of *Smerinthus ocellatus* in London produced 25 healthy and two parasitized larvae in 1985 whereas, in 1986, only two larvae completed their development out of a total of six. In some years upward of 80 per cent of all the larvae of this species in London may be parasitized (Pittaway, pers. obs). This level of parasitism may have brought about a pattern of later emergence, enabling some sphingids to miss many potential larval parasites. *Sphinx ligustri* usually emerges in southern England during mid-June, even though suitable larval hostplants are available from the end of April. However, larvae of this species are prone to parasitism by ichneumonid wasps, which overwinter as adults. By emerging so late it is possible that many potential parasitoids may have already found other hosts and died before suitable larvae of *S. ligustri* are available.

A combination of good weather and few parasitoids can, in some years, result in a population explosion, which may lead to mass migrations and the colonization of new areas. Annual migrations can have an important influence on the numbers of some species recorded in certain areas. Many of the *Sphinx ligustri* caught in south-east England each year are migrants from continental Europe. All adults of *Agrius convolvuli* caught in the Netherlands and the British Isles are either migrants or the first-generation offspring of migrants (Text fig. 18).

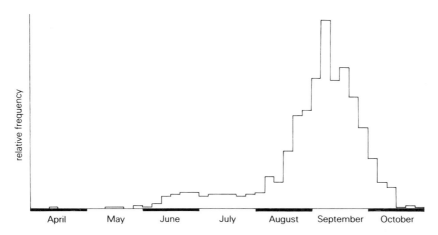

Text Figure 18 Typical seasonal appearance in the Netherlands of *Agrius c. convolvuli* (Linnaeus) (after Meerman, 1987)

Thus a species may undergo seasonal and/or periodical cycles of abundance and scarcity, and/ or show a steady local, area or even regional decline/increase in numbers in response to changing environmental conditions and migration levels. Density-dependent factors (predation, parasitism, disease) and density-independent factors (climate, weather, fires) together help regulate hawkmoth populations, preventing them from either exploding in numbers to plague proportions, or from totally dying out.

Migration

Many sphingids undertake 'migrations', but this term has come to mean many things. Is the single moth which leaves a colony on a long-distance flight and is discovered far outside its range, a migrant? Such individuals are best termed 'vagrants' – opportunistic travellers seeking to found new colonies or join old established ones. Indeed, much of the gene flow between isolated populations of the same species occurs in this manner (see p. 60).

True insect migration is best defined as the non-directional movement of a species, usually massed, from one area to another. However, unlike birds, few Lepidoptera make a return journey to their point of origin – their lifespan is too short. A species builds up numbers in an area and departs in response to certain, as yet poorly understood, environmental conditions. Thousands of individuals of both sexes can be involved. Most are sexually immature and develop only on reaching their destination. How this destination is determined is unknown, but swarms of migrating moths often pass apparently suitable breeding areas and may eventually alight at a very unfavourable site. For instance, every year small numbers of *Agrius convolvuli* arrive in Britain, but are unable to breed successfully as they cannot survive the winter here (the author was handed two full-grown larvae during September 1991, which produced moths during October). Such individuals probably originate in West Africa, from where mass migrations to the north have been well studied (Gatter & Gatter, 1990).

How and why migration began is open to conjecture. Species which evolved in hot, periodically dry semi-desert and steppe conditions may have developed migration systems either to take advantage of hostplant growth produced by localized rainstorms or avoid decimation if rains failed in that area in a given year. In this case, the populations become nomadic, shifting location every few years. *Hyles livornica* has adopted this system, as have many desert butterflies in North Africa and the Middle East (Pittaway, 1981).

A species may also be in an expansive phase, continually pushing at the parameters of its range in order to colonize new territory. This may explain the frequent presence of small numbers of typical Oriental species in Palaearctic areas of Afghanistan (Ebert, 1969). This has distinct evolutionary advantages. The last ice age rendered much of the Palaearctic region uninhabitable, forcing many species into Far Eastern and European refuges, as in the case of *Sphinx ligustri* and *Deilephila elpenor*. As the post-glaciation climate ameliorated, both species rapidly expanded their range once again.

With migration, a subtropical or tropical species can take advantage of unexploited niches in cooler temperate regions during the more favourable summer months. *Macroglossum stellatarum* appears to practice this in Europe and many individuals, the descendants of earlier migrants to the north, have been observed travelling south in autumn. Indeed, in winter, individuals may even migrate south from the Mediterranean and turn up in tropical West Africa (Rougeot, 1972). Whatever the reasons, many western Palaearctic sphingids migrate well beyond their normal ranges. Some, such as *Agrius convolvuli*, *Acherontia atropos* and *M. stellatarum*, have even occurred as far north as Iceland where no hawkmoths are endemic (Wolff, 1971).

Vagrants, whether self-propelled, wind-assisted or man-assisted (by boat, plane or accidental release) have been found throughout Europe. *Hyles hippophaes* has occurred at least once in Britain (Barrett, 1893), north-west Spain (Gómez Bustillo & Fernández-Rubio, 1976), the Crimea and Hungary. Some exotics, such as the American *Agrius cingulatus* and the Asian *Daphnis hypothous*, probably visit the western Palaearctic regularly, only to be confused with their resident counterparts.

The following exotic species have been captured within the western Palaearctic region, although some have almost certainly originated from breeders and cannot be considered as genuine vagrants:

SPECIES	ORIGIN
★Agrius cingulatus (Fabricius, 1775)	North America
Manduca sexta (Linnaeus, 1763)	North America
Manduca quinquemaculata (Haworth, 1803)	North America
Ceratomia undulosa (Walker, 1856)	North America
Sphinx drupiferarum Smith, 1797	North America
Polyptychus trisecta (Aurivillius, 1901)	West Africa
Ambulyx lahora (Butler, 1875)	North-West India, Pakistan
Pachysphinx modesta (Harris, 1839)	North America
Enyo lugubris (Linnaeus, 1771)	North America
Eumorpha anchemola (Cramer, [1780])	South America
★Daphnis hypothous (Cramer, 1780)	South Asia
Nephele didyma (Fabricius, 1775)	South-East Asia
Rhopalopsyche nycteris nycteris (Kollar, 1848)	North India, Pakistan
★Hippotion osiris (Dalman, 1832)	Africa
Hippotion eson (Cramer, 1779)	Africa
★Theretra boisduvalii (Bugnion, 1839)	South Asia
Theretra japonica (Boisduval, 1869)	East Asia

Several of these species, indicated by an asterisk (★), are dealt with in the general text. One very rare vagrant from West Africa, *Polyptychus trisecta*, is illustrated (Pl. F, fig. 6).

Types of biomes

Like all organisms, the Sphingidae are highly specialized, with each species adapted to live in a certain type of habitat or ecosystem (Plates A and B, pp. 67, 68). The suitability of each ecosystem will depend on the presence of suitable hostplants and such critical features as altitude, rainfall, sunshine and temperature. A number of similar ecosystems may share the same area and form a community. A collection of these is termed a 'biome', characterized by a particular 'climax community' that remains relatively stable in the typical climatic conditions of a given region. Worldwide there are many types of biome but in the western Palaearctic one finds only Tundra, Boreal Forest, Deciduous Temperate Forest, Mediterranean, Steppe and Desert (Text fig. 19). It should be remembered, however, that biomes are not uniform, but comprise a complex of communities within a given range of conditions; they are the result of an equilibrium established between the organisms of a region and its climate. Local differences in altitude and precipitation may also produce pockets of one biome within another. Additionally, certain ecosystems, for example riverine systems, may be found in several biomes. These often allow hawkmoths to penetrate biomes where they would not normally be found, such as the deserts of central Asia and North Africa.

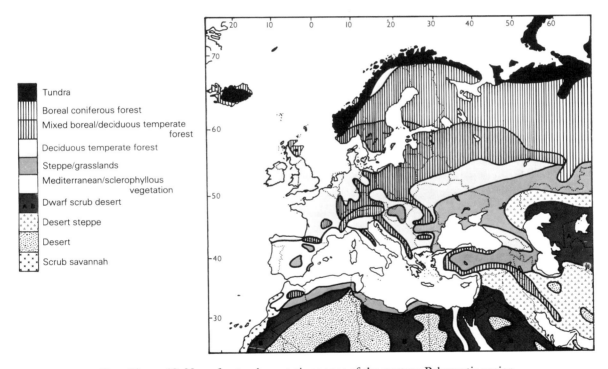

Legend:
- Tundra
- Boreal coniferous forest
- Mixed boreal/deciduous temperate forest
- Deciduous temperate forest
- Steppe/grasslands
- Mediterranean/sclerophyllous vegetation
- Dwarf scrub desert
- Desert steppe
- Desert
- Scrub savannah

Text Figure 19 Map of natural vegetation zones of the western Palaearctic region

(a) *Tundra* (Arctic)

This region lies between the northern tree-line and permanent ice, where the average summer temperature never exceeds 10°C. For two months there is continuous daylight with little or no frost, although the subsoil remains permanently frozen (permafrost). In winter everything freezes for months on end, there is permanent snow-cover and darkness for two months. Annual rainfall averages 250mm. Winds can be very strong.

The vegetation of this region is very distinctive, being composed of dwarf evergreen members of the Ericaceae (heaths) and of *Carex* (sedges), *Juncus* (rushes), mosses and lichens. No western Palaearctic Sphingidae are adapted to this zone although some, like *Agrius convolvuli*, *Hyles euphorbiae*, *H. gallii* and *H. livornica*, migrate into it.

(b) *Boreal forest* (Taiga)

This region forms an almost continuous band, some 1500km wide, right across the northern hemisphere (the Holarctic). To the north it merges with the tundra, to the south with mixed temperate forests. Its climate is profoundly influenced by cold polar air masses; only at its western and eastern extremities do moisture-laden sea winds have any ameliorating effect. Winters are cold to very cold, with average temperatures less than −7°C. Snow-cover can last up to six months during the short days of winter. In contrast, the short summers are relatively warm, with average temperatures reaching 15°C during the long days. Rainfall rarely exceeds 500mm. Thus the growing period for plants varies between 100 and 160 days.

The vegetation is dominated by conifers (*Picea*, *Abies* and *Pinus*, with some *Larix*). Often a single species of tree covers huge areas, *Picea abies* (Norway spruce) being the most important. In drier areas it cedes its dominance to *Pinus sylvestris* (Scots pine). In clearings and along rivers a number of broad-leaved trees occur, notably *Betula pubescens* (downy birch), *Sorbus aucuparia* (mountain ash), *Alnus incana* (grey alder) and *Populus tremula* (aspen). Beneath all of these there is a sparse shrub layer of various *Salix* species (sallows) and members of the Ericaceae (heaths).

In the western Palaearctic two species, *Laothoe amurensis* and *Smerinthus caecus*, are adapted specifically to this zone, penetrating south only a short way into the mixed temperate forests. In the same way, a number of deciduous temperate forest species (*Sphinx ligustri*, *S. pinastri*, *Smerinthus ocellatus*, *Laothoe populi*, *Hemaris fuciformis*, *H. tityus*, *Hyles gallii* and *Deilephila elpenor*) occur in the mixed forest zone and may even reach the boundaries of the true Taiga forests.

(c) *Deciduous temperate forest*

Farther south, especially where humid onshore sea-winds ameliorate the more severe continental climates, monotonous Boreal forest gradually gives way to species-rich deciduous forest. In the Palaearctic, this biome is unfortunately rather limited and irregular due to the effects of several Pleistocene glaciations and human deforestation. It reaches its greatest development in central Europe, the Caucasus/Alborz, and eastern Asia (Korea, Japan and northern China, with a relict area in the western Himalaya and Tian Shan (Shchetkin, 1956). In Europe two distinct sub-biomes are recognized – the Atlantic climatic region and the central European climatic zone. None of the Sphingidae is specifically adapted to the former; moreover, many continental species, such as *Proserpinus proserpina*, *Hyles euphorbiae* and *H. gallii*, are unable to tolerate the cool and wet conditions found there. It is also one of the cloudiest regions of the world with, for instance, only 1700 hours sunshine per annum in London. The average temperature difference between winter and summer is only 8–14°C: the July and January (in brackets) averages for London and Santiago de Compostella, northern Spain, are 16°C (3·8°C) and 18°C (7·3°C), respectively. Rainfall occurs throughout the year and ranges from 580mm in Paris to 1651mm in Santiago de Compostella.

Farther east, the high-pressure system of eastern Europe dominates the weather. It is less variable, less humid and more extreme. January mean temperatures vary between −3°C (Warsaw) and −1·7°C (Belgrade). Snow-cover may last up to three months, with continuous sub-zero temperatures. By contrast, the average mid-summer temperatures can be relatively hot – Warsaw 18·5°C and Belgrade 22·2°C. Thus the average temperature-difference between winter and summer usually exceeds 19°C. Rainfall rarely exceeds 635mm, but increases to 3000mm on some mountains. There is no distinct rainy season, rainfall being spread evenly throughout the year.

As the description suggests, the vegetation is dominated by deciduous broad-leaved trees, although certain *Pinus* species can form extensive mixed forests in drier areas. The woodlands are much more complex than those of the Taiga, with well-developed strata of herbaceous plants, shrubs, small trees and tall trees. The diversity of species is also much higher, resulting in a rich and abundant lepidopterous fauna. The dominant trees are various species of *Quercus* (oak), *Fagus* (beech), *Tilia* (lime), *Acer* (maple), *Ulmus* (elm), *Castanea* (chestnut), *Fraxinus* (ash) and *Carpinus* (hornbeam). Having escaped the repeated glaciations which decimated the European forests, those of the Caucasus and Alborz are also rich in genera formerly found in Europe, such as *Zelkova* (Ulmaceae), *Pterocarya* (Juglandaceae) and *Liquidamber* (Hamamelidaceae).

The greatest number of western Palaearctic sphingid species inhabit this biome. Besides those also found in the Boreal forest biome (p. 44) are *Mimas tiliae* and *Deilephila porcellus*. Some Mediterranean and steppe species (*Kentrochrysalis elegans*, *Marumba quercus*, *Proserpinus proserpina*, *Hyles euphorbiae* and *Hyles vespertilio*) penetrate this region from the south. They tend to be found in the driest, sunniest and warmest locations, in other words, in localities where the microclimate is similar to that of their ancestral homeland. Nearly all colonies of these species are confined to very warm south-facing banks and hillsides at low elevations in central Europe, where the larvae can bask in hot sunshine. Large caterpillars can increase their body temperature up to 10°C or more above that of their surroundings, as noted above (p. 22) and the level of opportunity for basking can make or break the survival of a colony and probably limits the spread of such species as *Hemaris croatica*. Conversely, most temperate-forest species can also occur farther south in favourable situations, such as along river valleys.

(d) *The Mediterranean (sclerophyllous) biome*

This is the most distinctive of those found in the western Palaearctic. The winters are wet and mild with an average temperature of 6°C. Summers are hot and sunny (2500 hours sunlight per annum), with temperatures averaging 21°C. Most of the rain falls as short storms between September and April, and mid-summer is marked by two or three rainless months. The amount of precipitation varies markedly and depends on proximity to the coast, mountain ranges and prevailing winds. Skopje in Macedonia receives 480mm per annum; just across the mountain ridge, on the southern coast of Montenegro, rainfall can exceed 4000mm.

The vegetation is dominated by sclerophyllous trees and shrubs with hard, evergreen leaves. Herbaceous plants which die back during the summer months and a rich diversity of spring and autumn annuals are also characteristic features.

A number of sphingids are resident in this biome, with many penetrating from here northwards into warmer regions of deciduous forest. Those characteristic of the Mediterranean region are *Hemaris dentata*, *H. croatica* and *Hyles nicaea*. Others, such as *Akbesia davidi*, *Smerinthus kindermanni*, *Clarina kotschyi* and *Rethera brandti*, are confined to the hotter and less humid eastern Mediterranean sub-region (Iranian and Anatolian plateaux) and penetrate only a short distance westwards. Although they all occur within the Mediterranean biome, each of these species inhabits a distinct ecosystem.

A few cold-tolerant tropical Sphingidae have also become established along the southern periphery of this region and others migrate into it. These include *Agrius convolvuli*, *Acherontia atropos*, *A. styx*, *Daphnis nerii*, *Sphingonaepiopsis nana*, *Hippotion osiris*, *H. celerio* and *Theretra alecto*. The survival of some of these, such as *A. convolvuli* and *H. celerio*, would be very unlikely were it not for continuous migrations from the tropics, many of which penetrate far to the north.

(e) *Steppe* (Prairie)

This is a temperate zone biome, also referred to as 'Pannonic Steppe' in central Europe or, in South America, as 'Pampas'. Rainfall is low (400–500mm) and insufficient to support trees away from sources of groundwater. Most of the rain falls between May and June (earlier in Anatolia), with July and August experiencing drought conditions. There is a large winter/summer temperature range of more than 19°C, the winters being cold and dry. For example, in Bucharest the average July and January temperatures are, respectively, 22·8°C and −3·9°C (see Text figs 14a,b, p. 34).

The vegetation is dominated by grasses interspersed with various species of Leguminosae and Compositae. Throughout this region, pockets of Mediterranean or temperate deciduous trees may also occur where there is sufficient water. In some areas, sparse *Quercus* (oak) woodland can dominate.

In Europe, Pannonic Steppe is found in central Hungary and southern Romania, and across the southern part of Russia, Moldova and the Ukraine bordering the Black and Caspian Seas. In the Middle East, much of the Anatolian and Iranian plateaux, northern Syria, Lebanon and Iraq have large areas of steppe intermixed with Mediterranean scrub (Pontic Steppe). Similar conditions also exist along the southern foothills of the Atlas Mountains in North Africa.

A number of sphingid species are very characteristic of this biome and often occur as isolated populations. These remained when the main steppe habitat retreated eastward as the climate grew wetter after the last ice-age. Among such are *Rethera komarovi*, *Sphingonaepiopsis gorgoniades*, *Hyles centralasiae*, *H. zygophylli* and *H. hippophaes*. *Hemaris croatica* and *Hyles nicaea* frequent both Pontic Steppe areas and Mediterranean regions.

(f) *Deserts*

Across the whole of the southern part of the western Palaearctic, the Mediterranean and Steppe biomes gradually merge into deserts (Sbordoni & Forestiero, 1985). Rainfall here is erratic, rarely exceeds 200mm per annum, and usually falls in the winter. The summers are rainless and often cloudless, with average July temperatures exceeding 30°C. The winds are strong and the evaporation rates so high that the unstable soils tend to be badly degraded.

Most plants are annuals which, after a downpour, grow rapidly, flower and set seed. Perennial plants tend to be xerophytic and widely spaced with extensive root systems.

The only true desert sphingid is *Hyles livornica*. Its main breeding cycle coincides with the winter and spring flush of annual flowers; its larva is polyphagous and the adult moth highly migratory – all adaptations to a desert life. One Mediterranean species, *H. tithymali*, has managed to penetrate the Saharan deserts where its larva feeds on xerophytic species of *Euphorbia*.

(g) *Mountain biomes*

These are not, strictly speaking, true biomes, but mountain ranges (Text fig. 20) do have a considerable effect on the climate, vegetation and fauna of the region. There tends to be a rapid succession of different species as the zones of vegetation change with altitude. Each of these zones often corresponds with a similar one found with increasing latitude. Thus in the Alps one can find both Boreal and Tundra biomes in what would otherwise be classified a temperate deciduous forest region.

Such a phenomenon has a marked effect on the Sphingidae of this region. *Mimas tiliae* does not occur above 1500m in the Alps, but in Iran it is rarely found below this altitude. Indeed, many species and subspecies in the southern part of the western Palaearctic are confined to mountain chains. It is very unlikely that *Hemaris fuciformis* or *H. tityus* would have survived the post ice-age drying-out and desiccation of North Africa were it not for the Atlas Mountains. Such has been the importance of the Tian Shan, Pamir and Hindu Kush in Central Asia as long-term refugia, that several distinct species and subspecies of Sphingidae have evolved in isolation there, e.g. *Dolbinopsis grisea*, *Laothoe philerema*, *Hemaris rubra*, *H. ducalis*, *Acosmeryx naga hissarica*, *Rethera afghanistana*, *Sphingonaepiopsis kuldjaensis*, *Hyles nervosa*, *H. salangensis* and *H. chamyla*.

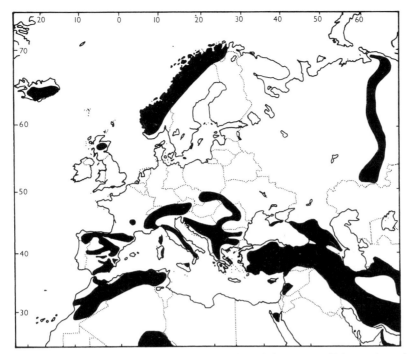

Text Figure 20 Map of principal mountain chains of the western Palaearctic region

Hostplants and sphingid preferences

All sphingids are primary consumers. The adults (with some few non-feeding exceptions) imbibe flower nectar, although some tropical species will drink eye-secretions (Bänziger, 1988) and some North American species have been seen probing decaying animal remains (Sbordoni & Forestiero, 1985). Larvae consume plant material – they are herbivores.

The most conspicuous herbivores in the region are mammals such as cattle, sheep, goats, rabbits and deer; however, by far the most important terrestrial plant-eaters are the insects. They have evolved a remarkably diverse variety of mouthparts to deal with their hostplants, eating leaves from both the outside and the inside, boring through stems and roots and devouring flowers, seeds and fruits. Considering the abundance of insects, their variety and appetites one wonders how plants still manage to flourish on earth. One reason is the production of defence structures, such as spines, thorns and spiky leaves. Remove these and many insects, or their larvae, will devour the resulting defenceless plant even though they would not normally eat it, e.g. larvae of *Sphinx ligustri* readily accept leaves of *Ilex aquifolium* (holly) once the spines have been removed. Another defence is nutritional deficiency. However, all these mechanisms are secondary to the plant world's main line of defence (Pittaway, 1983a).

(a) Chemistry of plants

Plants produce a wide range of secondary chemical compounds in vulnerable parts, which appear to perform no physiological function in the plants themselves but can act as potent repellents, growth retardants or even toxins to insects. Most widely known is pyrethrin, a *Tanacetum* (formerly included in *Chrysanthemum*) extract, which is relatively harmless to mammals but a potent insecticide.

One very large group comprises terpenes, or essential oils, which are simple aromatic carbon-compounds. Most plants contain one or more of these which give them their characteristic smell when crushed, such as menthol in mints.

Alkaloids are equally important. They abound in certain plant-families, such as the Solanaceae, and tend to be more widespread in tropical regions, examples being nicotine, strychnine, caffeine, solanin, quinine, tetrahydrocannabinol and opiates.

Another important group are the acetogenins, such as rotenone and the furanocoumarins. The latter are well represented in the Rutaceae and Umbelliferae families.

The Umbelliferae are also well endowed with certain types of a very important group of secondary chemicals, the glucosides. Depending on their action upon ingestion, these are classified as flavonoid, cardiac, phenolic, anthraquinone and thiocyanate (mustard) glucosides. Anthraquinones occur in a fairly small number of plant-families, for example, the Rhamnaceae, Polygonaceae and Rubiaceae. Cardiac glycosides are more widespread and are often sequestered by insects for their own defence, such as in the butterfly *Danaus plexippus* (Linnaeus). Similar compounds have also been isolated from adults of *Daphnis nerii* (Chang, 1982) and may perform a similar function.

Related to glucosides are saponins which, together with tannins, raphides (crystals of calcium oxalate), organic acids, and bitter compounds, in addition to those already mentioned, make up the chemical cocktails which protect plants from the attentions of insects (Fraenkel, 1959; Baldwin, 1988; Harborne, 1988; Spencer, 1988).

A gravid female moth intent on depositing eggs is likely to be found in an area where suitable hostplants are present because she herself was bred in that locality. She will home in on likely sites using as yet unidentified stimuli – probably a combination of chemical, microclimatic and visual. It is her responsibility to select the correct hostplant, that selection being dependent on what the larva can eat. At close range, individual hostplants are located by their chemical 'signature' – a characteristic mixture of volatile hydrocarbons and other secondary chemicals given off by that plant's metabolism, which are often the very same as those intended to deter insect attack. Before ovipositing, the moth may check a hostplant's identity by means of tarsal receptors, as some completely unrelated plants appear to have very similar signatures. If the signature is not correct, no eggs are deposited and that plant will escape the attentions of that particular insect.

Few if any herbivores (including lepidopterous larvae) have evolved the range of morphological and physiological adaptations which allow them to feed on all plants; most specialize to some degree. Caterpillars employ alimentary mixed-function oxidase enzymes to detoxify secondary plant compounds; the more polyphagous a species the more types and quantities of oxidase enzymes it will generally possess. Bernays & Janzen (1988) did show, however, that different taxa of moths can employ very different methods for dealing with secondary plant chemicals. They studied the Saturniidae and Sphingidae of Costa Rica and found that the large-headed saturniid larvae have short, simple mandibles which cut their food into large pieces of which only the edges are digested. The relatively small-headed sphingid larvae have long, ridged and toothed mandibles which tear and crush food into small pieces. The former generally feed on old, tough, tannin-rich leaves with low levels of secondary toxins, the latter on young, tannin-poor leaves containing high levels of secondary toxins to which they are resistant.

This selection implies that the larva has the ability to distinguish between edible and inedible plants. Caterpillars do bear a range of gustatory sensilla on their maxillae which respond to the variety of secondary and other compounds found in plant tissue (Remorov, 1984). If, during feeding, more positive than negative stimuli are received (all plants contain both beneficial and harmful substances in varying proportions), the plant is edible and feeding either commences or continues. As it is in the insect's interest to have positive stimuli supplied by secondary compounds it is capable of detoxifying, they become feeding- or even oviposition-stimulants for that species. However, if the sensilla are removed, many species can feed on what appear to be unsuitable hostplants (Remorov, 1984).

To protect themselves from the most important terrestrial herbivores, the insects, plants therefore produce a range of chemical compounds which vary, both in proportion and type, between species, genera and families. In order to feed, insects have found it necessary to evolve mechanisms to neutralize these compounds, but have been unable to breach the chemical defences of every plant group and so have become specialized, relying on stimuli produced by once-toxic compounds to identify suitable hostplants. However, the full role of plant secondary compounds is uncertain as conflicting results have been obtained from laboratory studies (Reavey & Gaston, 1991).

This situation is not static, however, but in a constant state of evolutionary flux for, to avoid being fed upon, a mutant plant which can produce a new deterrent will prosper at the expense of its fellows which cannot. But, as the resistant plant population builds up, a deterrent-factor pressure is put on the original insect population to adapt to this deterrent. Some will be able to respond; some may change to hostplants containing deterrents similar to the original hostplant. Thus, for those insects which 'crack' a new deterrent-code, competition will be reduced and often leads to speciation; this appears to have happened with *Hyles chamyla* in Central Asia, a species which almost certainly split off from *H. hippophaes* quite recently.

Similarly, were a polyphagous species to shift its range to a new area devoid of its original hostplants it might, through a drift in genotype, lose the ability to feed on those plants. Conversely, if there were no selection pressure against that factor, it might persist and express itself if the moth and its ancestral hostplant were reunited. This appears to have occurred in the European populations of *Sphinx ligustri*, which are thought to have originated from an ancestor in western North America (Hodges, 1971) and spread to Asia and then Europe. The larvae of this species normally feed on *Fraxinus*, *Ligustrum*, *Syringa* and *Spiraea* and, occasionally, *Lonicera*. Some of the related North American species prefer *Symphoricarpos*, a genus of the family Caprifoliaceae naturally confined to that continent, especially the western portion. However, with the introduction of *Symphoricarpos alba* into Europe, larvae of *Sphinx ligustri* are to be found on this plant in ever-increasing numbers (Meerman, 1987), as well as on the introduced *Forsythia* (from China) and *Physocarpus opulifolius* (from North America).

Most work on lepidopterous hostplants has been concerned with butterflies, so, in order to understand the present day pabulum of the European hawkmoths, it is worthwhile looking at the *Papilio machaon* complex of swallowtail butterflies. *P. machaon* Linnaeus is the world's most widely distributed swallowtail, ranging across Europe, North Africa, northern Asia and northern North America. In central and southern North America it is partly replaced by closely related species, such as *P. indra* Reakirt, *P. polyxenes* Fabricius, *P. brevicauda* Saunders, *P. zelicaon* Lucas and *P. joanae* Heitzman. In Europe and North Africa there are three other related species, namely *P. hospiton* Guenée, *P. alexanor* Esper and *P. saharae* Oberthür. All can feed in the larval stage on the taxonomically unrelated but chemically similar Rutaceae and Umbelliferae (Thompson, 1988). Rutaceae contain furanocoumarins, small amounts of tannin, a bitter compound, the glucoside rutin, and the monoterpenes methyl chavicol, anethole and anisic aldehyde. Although rutin may be absent, all the above are present also in the Umbelliferae. Differences in the selection of wild hostplants appear to be due mainly to adult behaviour for, although *P. machaon* and *P. polyxenes* will not normally oviposit on Compositae containing the same or similar secondary compounds, their larvae can be reared on *Tagetes* and *Cosmos*. However, two members of the complex, *P. machaon bairdii* Edwards and *P. machaon oregonius* Edwards, have overcome the as yet unidentified oviposition deterrent and adopted *Artemisia dracunculus* as their natural hostplant. Additionally, certain populations of *P. machaon* in Afghanistan and Alaska have also transferred to the genus *Artemisia*.

Hawkmoths show similar preferences in their selection of hostplants. The chemical relationships are as important as the evolutionary relationships between plant species. Indeed, the ability of most sphingid subfamilies and tribes to detoxify only certain secondary compounds is so characteristic that one could almost classify them according to these abilities (Pittaway, 1983a).

The orders Myrtales, Euphorbiales, Rhamnales and Rubiales must contain similar secondary compounds to enable the Macroglossinae to feed widely amongst them, or to specialize on one particular family in an order (see list of principal hostplant families below).

PROTEALES	EUPHORBIALES	RUBIALES
Elaeagnaceae	Euphorbiaceae	Rubiaceae
MYRTALES	RHAMNALES	DIPSACALES
Lythraceae	Vitaceae	Caprifoliaceae
Thymelaeaceae	SAPINDALES	Valerianaceae
Onagraceae	Zygophyllaceae	Dipsacaceae

Many botanists include the Elaeagnaceae in the order Myrtales, a view supported by its adoption by *Hyles hippophaes* for its hostplants.

(b) *Nutritional factors*

Why do several female moths oviposit on the same plant when apparently equally satisfactory, if not better, plants of the same species are all around? Why are some species of hostplant preferred to others? Why do larvae of *Marumba quercus* reared on *Quercus cerris* thrive better than those on *Q. robur* (pers. obs.)? Why do many of the Macroglossini prefer the flowers of their hostplants rather than the leaves? That the biochemical composition of leaves varies, both between and within species, is not in doubt (Reavey & Gaston, 1991). However, although considerable research has been undertaken regarding the chemical interactions between plants and insects (see p. 47), very little is known of the nutritional quality of known hostplants.

Recently, various American researchers have produced some interesting results from studies of the survival rate of the larvae of *Papilio (Pterourus) glaucus* (Linnaeus) (tiger swallowtail) on a variety of hostplants under conditions of an average day-length of sixteen hours and a temperature range of 19·5–23·5°C. Not only were differences recorded in the results obtained with different hostplants (Table 1), but further research by Scriber & Lederhouse (1983) demonstrated differences within that one species at different times of the year (Table 2). *P. glaucus* is usually bivoltine, or univoltine with a bimodal emergence.

Research into the hostplant preferences of the various western Palaearctic sphingids would

Table 1 Survival and development potential of larvae of *Papilio (Pterourus) glaucus* (Linnaeus), studied at Wisconsin, U.S.A. in 1979 (after Scriber, Lintereur & Evans, 1982)

Hostplant	% survival rate (to 3rd instar)	No. days as larva	Pupal wt. (mg dry)
Magnolia acuminata	67	35·4	249·0
Liriodendron tulipifera	63	30·6	301·1
Sassafras albidum	46	35·6	213·2
Lindera benzoin	0	–	–
Betula papyrifera	57	41·8	315·6
Populus tremuloides	0	–	–
Sorbus americana	80	41·5	272·9
Prunus virginiana	75	40·2	248·7
Prunus serotina	57	32·4	335·4
Ptelea trifoliata	0	–	–
Fraxinus americana	50	–	–

Table 2 Developmental duration of larvae of *Papilio (Pterourus) glaucus* (Linnaeus) reared in 1973–74 on two hostplants under field conditions at Ithaca, New York, U.S.A. (after Scriber & Lederhouse, 1983)

Hostplant	Number of days as a larva	
	June–July	Aug.–Sept.
Prunus serotina	36·3 (7810·8†)	47·0 (8030·4†)
Fraxinus americana	44·5 (9345·1†)	55·0 (7973·3†)

† actual accumulated thermal units

SPHINGID ECOLOGY

undoubtedly produce equally interesting results and to a limited extent has already done so for *Laothoe populi* (Table 3). It was found that larvae reared on *Salix fragilis* tended to pass through only four instars, whereas those on *Populus alba* underwent four or five. However, although more larvae survived on *S. fragilis* and fed up more rapidly, the adults from larvae reared on *P. alba* were larger and more fecund.

Table 3 Survival and development of larvae of *Laothoe p. populi* (Linnaeus) on two hostplants, given as the mean of three trials (after Grayson & Edmunds, 1989a)

| Hostplant | % larvae surviving to | | | No. of days as larva |
	7 days	21 days	pupation	
Salix fragilis	76·0	53·3	44·0	30·2
Populus alba	50·0	23·0	14·6	45·3

Ecological influences on speciation

One of the basic factors governing the interactions of species which make up an ecosystem is competition; two species cannot permanently occupy the same ecological niche. Thus competition leads either to the departure or extinction of one of the competitors or, more commonly, a shift in the genetic make-up of one (or both) species so that neither has identical ecological requirements.

The genus *Smerinthus* illustrates this well. All species utilize *Salix* as their main hostplant. As this plant genus is distributed across the entire western Palaearctic, this is the potential regional range of all local species of *Smerinthus*, a genus of Holarctic origin. However, in order to exploit separate niches and avoid competition, each has adapted to a certain range of climatic conditions: *S. caecus* requires short, cool summers and very cold, dry winters, hence its confinement to the northern boreal forests; *S. kindermanni* requires long, hot, dry summers and short, cold, dry winters, thus its south-easterly distribution; *S. ocellatus* can exploit intermediate ranges including cool, moist winters. There are areas where *S. ocellatus* does overlap in range with the other two species, but closer investigation reveals that local environmental conditions separate them. In Asia Minor *S. kindermanni* usually occupies hot valley floors, and *S. ocellatus* the cooler mountains (pers. obs.).

The same type of adaptation to different ecological conditions may be the main factor in bringing about the evolutionary separation in North Africa of *Hyles euphorbiae* from *H. tithymali* both of which feed on herbaceous *Euphorbia*. Gene flow decreases, following genetic linkage between ecologically important traits in sex chromosomes, resulting in species and sub-species being maintained primarily by ecological selection and not genetic barriers.

Parasites and predators

As moths have been on the earth for more than 190 million years, numerous other organisms have evolved to exploit this rich food source.

Very few adult moths are subject to lethal parasitism, probably due to their limited lifespan and inability to grow – they do not accumulate tissue-mass or food-reserves. Eggs and caterpillars, however, are subjected to severe attacks by parasitoids (see Table 4, pp. 52–55) in addition to the numerous fungi, bacteria, viruses and nematodes (Siebold, 1842; Hagmeier, 1912; Fiedler, 1927; Mathur, 1959) which take their toll. Although nematodes have been found in several species, most notably *Hexamermis albicans* in *Mimas tiliae* (Hagmeier, 1912), little is known of the exact effects they have on the host. Parasitoids have been studied in greater detail (Askew, 1971; Shaw & Askew, 1976).

Tiny wasps (Trichogrammatidae) may lay their eggs through the micropyle of a hawkmoth egg and develop within. In Iran, many ova of *Rethera komarovi manifica* are lost in this way (pers. obs.). Indeed the Macroglossini seem particularly prone to this type of parasitism. As the

51

Table 4
Parasitoids recorded* from Western Palaearctic sphingid hosts

Parasitoid spp. — HYMENOPTERA: Ichneumonidae

Sphingidae	Alcima orbitalis (Gravenhorst)	Ambyloppa fuscipennis (Wesmael)	Ambyloppa proteus (Christ)	Apechthis compunctor (Linnaeus)	Apechthis rufata (Gmelin)	Aphanistes armatus (Wesmael)	Aphanistes bellicosus (Wesmael)	Aphanistes megasoma Heinrich	Aphanistes ruficornis (Gravenhorst)	Artiranis fugitivus (Gravenhorst)	Banchus falcatorius (Fabricius)	Banchus palpalis Ruthe	Banchus pictus Fabricius	Barylypa insidiator (Förster)	Callajoppa cirrogaster (Schrank)	Callajoppa exaltatoria (Panzer)	Catadelphus arrogator (Fabricius)	Coelichneumon deliratorius (Linnaeus)	Coelichneumon sinister (Wesmael)	Cratichneumon sicarius (Gravenhorst)	Diphyus longigena (Thomson)	Diphyus monitorius (Panzer)	Diphyus palliatorius (Gravenhorst)	Enicospilus undulatus (Gravenhorst)	Goedartia alboguttata (Gravenhorst)	Habronyx heros (Wesmael)	Heteropelma amictum (Fabricius)	Heteropelma calcator (Wesmael)	Heteropelma capitatum (Desvignes)	Hyposoter didymator (Thunberg)	Ichneumon cerinthius Gravenhorst	Ichneumon insidiosus Wesmael	Ischnus migrator (Fabricius)	Lymantrichneumon disparis (Poda)	Mesochorus disciergus (Say)	Metopius dentatus (Fabricius)	Microleptes splendidulus Gravenhorst	Netelia testacea (Gravenhorst)	Netelia vinulae (Scopoli)	Pimpla hypochondriaca (Retzius)	Pimpla illecebrator (Villers)	Pimpla turionellae (Linnaeus)	Protichneumon fusorius (Linnaeus)
Sphinginae																																											
Agrius cingulatus																																											
Agrius convolvuli convolvuli		●														●																									●		
Acherontia atropos	●	●													●	●		●									●																
Acherontia styx styx																																											
Sphinx ligustri ligustri		●		●	●										●	●									●			●														●	●
Sphinx pinastri pinastri			●				●	●	●	●								●	●	●			●				●	●								●						●	●
Sphinx pinastri maurorum																																											
Dolbinopsis grisea																																											
Kentrochrysalis elegans																																											
Akbesia davidi																																											
Marumba quercus																																											
Mimas tiliae														●																					●						●	●	
Smerinthus kindermanni kindermanni																																											
Smerinthus kindermanni orbatus																																											
Smerinthus caecus																																											
Smerinthus ocellatus ocellatus											●			●	●				●	●			●												●					●	●	●	●
Smerinthus ocellatus atlanticus																																											
Laothoe populi populi		●							●	●				●				●				●									●									●	●	●	●
Laothoe populi populeti																																											
Laothoe austauti																																											
Laothoe philerema																																											
Laothoe amurensis																																									●	●	
Macroglossinae																																											
Hemaris rubra																																											
Hemaris dentata																																											
Hemaris ducalis																																											
Hemaris fuciformis fuciformis		●																																									
Hemaris fuciformis jordani																																											
Hemaris croatica																																											
Hemaris tityus tityus																																											
Hemaris tityus aksana																																											
Daphnis nerii																																											
Daphnis hypothous hypothous																																											
Clarina kotschyi kotschyi																																											
Clarina kotschyi syriaca																																											
Acosmeryx naga hissarica																																											
Rethera afghanistana																																											
Rethera amseli																																											
Rethera komarovi komarovi																																											
Rethera komarovi drilon																																											
Rethera komarovi rjabovi																																											
Rethera komarovi manifica																																											
Rethera brandti brandti																																											
Rethera brandti euteles																																											
Sphingonaepiopsis gorgoniades gorgoniades																																											
Sphingonaepiopsis gorgoniades pfeifferi																																											
Sphingonaepiopsis gorgoniades chloroptera																																											
Sphingonaepiopsis kuldjaensis																																											
Sphingonaepiopsis nana																																											
Proserpinus proserpina proserpina																●																				●							
Proserpinus proserpina gigas																																											
Proserpinus proserpina japetus																																											
Macroglossum stellatarum		●																																●									

(* † see notes on page 54)

Column headers (read vertically, left to right):

- Protichneumon pisorius (Linnaeus)
- Pyramidophorus flavoguttatus Tischbein
- Scolobates auriculatus (Fabricius)
- Therion circumflexum (Linnaeus)

Braconidae

- Aleiodes praetor (Reinhard)
- Cotesia abjectus (Marshall)
- Cotesia coryphe (Nixon)
- Cotesia glomeratus (Linnaeus)
- Cotesia nigriventris (Nees)
- Cotesia saltator (Thunberg)
- Microplitis ocellatae Bouché
- Microplitis vidua (Ruthe)

Pteromalidae

- Erdoesina alboannulata (Ratzeburg)

Eupelmidae

- Anastatus bifasciatus (Geoffroy)

Encyrtidae

- †Ooencyrtus pityocampae (Mercet)

Eulophidae

- Eulophus smerinthicida Bouček

Trichogrammatidae

- Trichogramma evanescens Westwood

Scelionidae

- Telenomus punctatissimus (Ratzeburg)

DIPTERA: Tachinidae

- Blondelia nigripes (Fallén)
- Campylochaeta inepta (Meigen)
- Compsilura concinnata (Meigen)
- Drino galii (Brauer & Bergenstamm)
- Drino inconspicua (Meigen)
- Drino lota (Meigen)
- Drino vicina (Zetterstedt)
- Drino (Palexorista) imberbis (Wiedemann)
- Drino (Zygobothria) atropivora (Robineau-Desvoidy)
- Drino (Zygobothria) ciliata (Wulp)
- Exorista fasciata (Fallén)
- Exorista grandis (Zetterstedt)
- Exorista larvarum (Linnaeus)
- Exorista sorbillans (Wiedemann)
- Frontina laeta (Meigen)
- Huebneria affinis (Fallén)
- Masicera pavoniae (Robineau-Desvoidy)
- Masicera sphingivora (Robineau-Desvoidy)
- Nemoraea pellucida (Meigen)
- Nilea hortulana (Meigen)
- Osvaldia spectabilis (Meigen)
- Pales pavida (Meigen)
- Phryxe erythrostoma (Hartig)
- Phryxe nemea (Meigen)
- Phryxe vulgaris (Fallén)
- Tachina grossa (Linnaeus)
- Tachina praeceps (Meigen)
- Thelaira nigripes (Fabricius)
- Winthemia bohemani (Zetterstedt)
- Winthemia cruentata (Rondani)
- Winthemia rufiventris (Macquart)

(continued overleaf)

Table 4 (cont'd) Parasitoids recorded* from Western Palaearctic sphingid hosts

Parasitoid spp. — HYMENOPTERA: Ichneumonidae

Sphingidae	Alcima orbitalis (Gravenhorst)	Amblyjoppa fuscipennis (Wesmael)	Amblyjoppa proteus (Christ)	Apechthis compunctor (Linnaeus)	Apechthis rufata (Gmelin)	Aphanistes armatus (Wesmael)	Aphanistes bellicosus (Wesmael)	Aphanistes megasoma Heinrich	Aphanistes ruficornis (Gravenhorst)	Aritranis fugitivus (Gravenhorst)	Banchus falcatorius (Fabricius)	Banchus palpalis Ruthe	Banchus pictus Fabricius	Barylypa insidiator (Förster)	Callajoppa cirogaster (Schrank)	Callajoppa exaltatoria (Panzer)	Catadelphus arrogator (Fabricius)	Coelichneumon deliratorius (Linnaeus)	Coelichneumon sinister (Wesmael)	Cratichneumon sicarius (Gravenhorst)	Diphyus longigena (Thomson)	Diphyus monitorius (Panzer)	Diphyus palliatorius (Gravenhorst)	Encospilus undulatus (Gravenhorst)	Goedartia alboguttata (Gravenhorst)	Habronyx heros (Wesmael)	Heteropelma amictum (Fabricius)	Heteropelma calcator (Wesmael)	Heteropelma capitatum (Desvignes)	Hyposoter didymator (Thunberg)	Ichneumon cerinthius Gravenhorst	Ichneumon insidiosus Wesmael	Ischnus migrator (Fabricius)	Lymantrichneumon disparis (Poda)	Mesochorus disciergus (Say)	Metopius dentatus (Fabricius)	Microleptes splendidulus Gravenhorst	Netelia testacea (Gravenhorst)	Netelia vinulae (Scopoli)	Pimpla hypochondriaca (Retzius)	Pimpla illecebrator (Villers)	Pimpla turionellae (Linnaeus)	Protichneumon fusorius (Linnaeus)
Macroglossinae (cont'd)																																											
Hyles euphorbiae euphorbiae																																											
Hyles euphorbiae conspicua																																											
Hyles euphorbiae robertsi																																											
Hyles tithymali tithymali																																											
Hyles tithymali mauretanica																																											
Hyles tithymali deserticola																																											
Hyles tithymali gecki																																											
Hyles tithymali himyarensis																																											
Hyles dahlii																																											
Hyles nervosa																																											
Hyles salangensis																																											
Hyles centralasiae centralasiae																																											
Hyles centralasiae siehei																																											
Hyles gallii gallii	●	●													●	●	●									●																	
Hyles nicaea nicaea																																											
Hyles nicaea castissima																																											
Hyles nicaea libanotica																																											
Hyles nicaea crimaea																																											
Hyles zygophylli zygophylli																																											
Hyles vespertilio																																											
Hyles hippophaes hippophaes																																											
Hyles hippophaes bienerti																																											
Hyles hippophaes caucasica																																											
Hyles chamyla																																											
Hyles livornica livornica																																											
Deilephila elpenor elpenor		●	●																						●																	●	
Deilephila rivularis																																											
Deilephila porcellus porcellus		●	●								●	●		●	●	●										●																	
Deilephila porcellus colossus																																											
Deilephila porcellus suellus																																											
Hippotion osiris																																											
Hippotion celerio																																											
Theretra boisduvalii																																											
Theretra alecto																											●				●												

NOTES

* It is necessary to add a cautionary note. The table gives *recorded* parasitoid data. It must however be stressed that misidentification of both parasites and hosts has led to considerable confusion and erroneous recording. The destruction of the host by the parasitoid makes subsequent confirmation difficult; the taxonomic revision and splitting of parasitoid genera and species may render previous determinations unsound. There is also the problem of overlooked extraneous hosts and secondary parasitism which can lead to misidentification. For further discussion of these problems, see Shaw (1982; 1990).

† An egg parasite

(i) Parasitoids recorded in this table under the different sphingid species are not always found throughout the geographical range of their hosts but may be replaced by other parasitic species.

(ii) This is a subject providing a vast opportunity for research. The retention of all evidence with full contemporaneous notes is essential for the compilation of reliable data. If possible, reared parasitoids with appropriate data should be deposited in suitable research collections. One museum curator who specializes in such parasitoids is Dr M. R. Shaw, National Museums of Scotland, Chambers Street, Edinburgh EH1 1JF, Scotland. Another is Dr H.-P. Tschorsnig, Naturkundemuseum,

Column headings (left to right):

- Protichneumon pisorius (Linnaeus)
- Pyramidophorus flavoguttatus Tischbein
- Scolobates auriculatus (Fabricius)
- Therion circumflexum (Linnaeus)
- **Braconidae**
- Aleiodes praetor (Reinhard)
- Cotesia abjectus (Marshall)
- Cotesia coryphe (Nixon)
- Cotesia glomeratus (Linnaeus)
- Cotesia nigriventris (Nees)
- Cotesia saltator (Thunberg)
- Microplitis ocellatae Bouché
- Microplitis vidua (Ruthe)
- **Pteromalidae**
- Erdoesina alboannulata (Ratzeburg)
- **Eupelmidae**
- Anastatus bifasciatus (Geoffroy)
- **Encyrtidae**
- †Ooencyrtus pityocampae (Mercet)
- **Eulophidae**
- Eulophus smerinthicida Bouček
- **Trichogrammatidae**
- Trichogramma evanescens Westwood
- **Scelionidae**
- Telenomus punctatissimus (Ratzeburg)
- **DIPTERA: Tachinidae**
- Blondelia nigripes (Fallén)
- Campylochaeta inepta (Meigen)
- Compsilura concinnata (Meigen)
- Drino galii (Brauer & Bergenstamm)
- Drino inconspicua (Meigen)
- Drino lota (Meigen)
- Drino vicina (Zetterstedt)
- Drino (Palexorista) imberbis (Wiedemann)
- Drino (Zygobothria) atropivora (Robineau-Desvoidy)
- Drino (Zygobothria) ciliata (Wulp)
- Exorista fasciata (Fallén)
- Exorista grandis (Zetterstedt)
- Exorista larvarum (Linnaeus)
- Exorista sorbillans (Wiedemann)
- Frontina laeta (Meigen)
- Huebneria affinis (Fallén)
- Masicera pavoniae (Robineau-Desvoidy)
- Masicera sphingivora (Robineau-Desvoidy)
- Nemoraea pellucida (Meigen)
- Nilea hortulana (Meigen)
- Osvaldia spectabilis (Meigen)
- Pales pavida (Meigen)
- Phryxe erythrostoma (Hartig)
- Phryxe nemea (Meigen)
- Phryxe vulgaris (Fallén)
- Tachina grossa (Linnaeus)
- Tachina praeceps (Meigen)
- Thelaira nigripes (Fabricius)
- Winthemia bohemani (Zetterstedt)
- Winthemia cruentata (Rondani)
- Winthemia rufiventris (Macquart)

Rosenstein 1, 7000 Stuttgart 1, Germany, who specializes in Tachinidae.
(iii) Records have been drawn from a great many sources. Those for hymenopterous parasitoids include Atanasov *et al.* (1981), Hinz (1983), Eisenstein (1984). Those for dipterous parasitoids include Rungs (1981), Whitcombe & Erzinçlioglu (1989), Ford & Shaw (1991) and B. Herting and H.-P. Tschorsnig (pers. comm.).

surviving larvae grow, various members of the Tachinidae (Diptera) (Text fig. 21), Ichneumonidae and Braconidae (Hymenoptera) take an interest in them. Belshaw (in press) gives an account of the various, ovipositional and behavioural strategies employed by Tachinidae, and Gauld & Bolton (1988) give a good overview of the biology of parasitic Hymenoptera. Most of the tachinids deposit their eggs on the larval cuticle or the leaf on which hosts are feeding. These hatch in a few days and the parasitic larva burrows into the host's body to feed on its tissues. It is often only after the host pupates in the ground that the larval parasitoid emerges through the cuticle and itself pupates. In North Africa numerous larvae of *Acherontia atropos* fall victim to *Drino (Zygobothria) atropivora* (Robineau-Desvoidy), whilst *Drino vicina* (Zetterstedt) takes its toll of *Hippotion celerio* (Rungs, 1981).

The various parasitic Hymenoptera develop somewhat differently. Their eggs are usually laid through the host's skin, although in those few species which develop as ectoparasites, they are placed on the cuticle. Up to 80 per cent of all *Smerinthus ocellatus* larvae in Britain are lost every year to a gregarious braconid, *Microplitis ocellatae* Bouché (Text figs 22, 23). The mature endoparasitic larvae usually vacate the host in its final instar and spin individual but coalesced cocoons around the host remains, often after it has prepared its own subterranean pupal chamber. In China this parasitoid is equally devastating to populations of *S. planus* (Wei & Zheng, 1987).

Whereas many braconid larvae may develop in a single host caterpillar, in the Ichneumonidae one parasitic larva per host is the norm. Most of the species that attack sphingids pupate within the host's pupal skin and emerge through a hole in the cuticle. Many of the Ichneumonidae are host-specific or restricted to two or three genera (Text fig. 24); the Braconidae display a preference for the Smerinthini in their relations with sphingids.

Recent work in the U.S.A. has shown that the hostplant selected by a caterpillar may determine its susceptibility to attack by a given parasitoid (Barbosa *et al.*, 1986). It is possible

Text Figure 21 (top left) *Masicera sphingivora* (Robineau-Desvoidy) (Diptera: Tachinidae), the commonest tachinid parasitoid of European sphingid larvae

Text Figure 22 (top centre) Larvae of braconid parasitoid *Microplitis ocellatae* Bouché (Hymenoptera: Braconidae) emerging through the skin of larva of *Smerinthus o. ocellatus* (Linnaeus)

Text Figure 23 (top right) Adult of *Microplitis ocellatae* Bouché (Hymenoptera: Braconidae)

Text Figure 24 (bottom left) Adult of the ichneumonid *Amblyjoppa proteus* (Christ) (Hymenoptera: Ichneumonidae), a common parasitoid of larvae of *Deilephila e. elpenor* (Linnaeus)

that this factor has played as great an evolutionary role in the selection of hostplants by sphingid species as has hostplant resistance.

Being subterranean, no pupa of any western Palaearctic sphingid is known to be directly parasitized, although it is strongly suspected that *Protichneumon pisorius* (Linnaeus) is capable of digging down to the pupae of *Sphinx pinastri* and parasitizing them (Hinz, 1983).

Numerous animals, both invertebrate and vertebrate, find hawkmoths and their larvae tasty morsels. Wasps of the genus *Vespa* destroy countless shrub- and tree-feeding larvae every summer, such as those of *Sphinx ligustri* and *Laothoe populi*. Mice, shrews, birds, bats, ants, beetles and spiders also take their toll and, were it not for the fact that the Sphingidae have developed an almost incredible variety of sophisticated defensive and protective strategies (Evans & Schmidt, 1990), many species would become extinct.

Defence

Moths and their larvae are mainly the victims of predators using visual skills, and many defensive mechanisms have evolved to counteract these. However, bats are unique among aerial predators in hunting moths by echo-location and not sight. Adult moths of the families Arctiidae, Noctuidae and Geometridae can detect the ultrasonic emissions of a bat and take avoiding action. This may explain why these moth families are very successful and why, in the Sphingidae, the Choerocampina, with their 'ear', are one of the most successful sphingid subtribes. However, it is thought most deter or escape capture by their size (many hawkmoths are as large as the insectivorous bats themselves) or by means of their flying skills and speed (Rothschild, 1989; Evans & Schmidt, 1990).

A number of larvae exhibit cryptic coloration, blending in with their hostplant (see p. 21). Their outline can be further broken up by deportment, body stripes and countershading. The larva of *Smerinthus kindermanni* always rests upside down under a narrow leaf with the anterior of the body raised (Pl. C, fig. 3). To enhance its disguise as a leaf further, it has lateral stripes and a pale dorsal surface which counteracts any shade the body may cast. When this position is reversed and it is turned the other way up, this effect is lost and the caterpillar becomes very visible.

To avoid diurnal parasitoids and predators, many sphingid larvae have become nocturnal, hiding on the ground or low down in their hostplant by day. In such species the larva changes colour from a cryptic green when young to a mottled brown which blends in with dead leaves and stones (see p. 21). In *Hyles vespertilio*, such a strategy works well (Milyanovskiĭ, 1959), but it does not prevent the larvae of *Deilephila porcellus* being actively sought out by tachinid flies. However, the lower the larval density the more chance of escaping detection, so 'spacing out' helps reduce the odds of discovery (Tinbergen *et al.*, 1967; Meerman, 1988a).

Several adult moths also utilize cryptic mimicry, blending in with the vegetation among which they rest (Rothschild, 1989). *Laothoe populi* resembles a cluster of dead leaves (Pl. F, fig. 5), an illusion further enhanced by hindwings which project above the forewing margin. *Mimas tiliae* breaks up its outline with various blotches and has a leaf-like wing shape.

If discovered, some hawkmoths can suddenly display brightly coloured hindwing patches to surprise and disorientate would-be predators and enable them to drop to the ground unnoticed (Blest, 1957). Such patches – usually red, a known warning colour – remain concealed beneath cryptically-coloured forewings when the moth is at rest. All *Hyles* species are so endowed.

The three species of *Smerinthus* found in the western Palaearctic augment the warning effect of the red on their hindwings by bearing false eye-spots. These resemble vertebrate's eyes and, in conjunction with a beak-like abdomen, may imitate the face of a small owl and so intimidate predators. Many inexperienced young birds retreat when so confronted. Similarly, some hawkmoth caterpillars bear striking thoracic eye-spots which can be enlarged by expanding the thorax, as do the larvae of *Daphnis nerii*, *Rethera komarovi* (Pl. D, fig. 1), *Deilephila elpenor* and

Hippotion celerio. In these the anterior segments are trunk-like and can be contracted, giving the appearance of a small serpent. Such larvae often lash from side to side like a striking snake.

Many lepidopterous larvae can also assimilate toxins from their hostplants and sequester them in their tissues. Such larvae are often brightly coloured and feed quite openly yet unmolested. *Hyles euphorbiae*, *H. dahlii*, *H. tithymali* and *H. nicaea* all feed as larvae on species of *Euphorbia*, and are known to be poisonous to birds – the main sequestered toxin (a diterpene, ingenol ester) being highly inflammatory and carcinogenic. It is interesting to note, however, that the adult moths contain relatively low concentrations of this carcinogen (Rothschild, 1985). Ingestion of such larvae often induces vomiting in birds as large as a crow. Indeed, one such experience has such a lasting effect that other larvae of that species, or similar-looking species, are shunned. Larvae of *Daphnis nerii* sequester small quantities of cardiac glycosides in their tissues which are passed on to the adult stage (Chang, 1982). Larvae of *Acherontia atropos* seem to confine plant toxins to their large gut and so gain only partial protection from them; they rely more on cryptic coloration for protection, as do the larvae of *D. nerii* (Pl. C, fig. 5).

However, an increase in the uptake of toxins into the body tissues can be achieved by utilizing alternative hostplants. Larvae of the North American *Manduca sexta*, when fed on *Nicotiana* (tobacco), excrete the nicotine they ingest. When reared on *Atropa belladonna* (deadly nightshade), larvae sequester and store atropine and synthesize two metabolites, which are found in the frass. The resulting pupae (and probably adults) are toxic to chickens (Rothschild, 1985). This ability to sequester and store varying levels of toxin depending on the hostplant eaten can also greatly affect the success of parasitoids attacking larvae of a given species (Thorpe & Barbosa, 1986; Beckage *et al.*, 1988).

Few adults of western Palaearctic Sphingidae inherit the unpalatability of their larvae, apart from the partial protection acquired by the *Euphorbia*-feeding *Hyles* species and *Daphnis nerii*. However, high concentrations of histamine (75 μg/g) have been found in adults of *Smerinthus ocellatus*, and acetylcholine in the male accessory gland and ejaculatory duct of *Laothoe populi* (Rothschild, 1985).

Many perfectly edible butterflies and moths mimic poisonous species to gain protection, a feature known as Batesian mimicry. For example, adult *Hemaris* species mimic bumblebees (*Bombus* spp.) (Pl. G, figs 1, 2). The latter, with their painful sting, are studiously avoided by most avian predators. In addition, the adults of *H. fuciformis* are also distasteful, thus exhibiting a degree of Müllerian mimicry, where several equally poisonous species resemble one another for added protection. In this case, predators learn to avoid the warning characteristics – the toll taken before they do so being distributed among the co-mimics. What type of mimicry is involved in the close resemblance of the larvae of *Hyles nicaea* to those of *Hyles centralasiae* is not understood. The former is known to be poisonous; the latter may or may not be, although its hostplants (Liliaceae) are.

When attacked, some hawkmoths are capable of emitting sounds. The larva of *Acherontia atropos* audibly clicks its jaws together; the adult hops about and squeaks. Some adult male hawkmoths will also stridulate their genitalic valvae and emit volatile pyrazines from abdominal hair-pencils if attacked (Rothschild, 1989). What effect these noises and smells have on potential predators is not known.

Most sphingid larvae will regurgitate their sticky, sometimes toxic, foregut contents over the attacker. This may be accompanied by violent lashing movements from side to side and is extremely effective against parasitoids and ants. *Marumba quercus* rarely regurgitates but it employs this lashing action to great effect against parasitoids – tachinid flies can be impaled on the head spines (pers. obs.). If the lashing fails, the caterpillar of this species will aggressively bite its attacker, a habit it shares with the larva of *A. atropos*.

Finally, many of the larger sphingid adults are armed with formidable tibial spurs which can draw blood if the moth is handled or attacked.

VII. CLASSIFICATION

From Linnaeus until the present century, the criteria on which entomologists based the systematic division of Lepidoptera were numerous. Classification systems differed one from another and are difficult to compare with those in use today as they were based mainly on superficial characters, such as wing pattern and shape.

As his criteria, Linnaeus used wing positions of adults at rest. Latreille based his classification on the times of adult activity (i.e. Diurni, Nocturni) whereas Boisduval used antennal shape – the Rhopalocera (butterflies) having 'clubbed horns' (i.e. clubbed antennae); the Heterocera (moths) having 'horns otherwise' (i.e. antennae without clubs). None of these distinctions is valid, for even though Boisduval's Rhopalocera do form a monophyletic group, his Heterocera do not.

Another loose form of classification still in use today – Microlepidoptera and Macrolepidoptera – is based on adult size, although some 'Micros' have wingspans up to 200mm. Four modern suborders of Lepidoptera represented in the western Palaearctic (Zeugloptera, Dacnonypha, Exoporia and Monotrysia) contain only families of Microlepidoptera, but a fifth, the Ditrysia, includes families of both 'Micros' and 'Macros', among which is the Sphingidae.

Many other classifications have been proposed in addition to those referred to above, based on different morphological features (Comstock (1893), Packard (1895), Karsch (1898), Tillyard (1918, 1919), Börner (1925), Hinton (1946) and Common (1970)); none of these has contributed much to the present systematic arrangement of the sphingids themselves, but since 1900 three systems have been of sufficient merit to have had a significant impact on sphingid classification, namely Rothschild and Jordan (1903), Carcasson (1968) and Hodges (1971).

The position of the Sphingidae within the Lepidoptera is well reviewed by Scoble (1991) in his valuable summary of the classification of the Lepidoptera at higher taxonomic levels. He points out that this classification is in a state of flux especially between superfamily and suborder levels but he follows Minet (1986) and Holloway et al. (1987) in keeping the Sphingidae within the Bombycoidea rather than treating them as a separate superfamily, the Sphingoidea, as do Common (1970), Hodges (1971) and Kuznetsov & Stekolnikov (1985) although all these authors have recognized their close relationship to the Bombycoidea. In this work the following classification is adopted:

ORDER	Lepidoptera
SUBORDER	Glossata
INFRAORDER	Heteroneura
DIVISION	Ditrysia
SUPERFAMILY	Bombycoidea
FAMILY	Sphingidae

In their monumental work, Rothschild & Jordan (1903) recognized 772 species. Prior to this, the Sphingidae had been classified on purely superficial characters with the result that many unrelated species had been lumped together and many closely related species had been separated. However, Rothschild & Jordan admitted that they were very much hampered in their attempts by a lack of material and knowledge regarding the early stages. Only a few of the more common species had been bred or reared and the published descriptions of these were inadequate. Nevertheless, they were the first authors to adopt a natural, phylogenetic classification, using characters such as the structure of antennae, palpi, pilifer and feet, spination of the legs and abdomen, and dry genitalia preparations of a large number of species. The family Sphingidae was split into two divisions (Asemanophorae and Semanophorae), five subfamilies and seven tribes.

Carcasson (1968), in his excellent work on the African Sphingidae, proposed a revision to the classification of the Sphingidae of Rothschild & Jordan based on the genital armatures of both sexes and on the early stages. Because they were using dry genitalia preparations, Rothschild & Jordan were not enabled '. . . to take these structures into sufficient account when defining genera and following up their relationships. In the Philampelini and Choerocampini this did not matter much as the genitalial structure of these insects is extremely uniform, but it did lead to a number of misconceptions in the more advanced Ambulicini'. The names proposed by Rothschild & Jordan for supra-generic taxa were adopted by Carcasson, although several older and more appropriate names were substituted. The terms Sesiinae and Sesiicae were rejected because the type genus of these taxa is not a sphingid. The other groups were given the terminations recommended in the *International Code of Zoological Nomenclature* (1961). The term 'subfamily' was substituted for 'division', which has no nomenclatural status; all subsequent taxa above the rank of genus were demoted by one step. The 'tribes' of Rothschild & Jordan became 'subtribes', a useful rank which was not recognized in the *International Code* until 1985.

Hodges (1971), in his authoritative work on the Sphingidae of America north of Mexico, drew up a further revised system of classification, commenting that Rothschild & Jordan based theirs '. . . on a very large number of characters, mainly of the adults, but unfortunately they studied the male and female genitalia in a dry condition, so they were unable to discern characters of the female genitalia, and they did not figure (and thus probably did not take into full account) the full array of male genital characters. These two character systems in combination provide what seems to be a cohesive framework for a supergeneric [*sic*] classification within the Sphingidae. Many additional characters of the larvae and pupae fall into line if the associations of the genital systems are followed rather than those of the spination of the pilifer or of certain palpal characters'. Hodges further added that 'Rothschild & Jordan pointed out that the subfamilies and tribes within the Semanophorae were not equal and that several species and genera within each had characters which would tend to put them either one way or the other with the system. The higher categories could not be defined on the basis of just a few characters, but, perhaps in combination, several characters would suffice for determination'. Hodges agreed with Carcasson in redefining the Semanophorae and Asemanophorae as subfamilies but points out that Carcasson incorrectly used these names; the *International Code of Zoological Nomenclature* states that family-group names must be based on a valid or available generic name. The subfamilies Sphinginae and Macroglossinae were proposed as substitutes, respectively.

The subfamilies and tribes used by Hodges (1971) have been adopted here, as have the subtribes proposed by Carcasson (1968). The supraspecific categories for the Sphingidae of the western Palaearctic region described in this work are shown in the Check List (pp. 75–77).

Species concepts and population biology

A species can be described as an aggregate of populations which share similar genetic material and hence have a similar appearance. They are also capable of interbreeding to produce viable, self-perpetuating offspring. Many processes can lead to the formation of a species; all stop or restrict the flow of genetic material between portions of the range of a species.

Bearing in mind the amount of genetic variability within a species, if it has a small, continuous range, or is highly mobile, the results of any selection pressure exerted on part or all of that range will be rapidly transferred throughout the populations and all individuals will look similar, e.g. *Macroglossum stellatarum*. However, most species are not migratory and have, or may have had, a large range. Selection pressure at either end of this range, or on different parts of it, may vary, resulting in recognizably different populations – *Hyles euphorbiae* is a good example of this.

Where there is a gradual change in characters (markings or colour) from one end of a continuous range to another, this is termed a cline. The populations are still genetically interconnected and resemble one another. For instance, *Clarina kotschyi* gradually changes from a small, reddish moth with wavy wing-margins in Lebanon to a much larger, browner subspecies with straight wing-margins in Turkey and Iran. That the two subspecies are closely related with identical early stages has already been proven (Pittaway, 1982b). Each breeds true when reared under identical environmental conditions and, when cross-paired, produce perfectly viable offspring.

Where the differences between two populations are so marked and characteristic that they are distinguishable from one another yet are still capable of interbreeding, these populations are usually termed subspecies, although there are certain circumstances when this definition can be very subjective (see *Hyles euphorbiae* and *H. tithymali*). Subspecies need not be totally isolated from one another, e.g. if part of a cline. However, most subspecies are separated by barriers. Isolated thus, subspecies may continue to diverge and their ability to interbreed may decrease as successive mutations alter the genotype to suit local environmental conditions. When both populations have become so distinct that they can no longer produce viable, self-perpetuating offspring, each is regarded as a true species.

The time factor in this process of evolution is very important. If only a short period elapses between separation and reunification, the differences between subspecies would become submerged by an intermingling of their genes. This might take some time and, until completed, numerous hybrid populations would exist at various stages of introgression. Often secondary clines, with either a narrow or wide suture-zone containing many intermediate hybrids, are produced in this way, e.g. such as between *Sphinx p. pinastri* and *S. p. maurorum* in the Pyrenees and southern France. This process also appears to have happened with the various subspecies and races of *Clarina kotschyi*, *Hyles euphorbiae*, *H. tithymali* and *Deilephila porcellus* in the western Palaearctic.

When a much longer separation has occurred, reunification may no longer allow interbreeding due to the number of accumulated differences between the 'subspecies', such as between *Acherontia atropos* and *A. styx*, *Sphingonaepiopsis gorgoniades* and *S. kuldjaensis*, and *Hyles hippophaes* and *H. chamyla*.

Unfortunately, unless there is an overlap in ranges and no introgression, two physically isolated and visually distinct populations cannot be positively identified as true species. The western Palaearctic *Smerinthus ocellatus* and the Oriental *S. planus* appear, as phenotypes, to be two valid species. However, in Europe, a virgin female *S. planus* will usually attract some males of *S. ocellatus* during the right season. They pair successfully and the offspring are viable, although of reduced vigour. Is *S. planus* merely a subspecies of *S. ocellatus*?

The relationship between *Hyles lineata* and *H. livornica* was questioned for many years until Harbich (1980a, 1982) proved them to be distinct – or did he? Certainly, crosses between North African and American moths produce non-viable offspring over three generations, but what of hybrids between Oriental, Australian and American material? With two distinct species at either end, could not the *H. lineata/H. livornica* complex be a cline? Under certain circumstances the terms 'species' and 'subspecies' must therefore often be used in a subjective fashion. Such is also the case with the *Hyles euphorbiae/H. tithymali* complex.

To retain its genetic individuality, a species or 'sibling species' usually possesses mechanisms to prevent interbreeding with related species. This is known as reproductive isolation and many factors are involved. These can be divided into two groups, pre-copulatory and post-copulatory mechanisms.

Pre-copulatory mechanisms reduce the incidence of unviable matings and wastage of genetic material, and enhance reproductive capacity. Several are recognized:

(a) Spatial isolation. Two closely related species occupy two separate geographical regions and never meet. This appears to be the case with the eastern Palaearctic *Smerinthus planus* and *S. ocellatus* from the western Palaearctic.

(b) Temporal isolation. Individuals of two species emerge at different times of the year, e.g. *Hyles euphorbiae* and *H. vespertilio* (see p. 65).

(c) Sexual isolation. Females of each species have sex pheromones which vary in composition. The male genitalia in some genera are very similar, e.g. *Hyles*, and would in themselves prove an ineffective barrier.

(d) Mechanical isolation. This consists principally of differences in male genitalia. Whilst not guaranteeing reproductive isolation, when coupled with other mechanisms, this process is effective.

Post-copulatory mechanisms are what might be called 'mechanisms of last resort'. These include the following:

(a) Zygote mortality. The genetic material within the two contributing gametes is incompatible. The zygote dies.

(b) Lack of hybrid vigour. The offspring die before reaching adulthood. This is the fate of most hawkmoth hybrids.

(c) Adult sterility. Gametes produced by the hybrid imago may have either an incomplete structure, or one which is incompatible with either parent.

(d) Hybrid breakdown. The F_1 generation is fully or partially fertile, but each subsequent cross has successively less vitality and ultimately the offspring are sterile. Hybrids of the interspecific cross *Hyles lineata* × *H. livornica*, when paired with each other, are 95 per cent viable (Harbich, 1980a, 1982). Their offspring, the F_2 generation, produce ova of which only 1–6 per cent hatch. Genitalia may also atrophy, as in the cross *Laothoe p. populi* × *L. austauti*.

Although genetic mechanisms operate to produce and maintain species, the main factors that fragment or isolate populations leading to speciation are physical or environmental. Of these, climate is by far the most important (see p. 34)

The last quarter of a million years has demonstrated, from pollen and fossil records (Polunin & Walters, 1985), just how quickly climates can change. Biogeographical regions have shifted north, south, east and even west in response to global warming and cooling trends.

In response, the ranges of both hawkmoths and hostplants must have expanded and contracted. In many instances isolated populations of a species must have been left behind, cut off from the main distribution by intervening, unfavourable conditions. Many of these would have been reabsorbed when the climate improved and the parent population expanded its range to recolonize former territory. Others would have died out as environmental conditions worsened; some would have persisted.

Climatic barriers act in two ways: directly on the hawkmoths' ability to tolerate certain climatic conditions; and indirectly, by acting on the hostplant. Twenty thousand years ago much of the Arabian Peninsula had a Mediterranean flora and fauna. As the climate in Europe ameliorated at the end of the last ice-age, Arabia became arid. The wildlife either died out, retreated northwards or took refuge in cooler mountainous areas (Pittaway, 1987). Today, *Hyles euphorbiae conspicua* ranges from Turkey to Palestine and Iraq. Two thousand kilometres to the south a small, relict population of the same or a closely related subspecies persists in the cool, Mediterranean-like Asir Mountains of south-western Saudi Arabia. Along the border with Yemen, it has come into contact with another relict sphingid population, that of *Hyles tithymali himyarensis*.

Likewise, *Hyles hippophaes* occurs today in three isolated populations – one in western

Europe, one in south-eastern Europe and the Aegean and one stretching from Anatolia eastward to Mongolia (Chu & Wang, 1980a; Pittaway, 1982a). This species, as a larva, feeds on *Hippophae* and *Elaeagnus* (both shrubs in the Elaeagnaceae). The Palaearctic *Hippophae rhamnoides* is a small, drought-resistant bush adapted to poor, stony, unstable soils. It has its main distribution from Anatolia eastward to China. Here it often forms pure stands, covering whole mountain-sides and mountain river-banks. In Europe, although isolated patches of this shrub exist on coastal sand-dunes, its main distribution covers the riverine gravel-beds of southern Germany, Switzerland and south-eastern France, as well as sand-bars and gravel banks along mountain river-valleys of the Pyrenees, French Alps and northern Italy. Further east it is found in similar places in Romania. This shrub does not occur naturally in the Aegean, nor western Anatolia, but both these areas have been extensively planted in recent years with the ornamental *Elaeagnus angustifolia* from central and eastern Anatolia.

At the beginning of the interglacial periods, including the present one, retreating ice and scouring melt-water produced large areas favourable to *Hippophae rhamnoides* in Europe. For example, in the Holsteinian Interglacial (*c.* 250,000 years BP), this shrub was initially one of the dominant woody plants, even in Britain (Polunin & Walters, 1985). With improving soil conditions and a wetter climate other trees and shrubs succeeded this species, forcing it into refugia.

Hyles hippophaes may once have had a continuous distribution from Britain to China. This fragmented into two populations with the break-up of the distribution of its hostplants, giving rise to the European *H. h. hippophaes* and Asian *H. h. bienerti*. After the last ice-age the original European population must have not only contracted in range but also split into a western and an eastern population which, as yet, have had insufficient time to diverge morphologically or genetically. However, that this sphingid can thrive in areas from which it was previously excluded due to a lack of hostplants has been demonstrated by its recent colonization of the Aegean and western Anatolia, where subsp. *hippophaes* can now be commonly found on *Elaeagnus angustifolia* (Pittaway, 1982a). *Hippophae rhamnoides* ranges as far north as Scandinavia and Britain, but these regions are now too cold and wet for *Hyles hippophaes*. Similar mechanisms must also have acted on *Deilephila porcellus* to produce subsp. *suellus* and subsp. *porcellus*, only to reunite them in the recent past (see species accounts, pp. 151, 158).

Small, isolated European populations also occur of such essentially Asian species as *Rethera komarovi* and *Sphingonaepiopsis gorgoniades*. Even more extreme is the present-day distribution of *Sphinx ligustri*. Having originally spread to Europe from China during one of the last interglacial periods, it does not now occur between the Tian Shan and north-eastern China due to inhospitable ecological conditions. The two populations have diverged to produce two recognizable subspecies (subsp. *ligustri* and subsp. *amurensis*). *Sphinx pinastri* was affected in a very similar manner. It is more than likely that subsp. *maurorum* evolved over a period in isolation in North Africa and the Iberian Peninsula during the last ice age and that it is only since the end of that period that it and subsp. *pinastri*, which appears to have taken refuge in south-eastern Europe during the ice age, have reunited in southern France to produce many intermediate forms.

Isolated islands which are cut off from the mainland can also lead to speciation, but this very much depends on the mobility of a species and the extent of the marine barrier. On Sardinia, Corsica and the Balearic Islands, the Pontomediterranean *Hyles euphorbiae* is replaced by *Hyles dahlii*. Both these species are derived from a common ancestor, but when the initial island colonization took place and why *H. euphorbiae* has subsequently been unable to colonize these islands remain a mystery. In the analogous butterflies *Papilio machaon* Linnaeus and the endemic *P. hospiton* Guenée, the former has managed to establish itself on Corsica and Sardinia and both species co-exist with little interbreeding (Pierron, 1990). However, 'islands' need not necessarily be marine; small mountain chains can also act as 'islands' when the surrounding lowlands become too hot and/or dry. The Tian Shan appear to have undergone this phenomenon several times, resulting in the formation of several endemic species and subspecies of sphingid (see page 46).

With time, differential selection pressures may also produce two species in what could be termed an 'altitudinal cline'. Gene flow between a bivoltine lowland race and a univoltine upland population may be sufficiently reduced to initiate subspeciation as each population adapts to its ecological niche. Such a situation appears to be occurring with *Hyles centralasiae centralasiae* and *H. c. siehei* in the mountains of south-eastern Turkey.

It should be remembered, however, that complete speciation may not be achieved during the course of any one of the above isolation mechanisms if the selection pressures on each isolated population remain the same. Reunification could eventually either produce an intermixed and viable hybrid population, such as in *Deilephila porcellus/suellus*, or lead to complete speciation if the resulting hybrids were less vigorous. Selection pressures would force the two subspecies apart. During this process both populations would tend to remain allopatric and should best be regarded as 'sibling species', as perhaps are *Laothoe populi* and *L. austauti*, although the evidence indicates that they are good species.

Hybrids

Natural hybrids amongst certain species of wild sphingids are not uncommon, especially when these are closely related. This is hardly surprising considering that the female sex pheromones of all genera are similar in composition. Additionally, some genera have separated only comparatively recently into their present constituent species and hence still have many features in common, such as compatible sexual appendages, similar ranges, overlapping flight periods, similar hostplants (see p. 135). Most hybrids are to be found within *Smerinthus, Laothoe, Hyles* and *Deilephila*, with the first and last pairs of genera being sufficiently closely related to allow intergeneric, as opposed to interspecific, crosses.

Text Figure 25 Full-grown hybrid larva of (a) *Hyles l. livornica* (Esper) × *H. e. euphorbiae* (Linnaeus)
(b) *Hyles vespertilio* (Esper) × *H. e. euphorbiae* (Linnaeus)

However, unlike plant hybrids which exhibit hybrid vigour in being stronger and larger than their parents, hybrid hawkmoths are usually weaker and sterile, and often exhibit abnormal behaviour, e.g. pupae of *Hyles* × *Deilephila* rarely overwinter. When parent species are sufficiently closely related to give fertile offspring, the latter's fertility is greatly reduced and often ceases at the F_2 generation.

Adult hybrids tend to be a mixture of their parents, both in size and markings. In most, these markings are fused together, while in some they exist separately, side by side (see Plate 7, fig. 9). Full-grown hybrid larvae may display a mixture of parental coloration, or be so different as to give the impression that they are of a distinct species (Text figs 25a,b).

The main requisites for hybridization to occur have been listed above. In the cross, *Hyles e. euphorbiae* ♂ × *H. vespertilio* ♀, Europe's commonest sphingid hybrid (Pl. 12, fig. 12), it has been demonstrated that the emergence of one species should be earlier than, but overlap with, that of the other species. *H. vespertilio* emerges 8–14 days before *H. euphorbiae*, thus older *H. vespertilio* males, which emerge before their own females (as do the males of *H. euphorbiae*), are presented with receptive females of their own species but not of *H. euphorbiae*. Fresh *H. euphorbiae* males have an initial choice of receptive females of *H. vespertilio*, followed by females of their own species.

Most of the hybrids shown in the table (Table 5) have been produced artificially, either by confining females of two species with males of one of them (Dannenberg, 1942), or by siphoning body fluids and cell clusters from one pupa and injecting them into another (Meyer, 1953). Some, however, occur naturally in the wild and these have been indicated by an asterisk(*).

Table 5 Table of all known natural* or artificially-produced sphingid hybrids. The left column gives hybrids sorted by male, the right column hybrids sorted by female.

♂	♀	♀	♂
Smerinthini		**Smerinthini**	
Mimas tiliae	× *Smerinthus o. ocellatus*	*Mimas tiliae*	× *Smerinthus o. ocellatus*
Smerinthus k. kindermanni	× *Laothoe p. populi*	**Smerinthus k. kindermanni*	× *Laothoe p. populi*
S. o. ocellatus	× *Mimas tiliae*	*S. o. ocellatus*	× *Mimas tiliae*
S. o. ocellatus	× *S. o. atlanticus*	*S. o. ocellatus*	× *S. o. atlanticus*
*S. o. ocellatus	× *L. p. populi*	**S. o. ocellatus*	× *L. p. populi*
S. o. atlanticus	× *S. o. ocellatus*	*S. o. ocellatus*	× *L. austauti*
S. o. atlanticus	× *L. p. populi*	*S. o. atlanticus*	× *S. o. ocellatus*
S. o. atlanticus	× *L. austauti*	**S. o. atlanticus*	× *L. austauti*
*Laothoe p. populi	× *S. k. kindermanni*	*Laothoe p. populi*	× *S. k. kindermanni*
*L. p. populi	× *S. o. ocellatus*	**L. p. populi*	× *S. o. ocellatus*
L. p. populi	× *L. austauti*	*L. p. populi*	× *S. o. atlanticus*
*L. p. populi	× *L. amurensis*	*L. p. populi*	× *L. austauti*
L. austauti	× *S. o. ocellatus*	**L. p. populi*	× *L. amurensis*
*L. austauti	× *S. o. atlanticus*	*L. austauti*	× *S. o. atlanticus*
L. austauti	× *L. p. populi*	*L. austauti*	× *L. p. populi*
*L. amurensis	× *L. p. populi*	**L. amurensis*	× *L. p. populi*
Macroglossini		**Macroglossini**	
*Hyles e. euphorbiae	× *Hyles t. tithymali*	*Hyles e. euphorbiae*	× *Hyles t. tithymali*
H. e. euphorbiae	× *H. t. mauretanica*	*H. e. euphorbiae*	× *H. t. mauretanica*
H. e. euphorbiae	× *H. dahlii*	*H. e. euphorbiae*	× *H. t. deserticola*
H. e. euphorbiae	× *H. n. nicaea*	*H. e. euphorbiae*	× *H. dahlii*
*H. e. euphorbiae	× *H. g. gallii*	**H. e. euphorbiae*	× *H. g. gallii*
*H. e. euphorbiae	× *H. vespertilio*	*H. e. euphorbiae*	× *H. n. nicaea*
*H. e. euphorbiae	× *H. h. hippophaes*	**H. e. euphorbiae*	× *H. vespertilio*
*H. e. euphorbiae	× *Deilephila e. elpenor*	*H. e. euphorbiae*	× *H. h. hippophaes*
H. e. euphorbiae	× *D. p. porcellus*	*H. e. euphorbiae*	× *H. l. livornica*
H. t. tithymali	× *H. e. euphorbiae*	**H. e. euphorbiae*	× *Deilephila e. elpenor*
H. t. tithymali	× *H. g. gallii*	*H. e. euphorbiae*	× *D. p. porcellus*
H. t. mauretanica	× *H. e. euphorbiae*	**H. t. tithymali*	× *H. e. euphorbiae*
H. t. mauretanica	× *H. g. gallii*	*H. t. tithymali*	× *H. h. hippophaes*
H. t. mauretanica	× *D. e. elpenor*	*H. t. tithymali*	× *H. l. livornica*
H. t. deserticola	× *H. e. euphorbiae*	*H. t. tithymali*	× *D. e. elpenor*

(Table 5 cont'd overleaf)

Table 5 (cont'd) Table of all known natural* or artificially-produced sphingid hybrids.
The left column gives hybrids sorted by male, the right column hybrids sorted by female.

♂	♀	♀	♂
H. dahlii	× H. e. euphorbiae	H. t. mauretanica	× H. e. euphorbiae
H. dahlii	× D. e. elpenor	H. t. mauretanica	× H. g. gallii
H. c. centralasiae	× H. t. deserticola	*H. t. mauretanica	× H. n. castissima
*H. g. gallii	× H. e. euphorbiae	H. t. mauretanica	× D. e. elpenor
H. g. gallii	× H. t. mauretanica	H. t. deserticola	× H. c. centralasiae
H. g. gallii	× H. t. deserticola	H. t. deserticola	× H. g. gallii
H. g. gallii	× H. dahlii	H. t. deserticola	× H. h. hippophaes
H. g. gallii	× H. vespertilio	H. dahlii	× H. e. euphorbiae
H. g. gallii	× H. h. hippophaes	H. dahlii	× H. g. gallii
H. g. gallii	× H. l. livornica	*H. g. gallii	× H. e. euphorbiae
H. g. gallii	× D. e. elpenor	H. g. gallii	× H. t. tithymali
H. n. nicaea	× H. e. euphorbiae	H. g. gallii	× H. t. mauretanica
*H. n. castissima	× H. t. mauretanica	H. g. gallii	× H. vespertilio
*H. z. zygophylli	× H. l. livornica	H. g. gallii	× H. h. hippophaes
*H. vespertilio	× H. e. euphorbiae	H. g. gallii	× H. l. livornica
H. vespertilio	× H. g. gallii	H. g. gallii	× D. e. elpenor
*H. vespertilio	× H. h. hippophaes	H. n. nicaea	× H. e. euphorbiae
H. vespertilio	× D. e. elpenor	*H. z. zygophylli	× H. l. livornica
H. vespertilio	× D. p. porcellus	*H. vespertilio	× H. e. euphorbiae
H. h. hippophaes	× H. e. euphorbiae	H. vespertilio	× H. g. gallii
H. h. hippophaes	× H. t. tithymali	H. vespertilio	× H. h. hippophaes
H. h. hippophaes	× H. t. deserticola	*H. vespertilio	× D. e. elpenor
H. h. hippophaes	× H. g. gallii	*H. h. hippophaes	× H. e. euphorbiae
H. h. hippophaes	× H. vespertilio	H. h. hippophaes	× H. g. gallii
H. h. hippophaes	× D. e. elpenor	*H. h. hippophaes	× H. vespertilio
H. l. livornica	× H. e. euphorbiae	H. h. hippophaes	× D. e. elpenor
H. l. livornica	× H. t. tithymali	H. l. livornica	× H. g. gallii
H. l. livornica	× H. g. gallii	*H. l. livornica	× H. z. zygophylli
*H. l. livornica	× H. z. zygophylli	H. l. livornica	× D. e. elpenor
H. l. livornica	× D. e. elpenor	*Deilephila e. elpenor	× H. e. euphorbiae
*Deilephila e. elpenor	× H. e. euphorbiae	D. e. elpenor	× H. t. mauretanica
D. e. elpenor	× H. t. tithymali	D. e. elpenor	× H. dahlii
D. e. elpenor	× H. t. mauretanica	D. e. elpenor	× H. g. gallii
D. e. elpenor	× H. g. gallii	D. e. elpenor	× H. vespertilio
*D. e. elpenor	× H. vespertilio	D. e. elpenor	× H. h. hippophaes
D. e. elpenor	× H. h. hippophaes	D. e. elpenor	× H. l. livornica
D. e. elpenor	× H. l. livornica	D. e. elpenor	× D. p. porcellus
*D. e. elpenor	× D. p. porcellus	D. p. porcellus	× H. e. euphorbiae
D. p. porcellus	× H. e. euphorbiae	D. p. porcellus	× H. vespertilio
D. p. porcellus	× D. e. elpenor	*D. p. porcellus	× D. e. elpenor

1. Alpine habitat, Mariazell, Austria: mixed montane boreal and cool, wet, continental deciduous temperate forest. Recorded hawkmoth species include *Mimas tiliae, Smerinthus o. ocellatus, Laothoe p. populi, Hemaris t. tityus, Hyles g. gallii, Deilephila e. elpenor*

2. River valley in Alto Adige/Südtirol, northern Italy: warm, dry, continental deciduous temperate forest with riverine ecosystem. *Laothoe p. populi, Proserpinus p. proserpina, Hyles vespertilio*

3. Farmland, Weinburg, S.E. Austria: warm, wet, continental deciduous temperate forest, modified by agriculture. *Sphinx l. ligustri, Mimas tiliae, Smerinthus o. ocellatus, Hemaris t. tityus, Deilephila e. elpenor, D. p. porcellus*

4. Middle Atlas Mountains habitat, Morocco: relict mixed Mediterranean and deciduous temperate forest. *Smerinthus ocellatus atlanticus, Laothoe austauti, Hemaris tityus aksana, Hyles tithymali mauretanica, H. nicaea castissima, Deilephila porcellus colossus*

5. Rendlesham Forest, England: dry pine forest in maritime deciduous temperate forest. *Sphinx p. pinastri, Hemaris f. fuciformis, Deilephila p. porcellus*

6. Arid landscape north of Shiraz, Iran: mixed Mediterranean and steppe vegetation. *Akbesia davidi, Hemaris croatica, Rethera komarovi manifica, Sphingonaepiopsis gorgoniades pfeifferi, Macroglossum stellatarum, Hyles euphorbiae robertsi, H. nicaea libanotica, Deilephila porcellus suellus*

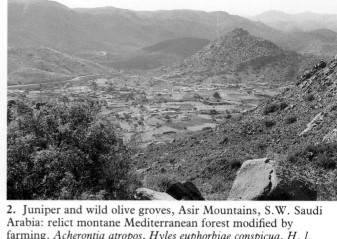

1. Northern foothills of Toros Mountains, southern Turkey: mixed Mediterranean and steppe vegetation with dry montane coniferous forests. *Kentrochrysalis elegans, Smerinthus k. kindermanni, Hemaris croatica, Clarina k. kotschyi, Rethera k. komarovi, Hyles c. centralasiae, H. n. nicaea, Theretra alecto*

2. Juniper and wild olive groves, Asir Mountains, S.W. Saudi Arabia: relict montane Mediterranean forest modified by farming. *Acherontia atropos, Hyles euphorbiae conspicua, H. l. livornica*

3. Edge of open plains with volcanic ash hills behind, Göreme, Turkey: steppe. *Hemaris croatica, Sphingonaepiopsis g. gorgoniades, Hyles hippophaes bienerti.* (The latter's main hostplant, *Elaeagnus angustifolia*, is in the foreground.)

4. Dadès Gorge, Morocco: mixed desert and steppe with riverine ecosystem. *Agrius c. convolvuli, Acherontia atropos, Smerinthus ocellatus atlanticus, Laothoe austauti, Macroglossum stellatarum, Hyles tithymali deserticola, H. nicaea castissima, H. l. livornica*

5. Landscaped gardens in Jeddah, Saudi Arabia: subtropical gardens in an otherwise desert region. *Agrius c. convolvuli, Acherontia s. styx, Daphnis nerii, Hyles l. livornica, Hippotion celerio*

6. Typical March landscape at Manifa, Saudi Arabia: desert. Flowering and other annual plants are very ephemeral in the desert, lasting little more than three months from February to April. *Hyles l. livornica*

Plate C – Ova; Larvae

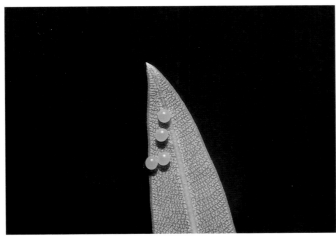

1. *Daphnis nerii*: ova on the underside of *Nerium oleander* leaf, laid by more than one captive female (Saudi Arabia). (*Page* 115)

2. *Marumba quercus*: full-grown larva on *Quercus cerris*. In a leaf cluster it is extremely well camouflaged (Spain). (*Page* 94)

3. *Smerinthus k. kindermanni*: camouflaged full-grown larva resting on *Salix* (Shiraz, Iran). (*Page* 98)

4. *Smerinthus ocellatus atlanticus*: countershading on full-grown larva in typical resting position provides effective camouflage (Morocco). (*Page* 101)

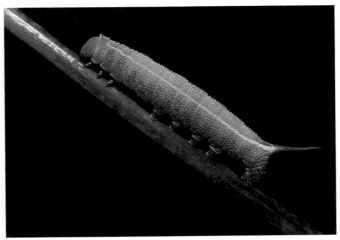

5. *Daphnis nerii*: half-grown final-instar larva – brown form on flower-head of *Nerium oleander* where it is usually well concealed (Saudi Arabia). (*Page* 115)

6. *Hemaris fuciformis jordani*: half-grown final-instar larva (Morocco). (*Page* 111)

1. *Rethera komarovi manifica*: alarmed full-grown larva on *Galium*, mimicking a small viper (Shiraz, Iran). (*Page* 125)

2. *Hyles tithymali himyarensis*: full-grown larva on herbaceous *Euphorbia* (northern Yemen). (*Page* 142)

3. *Hyles centralasiae siehei*: full-grown larva on cultivated *Eremurus* (Toros Mountains, Turkey). (*Page* 145)

4. *Hyles nicaea castissima*: full-grown larva on herbaceous *Euphorbia* – black form commonly found at high altitude in the Atlas Mountains (Algeria). (*Page* 148)

5. *Hyles vespertilio*: third-instar larva on *Epilobium dodonaei* (Austria). (*Page* 150)

6. *Hyles l. livornica*: full-grown larva – pale, eyed form on *Boerhavia* (Saudi Arabia). (*Page* 155)

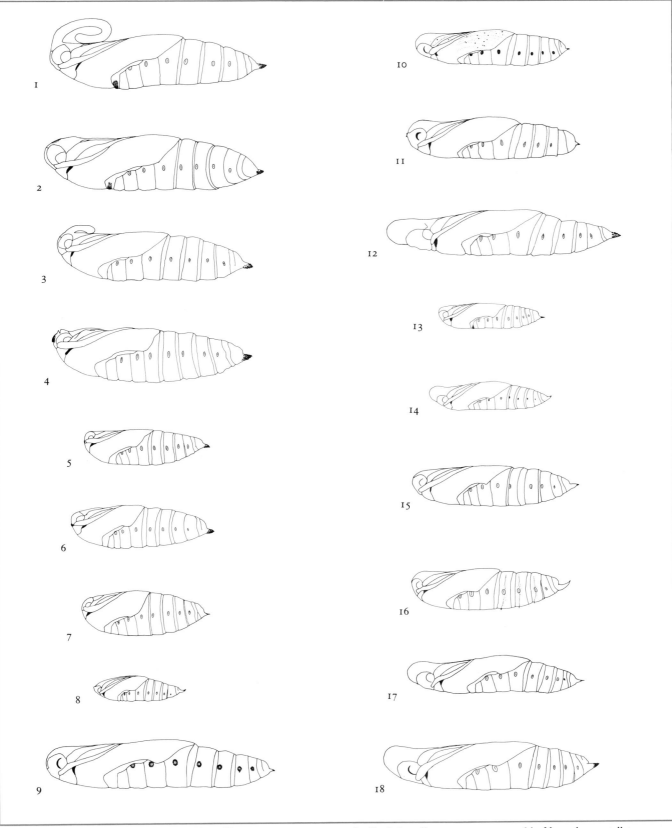

Sphinginae:
1. *Agrius c. convolvuli*
2. *Acherontia s. styx*
3. *Sphinx l. ligustri*
4. *Marumba quercus*

5. *Mimas tiliae*
6. *Smerinthus o. ocellatus*
7. *Laothoe p. populi*
Macroglossinae
8. *Hemaris f. fuciformis*

9. *Daphnis nerii*
10. *Clarina k. kotschyi*
11. *Acosmeryx n. naga*
12. *Rethera komarovi manifica*
13. *Proserpinus p. proserpina*

14. *Macroglossum stellatarum*
15. *Hyles h. hippophaes*
16. *Deilephila e. elpenor*
17. *Hippotion celerio*
18. *Theretra alecto*

1. *Acherontia atropos* (Kuwait). At rest the wings are held tent-like over the large body, the skull-mark on the thorax being prominent. (*Page 82*)

2. *Sphinx l. ligustri* ♀ (Austria). During the day it generally rests on vertical surfaces such as fence posts, its wings in a tectiform attitude. (*Page 85*)

3. *Smerinthus k. kindermanni* ♂ (Turkey). At rest, the cryptic pattern of the forewings provides good camouflage for hawkmoths of this genus, their eye-spots being concealed. (*Page 98*)

4. *Smerinthus o. ocellatus* ♂ (England). Suspended amongst foliage, this resting hawkmoth resembles dead leaves. Outer margin of forewing is less dentate than in the previous species. (*Page 100*)

5. *Laothoe p. populi* ♀ (England). When resting suspended, the hindwings protrude markedly above the sombre-coloured forewings, resembling a cluster of dead poplar leaves. (*Page 102*)

6. *Polyptychus trisecta* ♂ (Gibraltar). A very rare vagrant to the region from West Africa, captured at light in November, 1974. (*Page 42*)

1. *Hemaris f. fuciformis* (England). This fast-flying diurnal hawkmoth mimics bumblebees but is far more agile. It frequently visits nectar-bearing flowers in sunny woodland glades. (*Page 110*)

2. *Hemaris tityus aksana* (Morocco) Another diurnal bumblebee-mimic, this subspecies is found only in the Middle Atlas mountains, mainly in flower-rich meadows. (*Page 114*)

3. *Daphnis nerii* (Saudi Arabia). Against a background of foliage this large and attractively-patterned hawkmoth is well camouflaged by its cryptic coloration. (*Page 115*)

4. *Clarina k. kotschyi* ♀, (Shiraz, Iran). A grapevine-feeding hawkmoth, this subspecies is locally common on hillsides and in mountain valleys of Iran. (*Page 118*)

5. *Acosmeryx naga hissarica* (Afghanistan). On the bare ground of uncultivated mountain-ravines where it occurs, this brownish species is very well concealed. (*Page 120*)

6. *Proserpinus proserpina gigas* (Morocco). Active mainly at twilight, this large subspecies occurs in the Atlas Mountains alongside streams where *Epilobium* grows. (*Page 132*)

1. *Hyles e. euphorbiae* ♀ (Austria). The southern European and N. African subspecies of this widespread Palaearctic hawkmoth is commonly found in warm locations where herbaceous *Euphorbia* plants grow. (*Plate* 136)

2. *Hyles × tithymali* ♂ (Malta). This specimen, photographed in Malta, is almost certainly one of the natural hybrid populations of *H. euphorbiae* × *H. tithymali*; another is found in Crete. (*Page* 141)

3. *Hyles nicaea castissima* ♀ (Algeria). Confined to the Atlas Mountains of western N. Africa, this subspecies is distinguished by the straight distal margin to the median band of the forewing. (*Page* 148)

4. *Deilephila e. elpenor* (England). Despite its gaudy coloration, this hawkmoth is well concealed when settled on a colourful nectar-bearing flower such as this *Bougainvillea*. (*Page* 156)

5. *Hippotion celerio* (Saudi Arabia). When at rest during the day, the cryptic pattern of this species makes it very difficult to detect; at night it frequents tubular nectar-bearing flowers. (*Page* 162)

6. *Theretra alecto* ♂ (Cyprus). A partial migrant as far as Sicily and the Crimea, this species is a common resident in the south-east of the region where it can be a local pest of vineyards. (*Page* 164)

VIII. THE WESTERN PALAEARCTIC SPECIES OF SPHINGIDAE

The categories used by Hodges (1971) have been adopted as the classification system throughout this work for the various western Palaearctic genera (which have much in common with those of North America), but the subtribes proposed by Carcasson (1968) and as used by Derzhavets (1984) have also been incorporated. For the higher taxonomic categories, see p. 59 above.

CHECK LIST

SUPERFAMILY	**Bombycoidea**	
FAMILY	**SPHINGIDAE** Latreille, [1802]	
SUBFAMILY	**Sphinginae** Latreille, [1802]	
TRIBE	SPHINGINI Latreille, [1802]	
SUBTRIBE	Acherontiina Butler, 1876	
GENUS	*Agrius* Hübner, [1819]	
1	*cingulatus* (Fabricius, 1775)	Pink-spotted Hawkmoth
2	*convolvuli convolvuli* (Linnaeus, 1758)	Convolvulus Hawkmoth
GENUS	*Acherontia* [Laspeyres], 1809	
3	*atropos* (Linnaeus, 1758)	Death's Head Hawkmoth
4	*styx styx* (Westwood, 1848)	Eastern Death's Head Hawkmoth
SUBTRIBE	Sphingina Latreille, [1802]	
GENUS	*Sphinx* Linnaeus, 1758	
5	*ligustri ligustri* Linnaeus, 1758	Privet Hawkmoth
6	*pinastri pinastri* Linnaeus, 1758	Pine Hawkmoth
6a	*pinastri maurorum* (Jordan, 1931)	
SUBTRIBE	Sphingulina Rothschild & Jordan, 1903	
GENUS	*Dolbinopsis* Rothschild & Jordan, 1903	
7	*grisea* (Hampson, [1893])	Mountain Hawkmoth
GENUS	*Kentrochrysalis* Staudinger, 1887	
8	*elegans* (A. Bang-Haas, 1912)	Ash Hawkmoth
TRIBE	SMERINTHINI Hübner, [1819]	
SUBTRIBE	Ambulycina Rothschild & Jordan, 1903	
GENUS	*Akbesia* Rothschild & Jordan, 1903	
9	*davidi* (Oberthür, 1884)	David's Hawkmoth
SUBTRIBE	Smerinthina Hübner, [1819]	
GENUS	*Marumba* Moore, [1882]	
10	*quercus* ([Denis & Schiffermüller], 1775)	Oak Hawkmoth
GENUS	*Mimas* Hübner, [1819]	
11	*tiliae* (Linnaeus, 1758)	Lime Hawkmoth
GENUS	*Smerinthus* Latreille, [1802]	
12	*kindermanni kindermanni* Lederer, 1853	Southern Eyed Hawkmoth
12a	*kindermanni orbatus* Grum-Grshimailo, 1890	
13	*caecus* Ménétriés, 1857	Northern Eyed Hawkmoth
14	*ocellatus ocellatus* (Linnaeus, 1758)	Eyed Hawkmoth
14a	*ocellatus atlanticus* Austaut, 1890	

GENUS	*Laothoe* Fabricius, 1807	
15	*populi populi* (Linnaeus, 1758)	Poplar Hawkmoth
15a	*populi populeti* (Bienert, 1870)	
16	*austauti* (Staudinger, 1877)	Maghreb Poplar Hawkmoth
17	*philerema* (Djakonov, 1923)	Pamir Poplar Hawkmoth
18	*amurensis* (Staudinger, 1892)	Aspen Hawkmoth

SUBFAMILY **Macroglossinae** Harris, 1839

TRIBE DILOPHONOTINI Burmeister, 1856

SUBTRIBE Aellopodina Carcasson, 1968

GENUS	*Hemaris* Dalman, 1816	
19	*rubra* Hampson, [1893]	Kashmir Bee Hawkmoth
20	*dentata* (Staudinger, 1887)	Anatolian Bee Hawkmoth
21	*ducalis* (Staudinger, 1887)	Pamir Bee Hawkmoth
22	*fuciformis fuciformis* (Linnaeus, 1758)	Broad-bordered Bee Hawkmoth
22a	*fuciformis jordani* (Clark, 1927)	
23	*croatica* (Esper, 1800)	Olive Bee Hawkmoth
24	*tityus tityus* (Linnaeus, 1758)	Narrow-bordered Bee Hawkmoth
24a	*tityus aksana* (Le Cerf, 1923)	

TRIBE MACROGLOSSINI Harris, 1839

SUBTRIBE Macroglossina Harris, 1839

GENUS	*Daphnis* Hübner, [1819]	
25	*nerii* (Linnaeus, 1758)	Oleander Hawkmoth
26	*hypothous hypothous* (Cramer, 1780)	Jade Hawkmoth

GENUS	*Clarina* Tutt, 1903	
27	*kotschyi kotschyi* (Kollar, 1850)	Grapevine Hawkmoth
27a	*kotschyi syriaca* (Lederer, 1855)	

GENUS	*Acosmeryx* Boisduval, [1875]	
28	*naga hissarica* Shchetkin, 1949	Naga Hawkmoth

GENUS	*Rethera* Rothschild & Jordan, 1903	
29	*afghanistana* Daniel, 1958	Afghan Madder Hawkmoth
30	*amseli* Daniel, 1958	Amsel's Hawkmoth
31	*komarovi komarovi* (Christoph, 1885)	Madder Hawkmoth
31a	*komarovi drilon* (Rebel & Zerny, 1932)	
31b	*komarovi rjabovi* O. Bang-Haas, 1935	
31c	*komarovi manifica* Brandt, 1938	
32	*brandti brandti* O. Bang-Haas, 1937	Lesser Madder Hawkmoth
32a	*brandti euteles* Jordan, 1937	

GENUS	*Sphingonaepiopsis* Wallengren, 1858	
33	*gorgoniades gorgoniades* (Hübner, [1819])	Gorgon Hawkmoth
33a	*gorgoniades pfeifferi* Zerny, 1933	
33b	*gorgoniades chloroptera* Mentzer, 1974	
34	*kuldjaensis* (Graeser, 1892)	Kuldja Hawkmoth
35	*nana* (Walker), 1856	Savanna Hawkmoth

GENUS	*Proserpinus* Hübner, [1819]	
36	*proserpina proserpina* (Pallas, 1772)	Willowherb Hawkmoth
36a	*proserpina gigas* Oberthür, 1922	
36b	*proserpina japetus* (Grum-Grshimailo, 1890)	

GENUS	*Macroglossum* Scopoli, 1777	
37	*stellatarum* (Linnaeus, 1758)	Hummingbird Hawkmoth

SUBTRIBE	Choerocampina Grote & Robinson, 1865	
GENUS	*Hyles* Hübner [1819]	
38	*euphorbiae euphorbiae* (Linnaeus, 1758)	Spurge Hawkmoth
38a	*euphorbiae conspicua* (Rothschild & Jordan, 1903)	
38b	*euphorbiae robertsi* (Butler, 1880)	
39	*tithymali tithymali* (Boisduval, 1832)	Barbary Spurge Hawkmoth
39a	*tithymali mauretanica* (Staudinger, 1871)	
39b	*tithymali deserticola* (Staudinger, 1901)	
39c	*tithymali gecki* de Freina, 1991	
39d	*tithymali himyarensis* Meerman, 1988	
40	*dahlii* (Geyer, 1827)	Smoky Spurge Hawkmoth
41	*nervosa* (Rothschild & Jordan, 1903)	Ladakh Hawkmoth
42	*salangensis* (Ebert, 1969)	Salang Hawkmoth
43	*centralasiae centralasiae* (Staudinger, 1887)	Foxtail-lily Hawkmoth
43a	*centralasiae siehei* (Püngeler, 1903)	
44	*gallii gallii* (Rottemburg, 1775)	Bedstraw Hawkmoth
45	*nicaea nicaea* (de Prunner, 1798)	Greater Spurge Hawkmoth
45a	*nicaea castissima* (Austaut, 1883)	
45b	*nicaea libanotica* (Gehlen, 1932)	
45c	*nicaea crimaea* (A. Bang-Haas, 1906)	
46	*zygophylli zygophylli* (Ochsenheimer, 1808)	Bean-caper Hawkmoth
47	*vespertilio* (Esper, 1779)	Dusky Hawkmoth
48	*hippophaes hippophaes* (Esper, 1793)	Seathorn Hawkmoth
48a	*hippophaes bienerti* (Staudinger, 1874)	
48b	*hippophaes caucasica* (Denso, 1913)	
49	*chamyla* (Denso, 1913)	Dogbane Hawkmoth
50	*livornica livornica* (Esper, 1779)	Striped Hawkmoth
GENUS	*Deilephila* [Laspeyres], 1809	
51	*elpenor elpenor* (Linnaeus, 1758)	Elephant Hawkmoth
52	*rivularis* (Boisduval, 1875)	Chitral Elephant Hawkmoth
53	*porcellus porcellus* (Linnaeus, 1758)	Small Elephant Hawkmoth
53a	*porcellus colossus* (A. Bang-Haas, 1906)	
53b	*porcellus suellus* Staudinger, 1878	
GENUS	*Hippotion* Hübner, [1819]	
54	*osiris* (Dalman, 1823)	Large Silver-striped Hawkmoth
55	*celerio* (Linnaeus, 1758)	Silver-striped Hawkmoth
GENUS	*Theretra* Hübner, [1819]	
56	*boisduvalii* (Bugnion, 1839)	Boisduval's Hawkmoth
57	*alecto* (Linnaeus, 1758)	Levant Hawkmoth

NOTES. (i) In the systematic section following (pp. 78–165), some synonyms of genera and species are prefixed by the symbol ‡. This is to indicate that the name is not nomenclaturally available under the rules of the *International Code of Zoological Nomenclature* (1985).

(ii) Although only the English vernacular names are given in the check list above, those in eight additional languages, where known, are also included under species headings in the systematic section, designated by bold letters as follows:

English (**E**); French (**F**); German (**G**); Spanish (**Sp**); Dutch (**NL**); Swedish (**Sw**); Russian (**R**); Hungarian (**H**); and Czech (**C**).

FAMILY **SPHINGIDAE** Latreille, [1802]

Sphingides Latreille, [1802], *in* Sonnini's Buffon, *Hist. nat. gén. particulière Crustacés Insectes* **3**: 400.

Characterized by the combination of vein R_1 of the hindwing crossing to vein Sc from the middle of the cell, the absence of the base of vein M_1 and loss of vein 1A (Text figs 26a,b).

SUBFAMILY **Sphinginae** Latreille, [1802]

Sphingides Latreille, [1802], *in* Sonnini's Buffon, *Hist. nat. gén. particulière Crustacés Insectes* **3**: 400.
Type genus: *Sphinx* Linnaeus, 1758.

Distributed worldwide and represented by almost 400 species.

IMAGO: First segment of labial palpus medially without sensory hairs on the naked area; proboscis developed or vestigial; frenulum and retinaculum sometimes absent; eighth sternum evenly sclerotized; midtibia with one pair of spurs distally or with two pairs of spurs.

Genitalia. Symmetrical in the male, with the well-developed sacculus bearing complex excrescences; saccus also well developed. Both uncus and gnathos usually present with single process. Female with both ante- and postvaginal lamellae; base of ductus bursae heavily sclerotized.

LARVA: Head large and round or tapering dorsally. Horn usually present, as are oblique side stripes.

PUPA: Proboscis free or fused with body.

TRIBE SPHINGINI Latreille, [1802]

Type genus: *Sphinx* Linnaeus, 1758.

A tribe comprising more than 150 species distributed throughout the Old and New World.

IMAGO: Forewing ground colour brown or grey, with numerous undulating crossbands. Hindwing usually with dark submarginal bands; central area yellow, brown or pink. Inner surface of second segment of labial palpus depressed with a naked area, or scaled and lacking this depression. Proboscis normally well developed and functional. Abdominal spines of variable length and arranged in several overlapping rows.

Genitalia. Male with uncus broad basally and with uncinate extension or a flat tip; gnathos absent, but if present, bilobate or entire, then both uncus and gnathos heavily sclerotized apically; sacculus variable, but with two pointed, straight or curved processes, or with extended, pectinate apical processes. Female with short or long ductus bursae, which is evenly broadened.

LARVA: Usually with several oblique lateral stripes, and smooth in the final instar. Horn usually present (except in *Lapara*).

PUPA: In those species with adults bearing an elongate

proboscis, proboscis free from body. In species where the adult has a short proboscis, this is fused to body. Labrum displaced ventrally. Abdominal segments 5–7 bearing one or two spiracular furrows.

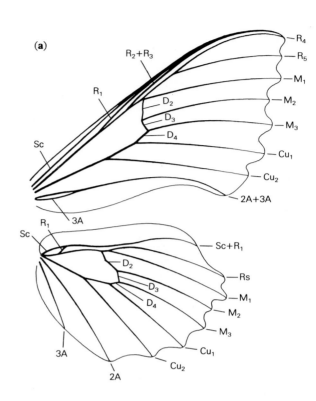

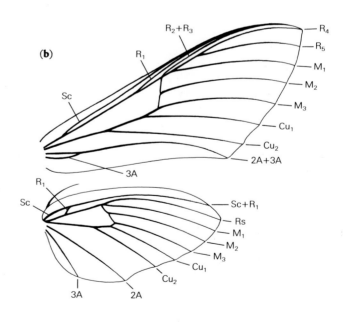

Text Figure 26 Wing venation of Sphingidae
(a) *Laothoe p. populi* (Linnaeus) (Sphinginae)
(b) *Hyles e. euphorbiae* (Linnaeus) (Macroglossinae)

SUBTRIBE Acherontiina Butler, 1876
Type genus: *Acherontia* [Laspeyres], 1809.

IMAGO: Segment 2 of labial palpus with a deep cavity on the inner side; terminal antennal segment long, filiform and covered with scales and setae; proboscis functional; posterior margin of abdominal segments with a few irregular rows of spinules.

LARVA: Head rounded; cuticle smooth in final instar. With several pronounced oblique lateral stripes.

PUPA: Glossy, with or without free proboscis. Mesonotum with transverse carina.

AGRIUS Hübner, [1819]
Agrius Hübner, [1819], *Verz. bekannter Schmett.*: 140.
Type species: *Sphinx cingulata* Fabricius, 1775.

   Timoria Kaye, 1919, *Ann. Mag. nat. Hist.* (9)**4**: 93.

(*Taxonomic note.* The species of this genus were included in *Herse* for many years; for the correct interpretation, see *Sphinx*, p. 84.)

Occurring throughout the tropical and subtropical regions, where it is represented by six species.

IMAGO: Less heavily built than *Acherontia*, with narrower forewings. Proboscis up to 2.5 times body-length, tapering towards tip. Inner lateral cavity of second segment of labial palpus deep and covered with long scales. Male antenna of almost equal thickness to terminal hook; that of female slightly club-shaped. Abdomen tapering. Tarsus slender and not compressed; mid- and hindtarsi with a basal brush on ventral side of first segment. Hindtibia with two pairs of spurs. Pulvillus very small. Paronychium with only one lobe.
Genitalia. Male valva with exterior patch of multi-dentate friction scales. Sacculus as in *Acherontia*, short, divided into two curved apical teeth, the dorsal one directed to base. Gnathos absent; uncus broad basally with a truncate apex; aedeagus stout, unarmed.

OVUM: Very small for size of moth, spherical. Bright, glossy blue-green when laid.

LARVA: As variable in ground colour as those of *Acherontia*, but horn smooth, erect and slightly curved. Head large, oval. Cuticle smooth in final instar; with oblique lateral stripes on abdominal segments or with longitudinal stripes.

PUPA (Text fig. 27): Smooth, glossy brown, with a prominent 'jug-handle' proboscis free from body. Cremaster large, with two closely approximated short spines at tip.

Text Figure 27 Pupa: *Agrius c. convolvuli* (Linnaeus)

HOSTPLANT FAMILIES: Mainly shrubs and herbaceous plants of the Convolvulaceae, Leguminosae and Boraginaceae.

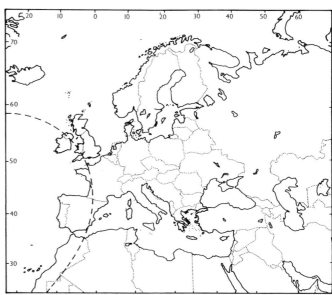

Map 1 – *Agrius cingulatus*

AGRIUS CINGULATUS (Fabricius, 1775)   Map 1
E – Pink-spotted Hawkmoth; Sweetpotato Hornworm (in U.S.A.)
Sphinx cingulata Fabricius, 1775, *Syst. Ent.*: 545.
Type locality: America.
   Sphinx affinis Goeze, 1780, *Ent. Beytr.* **3**(2): 215.
   Sphinx druraei Donovan, 1810, *The Natural History of British Insects* **14**: 1.
   Sphinx pungens Eschscholtz, 1821, *in* Kotzebue, *Entdeckungs-Reise . . .* **3**: 218.

ADULT DESCRIPTION AND VARIATION (Pl. 5, fig. 1)
Wingspan: 90–120mm. Differs from *A. convolvuli* (q.v.) in having bright pink abdominal 'ribs', as opposed to the reddish pink of *A. convolvuli*, and pink at base of hindwing. In f. *decolora* Edwards, the 'pink ribs' are pale pinkish white, almost white.

ADULT BIOLOGY
An avid visitor to flowers and a strong migrant.

FLIGHT-TIME
Migrant; all western Palaearctic records have been from mid-August to mid-October.

EARLY STAGES
OVUM: As for *A. convolvuli* (q.v., p. 80).

LARVA (not illustrated): Full-fed 90–100mm. Polymorphic: usually green, brown or yellow.
   Very similar to that of *A. convolvuli*, but has never been found in Europe.
Hostplants. Convolvulaceae, especially *Ipomoea batatas* (sweet potato) in the Americas.

PUPA: 55–60mm. Very like that of *A. convolvuli*, but with the proboscis reaching half-way down wings before reflexing under and ending in a 'bulb' which touches case near head.

PARASITOIDS
None recorded within the region.

BREEDING
As for *A. convolvuli*.

DISTRIBUTION
Occurs within the region as an autumn vagrant, though the frequency is not known due to its confusion with *A. convolvuli*. Specimens have been found in England (Barrett, 1893) and on ships off the French coast.

Extra-limital range. The tropics and subtropics of the New World, and the Galapagos and Hawaiian Islands. As a migrant, north to Canada, south to Patagonia and the Falkland Islands and, occasionally, to western Europe. *A. cingulatus* has recently established itself in the Cape Verde Islands west of Senegal, West Africa (Bauer & Traub, 1980), adults having arrived, presumably, from Brazil.

AGRIUS CONVOLVULI (Linnaeus, 1758)   Map 2
E – Convolvulus Hawkmoth, **F** – le Sphinx du liseron,
G – Windenschwärmer, **NL** – Windepijlstaart,
Sp – Esfinge de las correhuelas; esfinge de las enredaderas,
Sw – Åkervindesvärmare, **R** – Vyunkovyĭ Brazhnik,
H – folyófűszender, **C** – Lyšaj svlačcový

AGRIUS CONVOLVULI CONVOLVULI
(Linnaeus, 1758)
Sphinx convolvuli Linnaeus, 1758, *Syst. Nat.* (Edn 10) **1**: 490.
Type locality: [Europe].

> *Sphinx abadonna* Fabricius, 1798, *Ent. Syst.*, (Suppl.): 435.
> ‡*Sphinx patatas* Ménétriés, 1857, *Enum. corp. Anim. Mus. Petrop. Lepid.* **2**: 90.
> ‡*Sphinx roseafasciata* Koch, 1865, *Indo-Austr. Lepid.-Fauna*: 54.
> *Sphinx pseudoconvolvuli* Schaufuss, 1870, *Nunquam Otiosus* **1**: 15.
> *Protoparce distans* Butler, 1876, *Trans. zool. Soc. Lond.* **9**: 609.
> *Protoparce orientalis* Butler, 1876, *Trans. zool. Soc. Lond.* **9**: 609.

ADULT DESCRIPTION AND VARIATION (Pl. 5, figs 2, 3)
Wingspan: 95–130mm. Sexually dimorphic; female often considerably larger than male. Ground colour of forewing light to dark grey, with dark patches and markings which can be either extensive or absent. This large moth can be confused with no other sphingid of the region, except *A. cingulatus*. However, some specimens have yellow abdominal 'ribs' (f. *pseudoconvolvuli* Schaufuss), hence resembling North American *Manduca* species. Additionally, a very pale, almost white, high montane form occurs in the Tian Shan and Pamirs (f. *aksuensis* O. Bang-Haas).

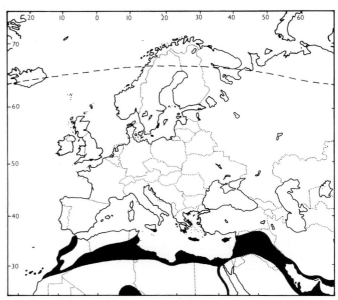

Map 2 – *Agrius convolvuli*

ADULT BIOLOGY
This species, being a great migrant, penetrates further north and in greater quantities than *Acherontia atropos*; in warm years this may involve huge numbers. However, as it is more sensitive to cold than *A. atropos*, far fewer pupae manage to survive European winters. It shows a preference for warm open areas, i.e. agricultural steppes, but can be found almost anywhere except dense forest. Outside its resident range, it is particularly attracted to potato fields and, in suburbs, to flower-beds and hedges overgrown with *Convolvulus*. Adults rest during the day on any solid surface, especially tree-trunks, fences, telegraph poles or bare earth. With wings folded roof-like over the body, they resemble a piece of weathered grey wood and are hence difficult to detect. Sometimes pairs can be found in such locations, but most, having paired towards midnight, part before dawn. At other times, *Agrius convolvuli* is an extremely active and powerful flier, frequenting beds of tubular flowers, such as *Nicotiana*, *Petunia*, *Lilium*, *Phlox* and *Jasminum*, hovering between one blossom and another while probing with its very long proboscis (up to 130mm). While on the wing between dusk and midnight, neither rain nor wind seem to deter this species. Light is also very attractive, most individuals arriving about two hours after dark.

FLIGHT-TIME
Migrant and multivoltine; in its resident range, from April to October, with up to four well-defined generations. Farther north, migrants start arriving from June to August, producing an autumn generation during August and September, individuals of which are considerably larger in size than African specimens.

EARLY STAGES
OVUM: Almost spherical, and very small for the size of the moth (1.30 × 1.15mm). Bright, glossy blue-green when laid, changing to yellowish green. Most are deposited singly on the upper and underside of the leaves of the hostplant, often in large numbers over a wide area. Each female can

oviposit up to 200 eggs which hatch within 10 to 15 days, partially collapsing before doing so.

LARVA (Pl. 1, fig. 9): Full-fed 100–110mm. Trimorphic: green, brown or yellow.

On hatching, the thin, 3–4mm long larva leaves its eggshell and immediately selects a resting place beneath a nearby leaf. Initially glaucous with a straight black-tipped horn, it gradually acquires a light green coloration through feeding. In the early stages, this involves nibbling holes through the leaf; little food is consumed and growth is slow. In the second and third instar, the ground colour deepens further, and pale yellow lateral stripes appear. By the fourth instar much more food is eaten, both by day and night, and growth is noticeably more rapid. Colour forms manifest themselves at this time, i.e. green, brown and, occasionally, yellow. When fully grown, most hide themselves by day, only emerging to feed at night, although some make no attempt at concealing themselves, but rest stretched out on the ground or on a stem, with the first two segments partly withdrawn into the third.

All are extremely sluggish, moving only enough to reach a new leaf after one has been consumed. By contrast, when a pupation site is sought, some larvae have been known to wander rapidly for several hundred metres. Prior to pupation, larvae anoint themselves (see also *Acherontia atropos*).

In its resident range, larvae are found from April to November; farther north during July, August and September.

Hostplants. Principally *Convolvulus* and *Ipomoea* species; also other Convolvulaceae and, occasionally, *Phaseolus*, *Chrysanthemum*, *Helianthus* and *Rumex*.

PUPA (Pl. E, fig. 1): 50–60mm. Rich, glossy, mahogany brown, with a large, distinctive, 'jug-handle' proboscis. Very sensitive and mobile, twitching violently if disturbed. Formed in a smooth-sided, hollow chamber in soft damp soil up to 10–20cm deep. The main overwintering stage throughout its resident range but rarely survives European winters.

PARASITOIDS

See Table 4, p. 52.

BREEDING

As for *A. atropos* (q.v., p. 83) except for the following:
  (i) adults must be force-fed on a honey and water mixture;
 (ii) strands of *Convolvulus* should be included in the oviposition cage;
(iii) larvae can be easily reared in plastic sandwich boxes, so long as they are not too crowded and kept very clean;
 (iv) pupae are very difficult to overwinter;
  (v) because emerging adults easily damage themselves due to their nervousness, pupae forming up together should be separated;
 (vi) several generations can be reared in a year (Harbich, 1980b).

DISTRIBUTION

Resident in only the warmest areas of the western Palaearctic, but a migrant to almost the entire region.

Extra-limital range. The entire tropical and subtropical Old World, with the exception of higher altitudes and more northern latitudes (to which it migrates), and a large

portion of Siberia. It is one of only three species of sphingid recorded from Iceland (Wolff, 1971).

OTHER SUBSPECIES

Marshall Islands, Pacific Ocean (170°E, 10°N) as subsp. *marshallensis* Clark, 1922. North-eastern China (Pei-tai-ho [Beidaihe, east of Beijing]) as subsp. *peitaihoensis* Clark, 1922; the validity of the latter subspecies is questionable.

ACHERONTIA [Laspeyres], 1809

Acherontia [Laspeyres], 1809, *Jena. allg. Lit.-Ztg* 4(240): 100.
Type species: *Sphinx atropos* Linnaeus, 1758.
  ‡*Manduca* Hübner, [1806], *Tentamen determinationis digestionis* . . . :[1].
  ‡*Atropos* Oken, 1815, *Okens Lehrbuch Naturgesch.* 3(1): 762.
  Brachyglossa Boisduval, 1828, *Eur. Lepid. Index meth.*: 33.
  Atropos Agassiz, 1846, *Nomencl. zool.* (Nom. syst. Generum Anim.) Lepid.: 9.

An Old World genus comprising three species, of which two occur in the region.

IMAGO: Large and heavily built, with relatively large dark forewing. Hindwing yellow with black submarginal lines. Dorsal surface of thorax with a skull-like marking. Proboscis shorter than thorax, stout and hairy, with a dorsal opening before tip. Labial palpi well separated. Antenna stout, of even thickness, with a fine terminal hook. Abdomen with yellow rib-like markings. Hindtibia with two pairs of spurs. Mid- and hindtarsi strongly compressed, without a ventral brush of long bristles at base. Pulvillus absent; paronychium reduced to form a short stumpy lobe. Squeaks if molested.

Genitalia. Male valva with a patch of large, multi-dentate friction scales exteriorly; sacculus short with two teeth.

OVUM: Oval, pale green or bluish, small for size of moth.

LARVA: Polymorphic, with a distinctly granulose yellow or light brown, downward-curving, S-shaped horn. Bears pronounced bi-coloured oblique lateral stripes and dorsal spots in the final instar. Clicks mandibles if molested.

PUPA (Text fig. 28): Smooth and glossy, with proboscis fused to body; transverse file-like ridges at base of proboscis. Cremaster broad, with two short apical spines.

HOSTPLANT FAMILIES: Trees, shrubs and herbaceous plants, mainly in the Solanaceae, Bignoniaceae, Verbenaceae and Oleaceae.

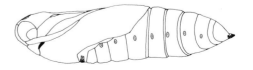

Text Figure 28 Pupa: *Acherontia s. styx* (Westwood)

ACHERONTIA ATROPOS (Linnaeus, 1758)
Map 3

E – Death's Head Hawkmoth, **F** – le Sphinx à tête de mort,
G – Totenkopf, **NL** – Doodshoofdvlinder,
Sp – Esfinge de la calavera; esfinge de la muerte,
Sw – Dödskallesvärmare, **R** – Mertvaya Golova,
H – halálfejes lepke, **C** – Smrtihlav obecný

Sphinx atropos Linnaeus, 1758, *Syst. Nat.* (Edn 10) **1**: 490.
Type locality: Europe.

‡*Atropos solani* Oken, 1815, *Okens Lehrbuch Naturgesch.*
3(1): 762.

Acherontia sculda Kirby, 1877, *Trans. ent. Soc. Lond.*
1877: 242.

ADULT DESCRIPTION AND VARIATION (Pl. 5, fig. 4; Pl. F, fig. 1)

Wingspan: 90–130mm. As illustrated. Colour somewhat variable, especially in intensity, clarity of markings and the presence of black hindwing bands. Individuals with no skull-like marking are occasionally found (f. *obsoleta* Tutt). Very difficult to confuse with any other species except *A. styx*. (See under *A. styx* for distinguishing features between the two species.)

ADULT BIOLOGY

The mottled forewing coloration of this species breaks up its outline when it rests by day on tree-trunks and walls or on leaves on the ground, with its wings held tent-like over the body (Pl. F, fig. 1). It is in such situations that mating pairs can sometimes be found, 'tail to tail' or 'side by side', from late evening to the following morning, although some separate after only a few hours.

Frequents cultivated areas where potato is grown and open scrub with solanaceous plants, tending to prefer drier and sunnier locations. Not as common as it used to be, due to the use of insecticides.

Whether in search of a mate, or feeding, *A. atropos* is active from dusk until after midnight. It is attracted to light and also to the nectar of potato, tobacco and sweet-william flowers, and orange blossom. Individuals have also been seen to frequent bee-hives where, upon entry, they feed on the honey, puncturing combs with their short, sharp proboscises. If disturbed while feeding, or for that matter at any other time, the adults raise their wings, run and hop around, while emitting high-pitched squeaks. This is especially noticeable in males that have recently arrived at a light source and have not yet settled down. Often this activity is accompanied by a mouldy smell given off by abdominal hairpencils (Birch, Poppy & Baker, 1990).

FLIGHT-TIME

Migrant and multivoltine; in its resident range, normally from March/April to August/September, although occasional specimens are on the wing in February and October. Continuous-brooded from North Africa southwards, where the winter is passed as a larva or pupa. Farther north, mainly during August, September and October as the offspring of immigrants in June and July.

EARLY STAGES

OVUM: Oval, 1.5 × 1.2mm, matt green or greyish blue with a slight polygonal network on its surface (evident only under magnification). Changes to golden buff just before emergence. Generally laid singly, low down underneath old leaves of the hostplant.

LARVA (Pl. 1, fig. 2): Full-fed 120–130mm. Trimorphic: green, brown or yellow.

On hatching, the 6mm-long larva consumes its egg-shell. Initially light, frosted green (the frostiness being imparted by small, closely packed, pale yellow tubercles and lines), it quickly darkens after feeding on leaf material. At this stage the horn is black and disproportionately long, with a bifurcate tip. In the second instar, thorn-like tubercles adorn the dorsal surface of the thoracic segments. In the third instar, blue or purple edging to the yellow lateral stripes appears and the still disproportionately large horn is now yellow with a black base. After the next moult the colours become more vivid: dark purple spots cover the dorsal surface from abdominal segment 1, with yellow replacing all the black pigmentation of the horn. However, it is only in the final instar that the horn assumes its characteristic down-curved shape, the dorsal tubercles vanish and yellow and brown colour-forms appear in addition to the more normal green.

Initially, the small larva clings to a vein on the lower surface of a leaf, nibbling small holes in its side between periods resting in a sphinx-like posture. Later, size and weight necessitate the use of a whole petiole or stem as a resting platform, with the well-camouflaged larva hanging from it. Whatever their size, larvae are extremely inactive, moving only in order to find a fresh leaf. This leads to severe and noticeable denudation over a small area. If disturbed, large larvae will repeatedly click their mandibles and may even bite an attacker severely.

Prior to pupation, the fully-grown larva darkens over a period of several hours, during which stage it anoints its whole body with 'saliva'; this appears to hasten the darkening process. This completed, a suitable location for pupation is sought.

Generally found from July to October in Europe, but also throughout the winter in North Africa.

Hostplants. Principally most Solanaceae, especially potato (*Solanum tuberosum*); also Verbenaceae, Oleaceae

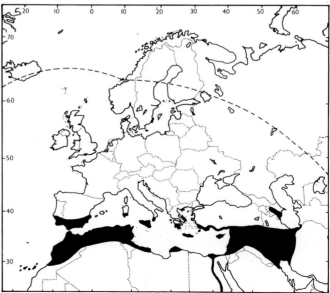

Map 3 – *Acherontia atropos*

(*Ligustrum, Olea* and *Fraxinus*), *Beta, Nerium, Buddleja* and many other plants.

PUPA: 75–80mm. Rich, glossy, mahogany-brown; very active. Proboscis fused to and flush with abdomen. Pupation normally takes place 15–40cm deep in a large, smooth-sided cavity. The main overwintering stage in its resident range; rarely survives European winters.

PARASITOIDS

See Table 4, p. 52.

BREEDING

Pairing. Not too difficult to achieve in a 130 × 130cm cylindrical, soft-netted cage at a temperature of 23–26°C. Although moths will pair in much smaller cages (40 × 40cm), success is limited and they damage themselves before many eggs are laid. To avoid the loss of tarsal claws by continuously climbing over the cage netting, a weak orange light should be suspended in the centre of the cage amongst a jar of long privet twigs. It is important for moths drawn to the centre to be able to crawl up this jar, so it must be covered with a small amount of netting. A number of males and females should be included together, all being hand-fed at least once a day on a 50 per cent honey solution followed by water only. Freshly emerged individuals refuse food until they are two to three days old.

Oviposition. Relatively easy to achieve when using the above recommendations. As the eggs are deposited on the twigs over a number of weeks, it is important that the moths be kept in as good a condition as possible. The eggs are best removed to a plastic box with no plant material.

Larvae. Very easy to rear at a temperature of between 23–26°C, with good ventilation. The pairing cage can be used again with an infra-red chick-rearing bulb for warmth. Sprigs of hostplant placed in water should be changed every three or four days. Do not rear in enclosed containers. In captivity, excellent alternative hostplants are privet (*Ligustrum*), summer jasmine (*Jasminum officinale*) and ash (*Fraxinus*), which can be interchanged.

Pupation. A deep, cardboard box with up to 40cm of soft, damp, loamy soil should be used for pupation. Alternatively, individual containers lined with damp absorbent paper are equally satisfactory. These should be placed in the dark and some ventilation is essential. Newly formed pupae, if placed on damp soil in an open aquarium and kept at above 20°C, will normally hatch within a few weeks. If they are to be overwintered, they should be placed immediately in an unheated room or outhouse and kept at a temperature of 4–8°C. Although they are capable of surviving freezing temperatures for a few days, excessive moisture is fatal; keep moderately dry in a dry environment and completely dry in a damp one. Under these conditions, most will enter diapause and overwinter, to hatch in May.

(For further details see Friedrich, 1986; Harbich, 1978a, 1978b, 1980c, 1981.)

DISTRIBUTION

An Afrotropical species which extends north to the Mediterranean (including the whole of North Africa and the Middle East) and across Turkey (Daniel, 1932; de Freina, 1979) to north-eastern Iran (Bienert, 1870; Sutton, 1963), Mesopotamia (Wiltshire, 1957), Kuwait (pers. obs.), western Saudi Arabia (Wiltshire, 1986), the Canary Islands and the Azores (Meyer, 1991). Throughout Europe as a migrant, including Iceland (Wolff, 1971).

Although resident in parts of southern Europe, a few individuals are found each year in central and northern Europe and, in some seasons, there are marked invasions, with larvae occurring in abundance during September and October. A few occasionally survive the winter as pupae, producing adults in the spring.

Extra-limital range. Tropical Africa to Ascension Island (Robinson & Kirke, 1990).

ACHERONTIA STYX (Westwood, 1848) Map 4
E – Eastern Death's Head Hawkmoth

ACHERONTIA STYX STYX (Westwood, 1848)
Sphinx styx Westwood, 1848, *Cabinet oriental Ent.*: 88.
Type locality: East Indies.

ADULT DESCRIPTION AND VARIATION (Pl. 5, fig. 5)

Wingspan 90–120mm. Similar to *A. atropos* but differs in having two medial bands on the underside of the forewing, instead of one, and usually no dark bands across the ventral surface of the abdomen. The skull-like marking is darker and there is a faint blue tornal dot enclosed by a black submarginal band on the hindwing upperside. The forewing discal spot (stigma) is orange; in *A. atropos* it is white. Variation as in *A. atropos*.

ADULT BIOLOGY

In behaviour, very similar to that of *A. atropos*. An avid robber of honey in bee hives in Oman (pers. obs.). Although not a regular migrant like *A. atropos*, limited numbers are found each year throughout the Middle East. It favours hot, open, low-lying country, either under semi-cultivation, or with oases.

FLIGHT-TIME

February, May, July and September, occurring in four

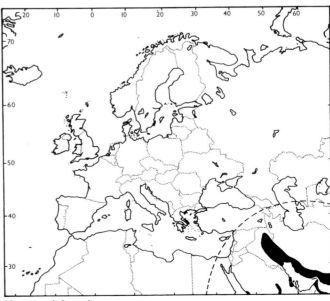

Map 4 – *Acherontia styx*

generations, although in Saudi Arabia, adults and larvae may be found until November in small numbers (Pittaway, 1979b).

EARLY STAGES

OVUM: Oval, 1.5 × 1.2mm, pale glossy green, changing to yellowish green just before hatching. Laid singly on the under and upper surfaces of leaves on peripheral twigs, usually hatching three to five days later.

LARVA (Pl. 1, fig. 8): Full-fed 90–120mm. Trimorphic: green, yellow or brown.

As in *A. atropos*, the newly-hatched, 5mm-long larvae immediately consume their eggshells. At this stage they are 'frosty' yellowish green with a long, black, fork-tipped horn. In the second instar white lateral stripes and numerous small white tubercles appear, many more than in *A. atropos*. Additionally, young larvae of *A. styx* never bear the dorsal, thoracic, thorn-like tubercles so pronounced in that species. As in *A. atropos*, it is not until after the third moult that the final colours become apparent. Fully-grown larvae closely resemble those of *A. atropos* except that the dark blue dorsal speckling is more pronounced on the anterior half of each abdominal segment, and the horn is less curved and lacks a reflexed tip.

In colour forms and behaviour, *A. styx* also resembles the previous species, being, if anything, more lethargic and confining its strip-feeding to two or three shoots. At 40°C, growth is exceedingly rapid, ovum to pupa taking as little as 10 days. Usually found from March to mid-November.

Hostplants. Principally Solanaceae, various Oleaceae and some Verbenaceae, such as *Clerodendrum inerme*, a common hedging plant in the Arabian Peninsula (Pittaway, 1981); also Cucurbitaceae, Leguminosae and *Sesamum indicum*, and, in Saudi Arabia, *Spathodea*, *Tecoma*, *Lantana*, *Vitex* and *Duranta* in gardens (pers. obs.).

PUPA (Pl. E, fig. 2): 50–60mm. Similar to, but paler than that of *A. atropos*. However, unlike that species, rarely formed more than 10cm below the surface of the soil, in a smooth-sided chamber. The main overwintering stage. Very tolerant of extremes of moisture or dryness.

PARASITOIDS

See Table 4, p. 52.

BREEDING

Conditions very similar to those for *A. atropos* are suitable for pairing and oviposition, but for the larvae more light and better ventilation are required. *Ligustrum* is a good alternative hostplant but, as it is almost impossible to change hostplants successfully, it is essential that the larva of this species be fed on the same species of plant throughout its lifespan. Wet plants should never be offered as they usually cause gastric problems. Pupation is also easy to achieve in damp sand at 20–30°C. Pupae overwinter very well at 10–15°C in an unheated room (November–April).

DISTRIBUTION

Lower Mesopotamia (Wiltshire, 1957), Saudi Arabia (Pittaway, 1979b; Wiltshire, 1980), eastern Oman (Wiltshire, 1975a,b), southern Iran and eastern Afghanistan (Ebert, 1969); as a migrant, it has been found in Turkey, Syria and Jordan (pers. obs.). *A. styx* has recently (1982) spread right

across Saudi Arabia to Jeddah (Wiltshire, 1986), and may well colonize the African mainland in the near future.

Extra-limital range. Pakistan, India, Sri Lanka, Burma and Thailand.

OTHER SUBSPECIES

Japan, China, Indo-China, Philippines, Malaysia and western Indonesia as subsp. *medusa* Butler, 1877.

SUBTRIBE Sphingina Latreille, [1802]

Type genus: *Sphinx* Linnaeus, 1758

IMAGO: Inner side of second segment of labial palpus flat (without a cavity); terminal antennal segment long, slender, sparsely covered with scales and setae; proboscis functional; posterior margin of abdominal segments with a few rows of spines.

LARVA: Head round; cuticle smooth in final instar or with tubercles on the thoracic segments.

PUPA: Smooth, glossy, with a small to large free proboscis.

SPHINX Linnaeus, 1758

Sphinx Linnaeus, 1758 *Syst. Nat.* (Edn 10) **1**: 489.
Type species: *Sphinx ligustri* Linnaeus, 1758.

    Spectrum Scopoli, 1777, *Introd. Hist. nat.*: 413.
‡*Herse* Oken, 1815, *Okens Lehrbuch Naturgesch.* 3(1): 762.
    Hyloicus Hübner,[1819], *Verz. bekannter Schmett.*: 139.
    Lethia Hübner, [1819], *Verz. bekannter Schmett.*: 141.
    Herse Agassiz, 1846, *Nomencl. zool.* (Nom. syst. Generum Anim.) Lepid.: 35.
    Lintneria Butler, 1876, *Trans. zool. Soc. Lond.* **9**: 620.
    Gargantua Kirby, 1892, *Synonymic Cat. Lepid. Heterocera* **1**: 692.
    Mesosphinx Cockerell, 1920, *Can. Ent.* **52**: 33.

A large genus found mainly in the New World, especially in the Nearctic region, but with several species occurring in the Palaearctic and two in the western part of the region.

OVUM: Ovoid, pale green.

IMAGO: Hindwing with two dark submarginal bands. Proboscis approximately equal to length of body. Antenna of even thickness in both sexes and with a very weak terminal hook. Eye small and lashed. Abdomen with grey, white, pink or yellow and black 'rib' markings. Hindtibia with two pairs of spurs, the proximal much longer than the distal. Distinguished from all other Sphingidae by the spined foretibia and lack of a pulvillus.

Text Figure 29 Pupa: *Sphinx l. ligustri* Linnaeus

Genitalia. According to Hodges (1971), '. . . the male and female genitalia seem to offer fairly consistent points of difference to separate the species from those of other genera . . . The male genitalia are somewhat diverse; however . . . the saccus is relatively broad with the apex rounded; the saccular margin is heavily sclerotized, and the apex may be either bilobate or with a single lobe; the gnathos is broadly rounded to slightly spatulate; the uncus is broad at the base, tapering to a point, in some there is a ventral development; the aedeagus is unarmed, and one side tends to project beyond the opening for the vesica. In the female genitalia, the genital plate, consisting of the joined lamellae ante- and postvaginalis, often is a broad, heavily sclerotized unit; the ostium bursae is often centered on this plate; the base of the ductus bursae may be heavily sclerotized or not; the corpus bursae is membranous, and a slender, U-shaped signum with a series of short, pointed projections is present'.

LARVA: Head slightly narrowed above, almost round, never pyramidal; body cylindrical. Colour usually some shade of green; typically sphingiform, with seven oblique lateral streaks of white or white and a dark colour. In the final instar rarely granulose. Horn erect, slightly curved and glossy in the final instar.

PUPA (Text fig. 29): Glossy reddish brown, with a small, free proboscis lying parallel to the body.

HOSTPLANT FAMILIES: Numerous, but mainly shrubs and trees of Oleaceae, Rosaceae, Caprifoliaceae, Ericaceae and Labiatae. A few species feed on Cupressaceae or Pinaceae.

SPHINX LIGUSTRI Linnaeus, 1758   Map 5

E – Privet Hawkmoth, **F** – le Sphinx du troène,
G – Ligusterschwärmer, **Sp** – Esfinge del aligustre,
NL – Ligusterpijlstaart, **Sw** – Ligustersvärmare,
R – Sirenevyĭ Brazhnik, **H** – fagyalszender,
C – Lyšaj šeříkový

SPHINX LIGUSTRI LIGUSTRI Linnaeus, 1758

Sphinx ligustri Linnaeus, 1758, *Syst. Nat.* (Edn 10) **1**: 490. Type locality: Unspecified [Europe].

    Sphinx spiraeae Esper, [1806], *Die Schmett.* (Suppl.), (Abschnitt 2): 21.
    Sphinx ligustri nisseni Rothschild & Jordan, 1916, *Novit. zool.* **23**: 253.
    Sphinx ligustri weryi Rungs, 1977, *Alexanor* **10**: 187. **Syn. rev.**
    Sphinx ligustri eichleri Eitschberger, Danner & Surholt, 1992, *Atalanta, Würzburg* **23**: 245–247. **Syn. nov.**

(*Taxonomic notes.* (i) de Freina & Witt (1984) synonymized subsp. *nisseni* and subsp. *weryi* with subsp. *ligustri* as there were insufficient features separating them. This synonymy is accepted here. For the same reasons, the resurrection of subsp. *weryi* by Eitschberger *et al.* (1992) cannot be accepted, and this subspecies is hereby re-synonymized with subsp. *ligustri*.
(ii) The creation of subsp. *eichleri* for individuals from Kars Province, Turkey, cannot be justified. Many regional populations of subsp. *ligustri* show distinct characters, all of which fall within the normal range of variation found within the nominate subspecies.)

ADULT DESCRIPTION AND VARIATION (Pl. 5, figs 6, 7; Pl. F, fig. 2)

Wingspan: 90–120mm. Difficult to confuse with any other European hawkmoth, although small, pale examples resemble the North American *Sphinx drupiferarum* J. E. Smith, which is sometimes found in this region as an 'escapee'. Exhibits great variation: pink entirely replaced by grey (f. *grisea* Closs); pink replaced by white on the hindwings and abdomen only (f. *albescens* Tutt); abdominal 'ribs' yellow (f. *lutescens* Tutt). Numerous other minor variations occur in which pink, brown or black pigments are extensive. In the Hungarian Depression, a pale form occurs which was once thought to be a distinct species (*S. spiraeae* Esper, [1806]), the larva feeding on *Spiraea*. There is no sexual dimorphism, although most females are generally larger than males.

ADULT BIOLOGY

An open-scrub and woodland-edge insect, showing a preference for limestone hills where *Ligustrum vulgare* is common, town suburbs with *L. ovalifolium* as a hedging plant, and river valleys rich in *Fraxinus* and *Spiraea*. Occurs up to 1500m in the Alps (Forster & Wohlfahrt, 1960) but only on the north-facing slopes of the Atlas Mountains in North Africa.

Rests by day on vertical surfaces such as fence posts, tree trunks and walls (Pl. F, fig. 2); rarely suspended from a twig. It is in such locations that mating pairs can be found 'tail to tail', although many separate in the afternoon and move two or three centimetres apart. Adults are active all night and can often be seen visiting privet (*Ligustrum*), honeysuckle (*Lonicera*) and many other sweet-scented flowers, with maximum activity about two hours after sunset. Both sexes are attracted to light.

FLIGHT-TIME

Univoltine; June in northern Europe. Farther south bivoltine, April/May, and again in August as a partial to full second brood.

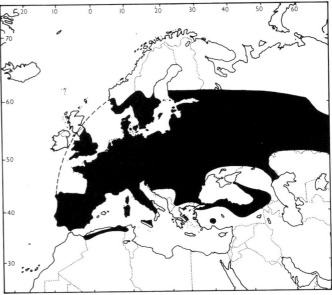

Map 5 – *Sphinx ligustri*

EARLY STAGES

OVUM: Pale green, elliptical, not very glossy (2.08 × 1.50mm). Up to 200 eggs are laid by each female on the underside of the leaves of the hostplant, usually singly, although two or three together are not uncommon. This stage lasts from 9–20 days, depending on temperature.

LARVA (Pl. 1, fig. 4): Full-fed 90–100mm. Dimorphic: usually green; also a rare purple form.

On hatching, the larva measures approximately 5mm and is pale yellow with a long dark horn. The eggshell is not eaten. With feeding, the primary colour changes to luminescent green, speckled with yellow tubercles. After the first moult, lateral streaks appear as a series of dots, their final white and purple colours not developing until the third instar; the yellow tubercles disappear in the last. Variation is not great, but some larvae have darker than normal side stripes, often complemented by a second, lower purple one. Others may have two or more horns in series, each successively smaller. In others, occasionally the primary body colour of green may be replaced by purple, but this form is very rare.

Young larvae rest beneath the midrib of a leaf, but when fully grown they assume a typical upside-down sphinx-like attitude, clinging to a petiole or stem by their anal and last two prolegs, with the thoracic segments hunched. Specimens feeding on *Ligustrum* or *Syringa* are mostly to be found clinging to stripped shoots within two metres of the ground, as are also many of those on *Fraxinus*, with saplings being preferred to mature trees.

The larval stage lasts between four and seven weeks, after which the green colour is replaced dorsally by purplish brown before the larva descends in search of a suitable pupation site.

Commonest during August and September, but with significant numbers in July and October, depending on the season and locality.

Hostplants. Principally *Ligustrum*, *Syringa*, *Fraxinus* and *Spiraea*, the last two mainly in central and south-eastern Europe; also *Ilex* (north-west Europe), *Physocarpus opulifolius*, *Viburnum*, *Euonymus*, *Cornus*, *Symphoricarpos*, *Ribes*, *Rubus*, *Pyrus*, *Sambucus* and *Lonicera* (occasionally reported on *Forsythia*).

PUPA (Pl. E, fig. 3): 50–55mm. Rich, glossy brown. Pupation normally takes place in soft, loamy soil up to 10cm deep, in a hollowed-out chamber lined with a few strands of silk. The overwintering stage.

PARASITOIDS

Albin (1720) describes and illustrates a parasitoid of this species which appears to be *Protichneumon pisorius*. See also Table 4, p. 52.

BREEDING

Pairing: Most successful in a rough wooden box (60 × 40 × 40cm high), covered with netting and placed outside but, contrary to what is recommended by Friedrich (1986), generally easy in any moderate-sized cage. Normally pairing takes place on the second night after emergence, pairs remaining coupled from dusk to mid-afternoon of the following day. Adults need not be fed, although more eggs will result if they are.

Oviposition: Takes place over the 6–8 days following separation, most eggs being deposited on the second and third nights. They are easy to obtain in any rough-sided cage; hostplants need not be included.

Larvae: Easily reared on cut or sleeved food in relatively crowded conditions. Up to the third instar, these can be kept in glass containers but, when larger, more ventilation is required. For their size, they consume large quantities of leaves.

Pupation: Easy to achieve if damp, light, loamy soil, up to 10cm deep, is provided. This should be placed in a wooden or cardboard box, as plastic receptacles often encourage disease. Hardened off, pupae should be kept on damp soil in an open tray and covered with damp peat or moss. The tray should then be placed in an outhouse, or unheated shed. If stored in tins, some succumb to a white fungus which grows within, causing stiffness and immobility of the abdominal segments.

DISTRIBUTION

Most of temperate Europe (including Sicily (Mariani, 1939)) to the Urals and lower Volga (Eversmann, 1844), central Asia (Kishida, 1987), northern and western Turkey (de Freina, 1979), northern Algeria and northern Tunisia (Rothschild & Jordan, 1903; Rougeot & Viette, 1978). Common in western Europe but rarer farther east. Although still quite numerous in northern Turkey (de Freina, 1979), it is absent from much of the Balkan Peninsula and northern Scandinavia. A partial migrant north-westwards.

Extra-limital range. See Other Subspecies.

OTHER SUBSPECIES

North-east China, Mongolia, northern Japan (Hokkaido), Kurile Islands and Amurland (Russia) as subsp. *amurensis* Oberthür, 1886. (*S. constricta* Butler, 1885, which is distributed throughout southern and central Japan (including the southern tip of Hokkaido), was formerly regarded as a subspecies of *S. ligustri* but is now considered by Kishida (1987) to be a distinct species.)

SPHINX PINASTRI Linnaeus, 1758   Map 6

E – Pine Hawkmoth, **F** – le Sphinx du pin,
G – Kiefernschwärmer, **Sp** – Esfinge del pino,
NL – Dennepijlstaart, **Sw** – Tallsvärmare,
R – Sosnovyĭ Brazhnik, **H** – fenyőszender,
C – Lyšaj borový

NOTE. In this work the following classification and synonymies have been adopted:

(i) *Sphinx pinastri pinastri* Linnaeus
         cenisius Jordan★
         medialis Jordan★
         euxinus Derzhavets

(ii) *Sphinx pinastri maurorum* Jordan
         massiliensis Jordan★

(iii) *Sphinx morio morio* Rothschild & Jordan

(iv) *Sphinx morio arestus* Jordan

The subspecies marked with an asterisk (★) appear to be the result of hybridization between several races and forms which evolved in separate refugia during the last ice age (see p. 63), a conclusion also reached by Kernbach (1958).

S. morio is now recognized as a species distinct from *S. pinastri*. Subsp. *morio* inhabits eastern Russia and northern

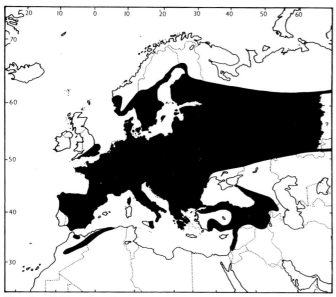

Map 6 – *Sphinx pinastri*

China (Derzhavets, 1979b; Chu & Wang, 1980b); subsp. *arestus* (Jordan, 1931) occurs in the Ussuri region (Russia), Hokkaido (Japan) and Korea. (Confusingly, *S. morio* was incorrectly reinstated as a subspecies of *S. pinastri* by Inoue in Inoue *et al.* (1982: 318).)

SPHINX PINASTRI PINASTRI Linnaeus, 1758

Sphinx pinastri Linnaeus, 1758, *Syst. Nat.* (Edn 10) 1: 492. Type locality: Unspecified [Europe].

    Sphinx piceae Gleditsch, 1775, *Forstwissenschaft* 1: 501.
    Sphinx saniptri Stricker, 1876, *Lepidoptera . . .* : 118.
    Hyloicus asiaticus Butler, 1876, *Proc. zool. Soc. Lond.* **1875**: 260.
    Hyloicus pinastri cenisius Jordan, 1931, *Novit. zool.* **36**: 243. **Syn. nov.**
    Hyloicus pinastri medialis Jordan, 1931, *Novit. zool.* **36**: 243. **Syn. nov.**
    Hyloicus pinastri euxinus Derzhavets, 1979, *Ent. Obozr.* **58**: 112–115. **Syn. nov.**

(*Taxonomic notes.* (i) *S. pinastri euxinus* cannot be considered a distinct subspecies as the genitalic differences used to separate it from *S. p. pinastri* are not marked; in these respects it differs as little from the nominate race as do 'subsp.' *cenisius* Jordan or 'subsp.' *medialis* Jordan, which are synonymized with subsp. *pinastri* (for the reasons given above).
(ii) The oriental *Sphinx morio* Rothschild & Jordan, 1903, long considered a subspecies of *S. pinastri*, is most certainly a separate species with distinct genitalic differences (Derzhavets, 1979; Litvinchuk, 1986).)

ADULT DESCRIPTION AND VARIATION (Pl. 5, fig. 8)

Wingspan: 70–96mm. Shows little resemblance to any other western Palaearctic hawkmoth, except to some extreme forms of *S. ligustri*. Ground colour very variable, ranging from dark brown (f. *brunnea* Spuler) to cream (f. *albescens* Cockayne). The normal grey colour may be of almost any shade, while the dark grey, discal bands on the forewing may be absent, or so heavy as to coalesce, forming one solid band (f. *semilugens* Andreas).

In the male genitalia, the upper sacculus branch is much longer than the lower, more or less heavily curved, almost always cylindrical in its apical half and with only rudimentary prongs. This is markedly different from that of subsp. *maurorum* (cf. Text figs 30a,b, p. 89). The apical apophysis and aedeagus are long.

ADULT BIOLOGY

Found in open or mixed pine forests, especially on dry heaths; also in mountain conifer forests up to 1600m in the Alps (Forster & Wohlfahrt, 1960), but at around 2000m in the Lebanon (Zerny, 1933; Ellison & Wiltshire, 1939).

Rests by day on a solid surface, usually the trunk of a pine tree, where it is extremely well camouflaged. Most often, single or isolated groups of trees are selected and it is in such situations that mating pairs are often found, for females are reluctant to fly until paired and remain on the trunk near where they emerged. Although some will separate in mid-afternoon, pairs normally remain *in copula* until dusk, when males fly off to feed, and possibly mate a second time, while females commence egg laying. Most sweet-smelling flowers, especially *Lonicera*, are avidly visited, with no time preference being shown. Both sexes are regularly attracted to light.

FLIGHT-TIME

Univoltine; in northern latitudes, June or even July; farther south bivoltine, during May/June, and again in August as a partial to full second brood.

EARLY STAGES

OVUM: Oval and slightly dorso-ventrally flattened (2.00 × 1.75mm); shiny pale green at first, changing to reddish yellow. Each female lays approximately 100 eggs, usually singly in close proximity to each other, on the needles or young twigs of isolated or small groups of trees. Development takes 14–20 days. Just before hatching the dark head of the larva becomes visible through the now transparent shell.

LARVA (Pl. 1, fig. 6): Full-fed 75–80mm. Dimorphic: green or brown.

On hatching, the larva consumes part of its eggshell. At this stage it is approximately 5mm long and dull yellow, with a disproportionately large head marked with brown, and a dark-tipped, forked horn. With feeding, the body colour changes to green and, after the first moult, six longitudinal creamy yellow lines appear. Still disproportionately large, the head is a paler green than the body, and has dark cheeks. During this stage, the larva sits lengthways on a pine needle, with which it blends very well. Apart from the longitudinal lines becoming broader and the horn and legs acquiring a reddish tint, little change takes place to its appearance over the next two moults. Fully grown, larvae may be either green with a marked brown dorsal band and three cream or white lateral stripes, or mainly brown, blending well with surrounding twigs. In both colour forms the whole body, which is slender and of even thickness, has dark, sunken lines, apparently dividing it up into narrow rings behind a large, prominent head. The body is smooth, but not glossy, unlike the head, underside, legs and shield. However, an oily appearance extends over the entire body as the larva darkens prior to pupation.

Initially, feeding consists of nibbling at the surface of a needle but, with increasing size, whole needles are consumed, being grasped between the legs and eaten from the tip to the base; older needles are preferred to younger ones. It is a sluggish diurnal feeder, never moving more than necessary, and comparatively little plant matter is consumed during this 4–8 week period. However, just before pupation, the larva becomes very restless and, after descending from the hostplant, it will wander some distance in search of a pupation site.

Feeds during July, August and September in the north, and June/July and September/October in southern regions.

Hostplants. Principally various *Pinus* spp. (especially *P. sylvestris*) and *Picea* spp. (especially in the Alps); also *Larix* (northern Europe), *Cedrus* (southern Europe), and the cultivated Douglas Fir (*Pseudotsuga menziesii*).

PUPA: 35–40mm. Very similar to that of *S. ligustri* and usually formed under moss or the needle mat found at the base of trees. The overwintering stage; may overwinter twice.

PARASITOIDS
See Table 4, p. 52.

BREEDING
Pairing. As for *S. ligustri* (q.v., p. 86).

Oviposition. As for *S. ligustri*, although inclusion of twigs of the hostplant gives better results.

Larvae. Easy to rear if sleeved outside on growing hostplant and not crowded. Glass containers tend to give poor results. Twigs can be placed in a well-ventilated cardboard or wooden box and replaced every day; heavy losses usually result, however, with cut food in water. Great care should be taken, if using cut hostplant, not to give damaged needles to young larvae as these ooze resin to which they adhere. As many wild-collected larvae are parasitized, it is best to rear this species from ova.

Pupation. Under moss on damp, shaded soil. As the pupae dry out easily, they should be stored in sealed tins or in damp moss. Forcing out of season is not very successful (Friedrich, 1986).

DISTRIBUTION
Europe (except Iberia, Ireland, northern Scandinavia and Arctic Russia), north-western Siberia (Eversmann, 1844), the Caucasus (Derzhavets, 1979b) to Lebanon (Zerny, 1933; Ellison & Wiltshire, 1939) and southern Turkey (Daniel, 1932, 1939; Kernbach, 1958). In Europe, found to the north-east of a line from Marseilles across south-west France and the Pyrenees (Pittaway, 1983b). There are some local European populations showing small differences in genitalia which were separated into individual subspecies by Jordan (1931), but this is unwarranted. Also found on Corsica (Bretherton & de Worms, 1963) and in the northern Aegean on Thassos (Koutsaftikis, 1970).

This species was formerly described as being a very rare vagrant to Britain (Drury, 1837) but, with the extensive cultivation of pine plantations over the last 100 years, it has become well established and has spread over most of southern England where pine trees occur. A good account of its early status in England is given by Barrett (1893) and South (1907).

Extra-limital range. Only to central Siberia.

S. pinastri appears to be resident in certain parts of the Canadian Rocky Mountains, and specimens have been found in the eastern U.S.A. (Hodges, 1971). Whether the former are native or escapees is not known. It is possible that they are the last remnants of the original North American population of this species which went on to colonize the Palaearctic, but this is pure speculation.

SPHINX PINASTRI MAURORUM (Jordan, 1931)
Stat. rev.
Hyloicus pinastri maurorum Jordan, 1931, *Novit. zool.* **36**: 243.

Type locality: Hammam R'Irha, Algeria.

Hyloicus pinastri massiliensis Jordan, 1931, *Novit. zool.* **36**: 243. **Syn. nov.**

(*Taxonomic notes.* (i) Eitschberger *et al.* (1989) raised subsp. *maurorum* to specific rank but gave no reasons for doing so. This suggestion is rejected, especially as subsp. *maurorum* and subsp. *pinastri* interbreed quite freely over several generations and have formed natural intermediate populations in central and southern France. It is probable that subsp. *maurorum* evolved for some time in isolation in North Africa and the Iberian Peninsula and that only after the end of the last ice age did it reunite with subsp. *pinastri*, which appears to have retreated to south-east Europe during that ice age. Kernbach (1958) reached similar conclusions.

(ii) 'Subsp.' *massiliensis* is synonymized with subsp. *maurorum* for the same reasons as given above for the other subspecies of *pinastri* described by Jordan, namely, that it is a hybrid race between subsp. *maurorum* and the nominate subsp. *pinastri*, but one which is closer to the former.)

ADULT DESCRIPTION AND VARIATION (Pl. 5, fig. 9)
Wingspan: first generation 70–80mm; second generation smaller. First generation generally indistinguishable superficially from central European examples. Second generation often very pale, as the specimen illustrated. In the male genitalia, unlike those of subsp. *pinastri*, both branches of the sacculus are short, the upper one flat, pronged, triangularly elongate and pointed (cf. Text figs 30a,b, p. 89). However, as in subsp. *pinastri*, the apical apophysis and aedeagus are long.

ADULT BIOLOGY
Apart from its preference for dry pine forests at higher altitudes, nothing is known of the adult biology of this subspecies.

FLIGHT TIME
Bivoltine; May/June and August in two broods.

EARLY STAGES
OVUM: As subsp. *pinastri*.

LARVA: As subsp. *pinastri*.

Hostplants. Principally *Pinus* spp., especially *P. halepensis* in North Africa; also *Cedrus*.

PUPA: As subsp. *pinastri*.

PARASITOIDS
See Table 4, p. 52.

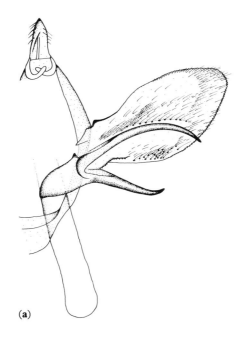

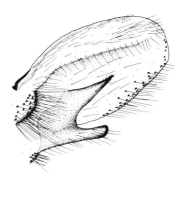

(b)

Text Figure 30 Male genitalia
(a) *Sphinx p. pinastri* Linnaeus
(b) *Sphinx p. maurorum* (Jordan)

BREEDING
As for subsp. *pinastri*.

DISTRIBUTION
The Iberian Peninsula, including both the north and south slopes of the Pyrenees; the extreme south of France bordering the Mediterranean, and the Atlas and Rif Mountains of North Africa (Rungs, 1981). Its discontinuous distribution is due to a preference for dry pine forests at higher altitudes. *Extra-limital range.* None.

OTHER SUBSPECIES
None.

SUBTRIBE Sphingulina
Rothschild & Jordan, 1903
Type genus: *Sphingulus* Staudinger, 1887.

IMAGO: Antenna of male with well-developed setae, those of the female weakly ciliate; terminal segment short. Proboscis shorter than thorax. Posterior margin of abdominal segments with weak spines, which are sometimes lacking on the sternites.

LARVA: Head large and triangular, tapering dorsally. Cuticle rough, covered with minute tubercles in final instar. Horn straight. Several oblique lateral stripes present on abdominal segments.

PUPA: Known only for *Dolbina*. Smerinthine, similar to *Marumba*, with no dorsal sculpturing on thoracic segment 3 but with indications of antespiracular ridges on abdominal segments 5–7.

DOLBINOPSIS Rothschild & Jordan, 1903

Dolbinopsis Rothschild & Jordan, 1903, *Novit. zool.* **9** (Suppl.): 156 (key), 159.
Type species: *Pseudosphinx grisea* Hampson, [1893].

This genus is very closely related to *Sphingulus* and *Kentrochrysalis* and contains only one central-southern Palaearctic species.

IMAGO: Ground colour of head, thorax and abdomen greyish with markings similar to those of the Indian *Dolbina inexacta* (Walker, 1856), but without white on the thorax. Thorax less stout than in *Dolbina*. Antennal segments not quite touching one another ventrally; penultimate segment longer than high. Proboscis non-functional, labial palpi reduced. Foretibia ending in a naked thorn; tibial spurs very short, with two pairs on hindtibia. Pulvillus absent, paronychium scarcely indicated. Rs and M_1 of hindwing separate; M_2 from centre of cell.

Genitalia. According to Rothschild & Jordan (1903), in the male, the valva lacks a patch of modified friction-scales, and is broadly rounded, broadest beyond middle. Sacculus with a finger-like ventro-distal process, and two dorsal processes, of which the proximal one is the narrower. Juxta obliquely truncate; aedeagus armed with a horizontal cornutus (tooth) pointing sinistrad. On the inner side of the aedeagus opposite the cornutus there is a patch of spines, which are visible from the outside as fine dots.

OVUM: Unknown.

LARVA: Unknown.

PUPA: Unknown. Presumably the overwintering stage.

HOSTPLANT FAMILY: Reputed to be Oleaceae (Derzhavets, 1984).

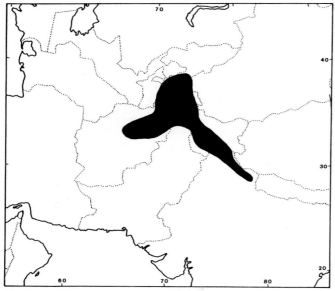

Map 7 – *Dolbinopsis grisea*

DOLBINOPSIS GRISEA (Hampson, [1893]) Map 7

E – Mountain Hawkmoth

Pseudosphinx grisea Hampson, [1893], *Fauna Brit. India (Moths)* 1: 104.
Type locality: Kulu [Himachal Pradesh, N.W. India].

ADULT DESCRIPTION AND VARIATION (Pl. 6, fig. 1)

Wingspan: 52–64mm. Very few specimens are known, although Ebert (1969) states that it is not uncommon in eastern Afghanistan. The specimen illustrated is a good example showing the normal coloration and pattern.

ADULT BIOLOGY

All that is known is that it is an inhabitant of light, temperate montane forest, including juniper (*Juniperus*) woodland. Very few females have come to light.

FLIGHT-TIME

Early April, early June to August, and early October, depending on altitude. Derzhavets (1984) states that this species is bi- or trivoltine.

EARLY STAGES

OVUM: Similar to that of *Sphinx pinastri*, but pale green (Ebert, 1969).

LARVA: Apparently unknown, although Derzhavets (1984) states that the hostplant is *Fraxinus*.

PUPA: Unknown. Presumably the overwintering stage.

PARASITOIDS

Unknown.

BREEDING

Unknown.

DISTRIBUTION

At present known only from Kashmir, eastern Afghanistan (Ebert, 1969) and eastern Tajikistan (Derzhavets, 1984). *Extra-limital range.* The temperate Himalayan foothills of Himachal Pradesh, India.

KENTROCHRYSALIS Staudinger, 1887

Kentrochrysalis Staudinger, 1887, *in* Romanoff, *Mém. Lépid.* 3: 157.
Type species: *Sphinx streckeri* Staudinger, 1880.

(*Taxonomic note.* de Freina & Witt (1987) placed *Dolbina elegans* in the genus *Kentrochrysalis* following suggestions by Kernbach (1959), who demonstrated that the male genitalia of *D. elegans* diverged markedly from what was typical of *Dolbina* and that *D. elegans* was clearly much closer to *Dolbinopsis*, *Sphingulus* and *Kentrochrysalis*, particularly the latter. Only when the early stages of this species are reared will its present generic position be confirmed.)

A Palaearctic genus of four species, one of which occurs in the western half of the region.

IMAGO: Head small, scaling prolonged to an inter-antennal tuft. Pilifer reduced to a tubercle, clothed with scales and bristles; genal process short. Proboscis short and non-functional. Labial palpi reduced, with the joint between segments 1 and 2 naked; inner surface of segment 1 scaled, except at base. Antenna in male with pronounced setae; terminal segment two or three times as long as high basally, triangular. Tibiae not spinose; spurs short, hindtibia with two pairs, but proximal pair often concealed under the scaling. First tarsal segment almost exactly as long as the four other segments together, and lacking prolonged spines. Mesotarsal comb absent, hindtarsus not longer than cell of hindwing. Pulvillus and paronychium present, but paronychial lobes very slender.

Genitalia (Text figs 31a,b). In the male, sacculus with a long ventro-distal process. Uncus scaled medially; gnathos with a broad medial lobe. Aedeagus produced into a long hook, which is curved proximally. In the female, ostium bursae covered by a broad lobe.

OVUM: Ovoid, pale yellowish green.

LARVA: In those species in which it is known, very similar to *Smerinthus* but with a long, straight, granulose horn and a pale suffusion below each lateral stripe.

PUPA: Similar to *Smerinthus* (Text fig. 35, p. 97). Proboscis fused to body. According to Staudinger (1887), that of *K. streckeri* with a pair of long, narrow pointed tubercles laterally on abdominal segments 1 and 2 similar to those of *Phyllosphingia dissimilis* (Bremer, 1861), but these are not found in the other species. Cremaster elongate, with a pair of small apical spines and lateral spines.

HOSTPLANT FAMILY: Usually trees and shrubs of the Oleaceae.

KENTROCHRYSALIS ELEGANS
(A. Bang-Haas, 1912) Map 8

E – Ash Hawkmoth

Dolbina elegans elegans A. Bang-Haas, 1912, *Dt. ent. Z. Iris* 26: 229.
Type locality: İskenderun, 'northern Syria' [southern Turkey].

    Dolbina elegans steffensi Popescu-Gorj, 1971, *Trav. Mus. Hist. nat. "Gr. Antipa"* 11: 222.

(*Taxonomic note.* There are no detectable differences be-

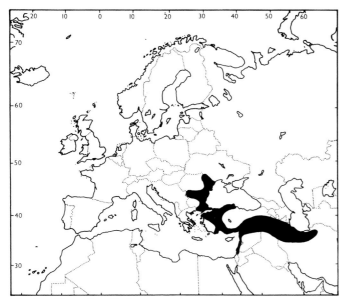

Map 8 – *Kentrochrysalis elegans*

tween the genitalia and morphology of European, Turkish and Syrian examples. Thus subspecific status is not justified for the European population, previously referred to as subsp. *steffensi*.)

ADULT DESCRIPTION AND VARIATION (Pl. 6, figs 2, 3)

Wingspan: 40–53mm. As illustrated. Very easily overlooked or confused with various other small moths when captured in a light-trap.

ADULT BIOLOGY

Found in light forest and scrubland in valleys with cultivation. The first European specimens were taken on the balcony of a tourist hotel at the Black Sea coastal resort of 'Sunny Beach', near Nessebar, Bulgaria (Soffner, 1959)!

FLIGHT-TIME

April/May to early September in two or sometimes three broods. In Europe, mainly in July (Popescu-Gorj, 1971; Ganev, 1984), although in southern Bulgaria it is regularly taken in April (S. V. Beschkow, pers. comm.) as well as in July.

EARLY STAGES

OVUM: Undescribed.

LARVA (not illustrated): Undescribed. All that appears to be known is that it is yellowish on hatching (Soffner, 1959). *Hostplants*. Probably species of *Fraxinus*, but possibly also *Olea*, *Syringa* and *Ligustrum*.

PUPA: Undescribed. Presumably the overwintering stage.

PARASITOIDS

Unknown.

BREEDING

Nothing known.

DISTRIBUTION

From Moldova (Plugaru, 1971), through eastern Romania (Popescu-Gorj, 1971), eastern and southern Bulgaria (Soffner, 1959; Ganev, 1984; S. V. Beschkow, pers. comm.), northern Greece, western and southern Turkey (de Freina, 1979) to northern Syria (Bang-Haas, 1912; Daniel, 1939), Israel (Kernbach, 1959), northern Iraq (Wiltshire, 1957) and northern Iran. The distribution indicated by the map may represent ease of access to localities rather than actual distribution and the species may also occur in other areas.

Extra-limital range. None.

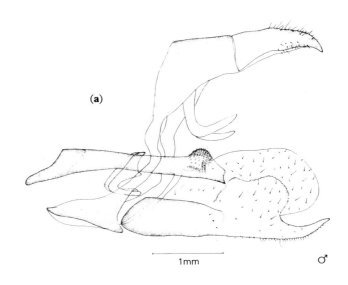

Text Figure 31

Genitalia of *Kentrochrysalis elegans* (A. Bang-Haas)

(a) male (b) female

TRIBE SMERINTHINI Hübner, [1819]
Type genus: *Smerinthus* Latreille, [1802].

A tribe of more than 200 species distributed throughout the Old World tropics, but with a significant number in the Holarctic region.

IMAGO: Forewing primarily yellow-brown, greyish lilac or green, with a few crossbands and spots. Hindwing usually with a brownish or reddish brown field; crossbands reduced to a pale or ocellate spot at anal angle. Proboscis often atrophied. Spines on caudal margins of abdominal terga generally diffuse.

Genitalia. Male with uncus variable, with a flat, entire or split tip; gnathos absent, bilobate or entire; sacculus with two simple or bilobate processes and pectens; saccus not well developed; aedeagus relatively short and often with a stout apical cornutus. Valva often with pouches on inner surface distally. Female with ductus bursa usually broad and short.

LARVA: Horn developed, except in *Pachysphinx*; cuticle rough with fine tubercles; oblique lateral stripes present, sometimes in combination with longitudinal lines. Body usually tapering forward from segment 8.

PUPA: Proboscis fused to body. Labrum displaced ventrally. Abdominal segments 5–7 with 1–4 interrupted, punctate furrows over each spiracle.

SUBTRIBE Ambulycina
Rothschild & Jordan, 1903
Type genus: *Ambulyx* Walker, 1856 (=*Protambulyx* Rothschild & Jordan, 1903)

IMAGO: Proboscis sometimes reaching abdomen but usually short and non-functional. Terminal antennal segment long or short. Forewing with extended or falcate tip.

LARVA: Head large, usually rounded. Body granulose with several oblique lateral stripes.

PUPA: Anterior end blunt.

AKBESIA Rothschild & Jordan, 1903
Akbesia Rothschild & Jordan, 1903, *Novit. zool.* **9** (Suppl.): 173 (key), 191.
Type species: *Smerinthus davidi* Oberthür, 1884.

A monotypic genus confined to the central-southern Palaearctic.

IMAGO: Apex of forewing pointed; outer margin sinuate, convex. Frenulum and retinaculum present. Proboscis extending to beyond middle of abdomen. Labial palpus short, not porrect. Pendant scales (lashes) present at upper edge of eye. Male antenna with long fasciculate setae; female antenna simply cylindrical without lateral grooves or prolonged setae; terminal segment in both sexes short. Transverse head crest between antennae. Tibiae without spines, but foretibia with long apical thorn; hindtibia with two pairs of spurs. Pulvillus small. Paronychium present, but ventral lobe vestigial. Veins Rs and M_1 of hindwing

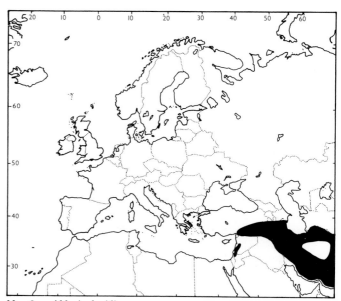

Map 9 – *Akbesia davidi*

stalked; M_2 from middle of cell.

Genitalia. In male, valva broadly sole-shaped without friction scales on outer side; apex rounded. Sacculus represented by a small sub-basal, broadly triangular ridge, which bears an obliquely vertical carina. Aedeagus without any special armature, but with apical edge produced on one side. In female, lamellae weakly chitinized with no special armature.

OVUM: Unknown.

LARVA: Unknown.

PUPA: Unknown.

HOSTPLANT FAMILIES: Unknown, but related genera on Anacardiaceae.

AKBESIA DAVIDI (Oberthür, 1884)   Map 9
E – David's Hawkmoth
Smerinthus davidi Oberthür, 1884, *Annls. Soc. ent. Fr.* (6)**4** (Bull.): xii.
Type locality: Asia Minor.

ADULT DESCRIPTION AND VARIATION (Pl. 6, fig. 4)
Wingspan: 60–70mm. A very striking species, as illustrated. Difficult to confuse with any other in the western Palaearctic. Shows little variation in coloration.

ADULT BIOLOGY
Although a local species, it often occurs in large numbers at certain sites in rocky, hilly areas supporting scattered trees and shrubs. Attracted to light, often in large numbers.

FLIGHT-TIME
Bivoltine; in Iran, April and late July/August, but across northern Iraq during May/June and August (Wiltshire, 1957).

EARLY STAGES
OVUM: Unknown.

LARVA: Unknown.

Hostplants. Not known, but possibly *Rhus* or *Pistacia*; see Hostplant Families under *Akbesia* above.

PUPA: Unknown. Presumably the overwintering stage.

PARASITOIDS
Unknown.

BREEDING
Nothing known.

DISTRIBUTION
From southern Turkey (Oberthür, 1884; Daniel, 1939) eastward across south-eastern Turkey, north-eastern Iraq (Wiltshire, 1957) to eastern Afghanistan (Sarobi) (Ebert, 1969), and southwards along the Iranian Plateau to the borders of Pakistan. Uncommon in its western range. There is an unconfirmed record of this species occurring in northern Israel and the Golan Heights (D. Benyamini, pers. comm.); it may thus occur in other parts of western and northern Syria, as well as eastern Lebanon.

Extra-limital range. None.

SUBTRIBE Smerinthina Hübner, [1819]
Type genus: *Smerinthus* Latreille, [1802].

IMAGO: Proboscis short and non-functional in most species. Terminal antennal segment short, with small or large setae. Forewing usually with undulate or serrate margin; frenulum sometimes reduced.

LARVA: Head large, usually pyramidal.

PUPA: Anterior end blunt.

MARUMBA Moore, [1882]
Marumba Moore, [1882], *Lepid. Ceylon* 2: 8.
Type species: *Smerinthus dyras* Walker, 1856.
  Burrowsia Tutt, 1902, *Nat. Hist. Br. Lepid.* 3: 386.
  Kayeia Tutt, 1902, *Nat. Hist. Br. Lepid.* 3: 386.
  Sichia Tutt, 1902, *Nat. Hist. Br. Lepid.* 3: 386.

A genus of 23 species occurring in the Palaearctic and Oriental regions, with one endemic to western Palaearctic.

IMAGO: Forewing buff, brown, or grey, with transverse lines and dentated outer margins. Hindwing apex rounded. Proboscis reduced to two separate lobes, non-functional. Antenna with prolonged setae in male; terminal segment short. Abdomen without broad dorsal scales, but densely clothed in thin spines. Tibiae spinose, but hindtibia without median pair of spurs. Tarsus stout, the spines of one row of the underside more or less erect in a comb-shape. Proportional length of hindwing veins D_2 and D_3 variable, but D_2 never twice the length of D_3, sometimes even shorter than D_3.

Genitalia. In male, valva separated by a distal incision to form a ventral and dorsal lobe; modified friction scales absent; sacculus represented by a very strongly chitinized hook curving upwards. Aedeagus without processes, but

Text Figure 32 Pupa: *Marumba quercus* ([Denis & Schiffermüller])

more or less rugose or granulose at the tip.

OVUM: Very large for the size of the moth, oval, pale whitish green, sometimes with brownish markings.

LARVA: Typically sphingiform, green, rough with fine tubercles, with either yellow or white lateral stripes, resembling *Smerinthus* (see p. 97). Anal flap and parapodia covered with large granulations. Horn straight or slightly curved and granular. Young larvae usually have a pronounced apical point to the head.

PUPA (Text fig. 32): Rugose, glossy, with no pronounced proboscis. Very similar to *Smerinthus*, i.e. short and thick, but with two pronounced frontal ridges. Coxal piece and ante-spiracular ridges present. Cremaster broad, short, usually with a pair of apical spines.

HOSTPLANT FAMILIES: Trees and shrubs mainly of the Juglandaceae, Rosaceae, Lauraceae, Tiliaceae, Malvaceae and Fagaceae.

MARUMBA QUERCUS
([Denis & Schiffermüller], 1775) Map 10
E – Oak Hawkmoth, **F** – le Sphinx du chêne,
G – Eichenschwärmer, **Sp** – Esfinge del roble,
NL – Eikepijlstaart, **R** – Dubovyĭ Brazhnik,
H – tölgyfaszender, **C** – Lyšaj borový
Sphinx quercus [Denis & Schiffermüller], 1775, *Ankündung syst. Werkes Schmett. Wienergegend*: 41, 244.
Type locality: Vienna district, Austria.
  Marumba quercus mesopotamica O. Bang-Haas, 1938, *Ent. Z., Frankf. a. M.* 52: 180. **Syn. nov.**
  Marumba quercus schirasi O. Bang-Haas, 1938, *Ent. Z., Frankf. a. M.* 52: 180. **Syn. nov.**

(*Taxonomic note*. Although this is a variable species with a number of colour forms, some of which, such as f. *mesopotamica* O. Bang-Haas and f. *schirasi* O. Bang-Haas, show geographical tendencies, none warrants subspecific status.)

ADULT DESCRIPTION AND VARIATION (Pl. 6, fig. 5)
Wingspan: male 85–100mm; female up to 125mm. Similar to a large, pale, ochreous *Laothoe populi* (q.v., p. 102) although, when at rest, the hindwings do not project above the forewings. Ground colour very variable, ranging from very pale ochreous or buff to dark brown (f. *brunnescens* Rebel); the latter is the predominant colour form among females in Asia Minor and Mesopotamia. In some, smoky grey is the main colour, while individual brown females may have yellowish white transverse lines on the wings, or a complete absence of any line. Another form has bright pink basal areas on the hindwings with the ventral surface of a similar colour (f. *mesopotamica* Bang-Haas).

ADULT BIOLOGY

Dry, sunny, wooded hillsides with a preponderance of young, shrubby oaks are favoured, usually in areas where the soil is of a light, gritty nature. Occurs up to 1500m in Spain, but in Lebanon is restricted to around 1200m (the oak zone). Adults rest by day suspended amongst foliage where they resemble dead leaves. A few may be found on tree trunks, especially *Quercus suber* (cork oak), having climbed there after emergence. It is here that mated females can sometimes be discovered, having parted from the male before dawn. As neither sex feeds in the adult stage, flowers have no attraction, although both sexes come to light.

FLIGHT-TIME

Bivoltine; May/June, with a partial to full second emergence in August.

EARLY STAGES

OVUM: Large, oval, pale green; up to 100 laid per female. Deposited singly on the underside of leaves, preferably on somewhat isolated bushy saplings.

LARVA (Pl. 1, fig. 3; Pl. C, fig. 2): Full-fed 65–80mm. Dimorphic: green or bluish white.

On hatching, the 8mm-long, whitish green larva may or may not eat the eggshell before wandering off to find a leaf on which to moult. Not until the second instar does normal feeding commence from beneath that leaf. At this stage the final coloration is established: an apple- or blue-green ground colour speckled with very fine yellow tubercles. Laterally, the body is marked with seven oblique yellow stripes, alternately wide and narrow, the last of which merges with a blue horn. Head concolorous with body, with yellow or orange cheek lines, and heavily spined in the first four but not the final two of the six instars (Text figs 33a,b). When this species feeds on *Q. ilex*, the larva tends to be greyish blue with white stripes and tubercles. Whatever the colour form, a fully-grown larva is extremely well camouflaged as it sits in a leaf cluster. Prior to pupation, it

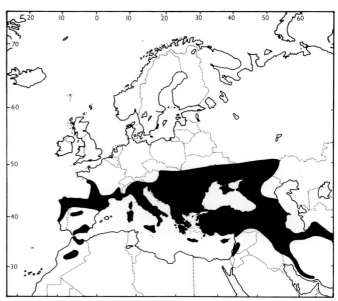

Map 10 – *Marumba quercus*

becomes reddish brown before descending from its host-plant during the night or early morning.

Occurs between June and September, depending on latitude, altitude and generation.

Hostplants. Quercus, especially dry-leaved species such as *Q. suber*, *Q. ilex* and *Q. cerris*. Shrubby bushes are preferred to large mature trees.

PUPA (Pl. E, fig. 4): 45–55mm. Rugose but glossy, dark reddish brown with a double-tipped cremaster. Pupation is noticeably deep in fine, gritty soil, where a large chamber is constructed. The overwintering stage.

PARASITOIDS

None recorded.

BREEDING

Pairing. Easy to obtain in any airy cage placed outside. Takes place around midnight, though most pairs separate before morning. Assembling can be very productive in areas where the species occurs. The moth does not feed.

Oviposition. Easy to obtain, even in a small box, as long as the female has something on which to cling. Most eggs are deposited on the second or third night after pairing.

Larvae. Best sleeved in a dry, very sunny position on *Quercus cerris* or *Q. suber*. It should be noted that North American oaks, such as *Q. borealis*, are unsuitable hostplants and usually result in heavy losses if used. Dry-leaved deciduous oaks are preferred. If in an area of little sunshine and moderate rainfall, rearing in a greenhouse on potted oak is the best solution. This species will succeed very well on cut twigs of *Q. cerris* placed in water, as long as the atmosphere is dry and warm.

The newly-emerged larva must be allowed access to the discarded eggshell, although it may completely ignore this and wander off to prepare for its first moult. It will not feed on any plant material in the first instar and must be allowed to find a leaf where it will be undisturbed. Thereafter it is difficult to start on cut food which, if used, should be clean, healthy and changed every three days. At this stage, larvae

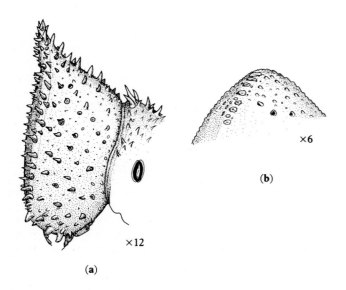

Text Figure 33 Larval heads of *Marumba quercus* ([Denis & Schiffermüller]): (a) third-instar; (b) final-instar

must not be crowded or they may kill each other. Rearing in glass containers is possible up to the third instar; thereafter good ventilation is essential. When changing the hostplant, larvae must be transferred to new leaves, for, if only placed nearby, many will starve to death rather than search for new food. Once on its hostplant, it rarely strays and can be kept until its final (sixth) instar without a cage. Losses are to be expected if larvae are disturbed during ecdysis; even so, some will still succumb before their seven-week duration is over.

Pupation. Each larva should have its own container with 10cm of light, gritty, soil/loam and sand in a 1:1 ratio. Hardened-off pupae are best kept on damp soil in a tray in a frost-free room (3°C). In April, they should be placed on dry peat in an emerging-cage. Losses are usually great if kept damp or with other pupae in a closed container.

DISTRIBUTION

From the Rif and Atlas Mountains of Morocco (Zerny, 1936; Rungs, 1981) across southern and central Europe (Herbulot, 1971; de Freina & Witt, 1987) to Turkey (Bang-Haas, 1938; Daniel, 1939; de Freina, 1979), Transcaucasia (Derzhavets, 1984), Lebanon (Zerny, 1933; Ellison & Wiltshire, 1939), Israel (Eisenstein, 1984), northern Iraq (Wiltshire, 1957) and south-west Iran (Shiraz) (Bang-Haas, 1938). Also from northern Iran (Brandt, 1938) to the lower Volga River (Eversmann, 1844) and southern Turkmenistan (Derzhavets, 1984).

This species occurs at low population densities so its presence may go undetected for long periods. Moreover, *M. quercus* is prone to wander beyond its normal range and vagrants have been discovered in northern France, the Netherlands (Meerman, 1987), Bavaria (Forster & Wohlfahrt, 1960) and Switzerland (Vorbrodt & Müller-Rutz, 1911).

Extra-limital range. None.

MIMAS Hübner, [1819]

Mimas Hübner, [1819], *Verz. bekannter Schmett.*: 142.
Type species: *Sphinx tiliae* Linnaeus, 1758.

    Lucena Rambur, [1840], *Faune ent. Andalousie* 2: 329.

A Palaearctic genus with one, or possibly two, species.

IMAGO: Very similar to *Marumba*, but forewing lobate distally. Proboscis very short and atrophied. Labial palpus with rough scaling and hair, the joint between segments 1 and 2 covered by the scaling and not visible from the outside as a naked spot; much smaller in female than male. Antenna triangular in cross-section in male, with lateral, seta-bordered grooves, the upper edge of which is laterally produced; cylindrical in female, without lateral grooves and prolonged setae. Apical segment short in both sexes. Abdominal spines restricted to segment edges, soft, sparse and partly modified into scales; ventral scales long. Tibiae with spurs; foretibia without apical thorn, epiphysis long, almost extending to apex of tibia; hindtibia with two pairs of spurs. Pulvillus, paronychium, retinaculum and frenulum present. Vein M_3 of hindwing before centre of cell.

Genitalia. In the male, valva rounded or subtriangular. Sacculus separated distally by a shallow, sinuate incision to give two small tooth-like apical lobes, which are occasion-

Text Figure 34 Pupa: *Mimas tiliae* (Linnaeus)

ally rough at edges with small notches. Aedeagus with a movable apical process, which is armed apically with two small sharp cornuti, these varying in size and position. In the female, lamellae rather large; proximal edge of ostium bursae raised, shallowly sinuate.

OVUM: Ovoid, pale green, with a brown tint developing after a few days.

LARVA: Resembles that of *Marumba*, but tapers anteriorly when fully grown, with a distinct, yellow, tuberculate structure on anal flap. Horn erect, slightly curved and granulose.

PUPA (Text fig. 34): Similar to that of *Marumba* and *Smerinthus* but, like the latter, lacks frontal ridges; also rough instead of glossy; more elongate. Cremaster curved dorsally, with a pair of spines on an extended tip and a few lateral ones.

HOSTPLANT FAMILIES: Mainly trees of the Tiliaceae, Fagaceae, Betulaceae, Ulmaceae and Rosaceae.

MIMAS TILIAE (Linnaeus, 1758)   Map 11

E – Lime Hawkmoth, **F** – le Sphinx du tilleul,
G – Lindenschwärmer, **Sp** – Esfinge del tilo,
NL – Lindepijlstaart, **Sw** – Lindsvärmare,
R – Lipovyï Brazhnik, **H** – hársfaszender,
C – Zubokřídlec lípový

Sphinx tiliae Linnaeus, 1758, *Syst. Nat.* (Edn 10) **1**: 489.
Type locality: Unspecified [Europe].

    Papilio ulmi Lucas, 1864, *Hist. nat. Lepid. Eur.* (Edn 2): 147.

    Mimas tiliae montana Daniel & Wolfsberger, 1955, *Z. wien. ent. Ges.* **40**: 63.

(*Taxonomic note.* At high altitudes, or when pupae are chilled prior to the emergence of the moth, a darker form, f. *montana* Daniel & Wolfsberger, 1955, is produced, which was regarded as a subspecies, but this is unwarranted.)

ADULT DESCRIPTION AND VARIATION (Pl. 6, figs 6, 7)
Wingspan: 60–80mm. Sexually dimorphic; a very distinct species, not confusable with any other sphingid of the area despite being highly variable. Ground colour brown (f. *brunnea* Bartel), grey (f. *pallida* Tutt), yellow (f. *lutescens* Tutt) or green (f. *virescens* Tutt). Sexual dimorphism marked: normally, the forewing ground colour is brownish in females and decidedly green in males, but there are many exceptions. The wing pattern is also very variable, the dark median band of the forewing being entire (f. *transversa* Tutt), narrowly interrupted (f. *tiliae*), or completely absent (f. *obsoleta* Clark). Gynandromorphs are also quite common. Sexual dimorphism is also evident in body-shape: the female abdomen is straight and fat because most eggs are already fully formed when the female emerges, as in all

species of Smerinthini. The male abdomen, on the other hand, is strongly curved and slender.

ADULT BIOLOGY

Rarely found outside deciduous woodland areas, where a preference is shown for open river-valley vegetation, rich in *Tilia* and *Ulmus* species; avenues of trees lining suburban and urban roads and parks; plantations and orchards of cherries (especially in central Europe); and warm, wet mountain sides overgrown with *Alnus viridis*. Up to 1500m in the Alps.

After emergence during the morning, many adults are to be found resting on the trunks of host-trees, where they have expanded their wings. Subsequently, most prefer to hang suspended amongst foliage. It is here that mating pairs are to be found – the male hanging below the female and remaining in this position for up to 20 hours before dropping to the ground and flying off at dusk. The female takes flight shortly after dusk and almost immediately commences egg-laying. However, they are active for only a short period and two hours after sunset few adults are still flying. As *M. tiliae* does not feed, the adults are not attracted to flowers, but large numbers of males come to light or a virgin female.

FLIGHT-TIME

Univoltine in northern Europe, from the end of May to early July; in southern regions bivoltine, in May and August.

EARLY STAGES

OVUM: Oval (1.75 × 1.40mm) and distinctly dorso-ventrally flattened, glossy, pale olive green (a tint imparted by its gum). Up to 130 may be deposited by each female, usually in pairs, on the underside of the leaves of its hostplant. If a tree is selected, the eggs are laid high in the crown.

LARVA (Pl. 1, fig. 1): Full-fed 55–65mm. Dimorphic: green or bluish white.

On hatching, pale green, about 6mm long and very slender, with a tail-horn about one-third of its body-length. Ignoring the eggshell, it wanders about for some time before settling on the underside of a leaf in a characteristic, erect, sphinx-like stance. Feeding normally takes place at night, with growth being confined to an increase in length over the first week; by the first moult most larvae measure *c.* 11mm. With further growth, lateral stripes become more and more prominent, changing from light to dark yellow. In the fourth instar small red streaks may appear on the anterior side of these, contrasting vividly with a yellowish or blue-green skin, which is covered in yellow tubercles. The fully-grown larva is noticeably more slender than those of related species, tapering anteriorly to a rather narrow, triangular head. The horn is blue or purple dorsally, red and yellow ventrally and situated above a reddish yellow, tuberculate anal flap.

Prior to pupation, the bright green coloration gradually changes to dull greenish pink on the underside; the red marks fade and the upper surface assumes a greyish brown hue in which the now cream tubercles stand out. The body also shrinks considerably. If touched during this stage, the larva will twitch violently from side to side.

Found throughout July, August and September, al-

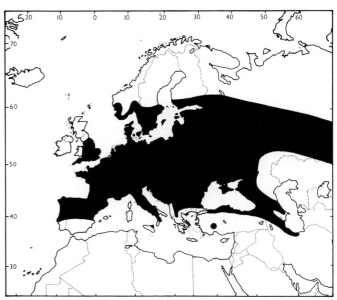

Map 11 – *Mimas tiliae*

though the second brood may feed until October. At times, larvae can be so abundant as to constitute a pest (Albin, 1720; Gninenko *et al.*, 1983).

Hostplants. Principally *Tilia*, *Ulmus*, *Alnus* and *Prunus* (cherry, cultivated or wild). Also, in northern and central Europe, *Betula*, *Quercus*, *Corylus*, *Acer*, *Sorbus* (all of which may, in some places, be major hostplants, e.g. *Acer pseudoplatanus* in Berlin and *Sorbus aucuparia* in the Alps), *Malus*, *Pyrus* and *Fraxinus*. In south-eastern areas of Europe also on *Juglans*, *Castanea* and *Aesculus*.

PUPA (Pl. E, fig. 5): 30–35mm. Very dark brown with a reddish tinge, rough and not glossy, unlike most sphingids. Pupation is normally at the base of the host-tree, under a loose mat of grass or, sometimes, damp, dead leaves and flat stones. Bare ground prompts most to wander off elsewhere, although light gritty soil will induce some to burrow 1–2 cm deep. Occasionally, pupae can be found high up in elm trees, wedged into cracks in the bark, or even under loose bark. The overwintering stage.

PARASITOIDS
See Table 4, p. 52.

BREEDING

Pairing. Very easy to achieve in a net cage at dusk, so long as the female can hang freely. Pairs remain *in copula* until the following evening. The moth does not feed.

Oviposition. Very easy to achieve in any rough-sided container.

Larvae. Very easy to rear if sleeved, or on cut food in water, or even in glass containers provided they are not crowded. If left without food for a short while or overcrowded, most will eat each other's horns.

Pupation. Just beneath damp moss placed over light, gritty soil. Pupae are easily stored in sealed tins but, in April, they should be placed on damp peat or soil and sprayed every two or three days, or daily in hot weather. In western Europe, most adults will emerge between 09.00 and 11.00 hours.

DISTRIBUTION

Across central and southern Europe (including Sicily (Mariani, 1939)) and northern and western Turkey (de Freina, 1979) to Transcaucasia, northern Iran (Sutton, 1963) and western Siberia (Eversmann, 1844; Derzhavets, 1984). Absent from Ireland (Lavery, 1991), Scotland (Gilchrist, 1979), northern Scandinavia and Arctic Russia. *Extra-limital range.* None.

OTHER SUBSPECIES

None. *M. christophi* (Staudinger, 1887) from eastern Russia (Derzhavets, 1984), Japan, northern China (Chu & Wang, 1980b) and Korea is considered to be a distinct species by Inoue (1982).

SMERINTHUS Latreille, [1802]

Smerinthus Latreille, [1802], *in* Sonnini's Buffon, *Hist. nat. gén. particulière Crustacés Insectes* 3: 401.
Type species: *Sphinx ocellata*, Linnaeus 1758.

> *Dilina* Dalman, 1816, *K. svenska VetenskAkad. Handl.* **1816**(2): 205.
> *Merinthus* Meigen, 1830, *Syst. Beschreibung eur. Schmett.* 2(4): 148.
> *Bebroptera* Sodoffsky, 1837, *Bull. Soc. imp. Nat. Moscou* **1837**(6): 84.
> *Eusmerinthus* Grote, 1877, *Can. Ent.* **9**: 132.
> *Copismerinthus* Grote, 1886, *N. Am. Lepid. Hawk Moths N. Am.*: 35.
> *Daddia* Tutt, 1902, *Nat. Hist. Br. Lepid.* 3: 386.
> *Bellia* Tutt, 1902, *Nat. Hist. Br. Lepid.* 3: 386 [homonym].
> *Bellinca* Strand, 1943, *Folia zool. hydrobiol.* **12**: 98 (objective replacement name for *Bellia* Tutt).

A Holarctic genus containing 10 species.

IMAGO: Most species are recognizable by having a blue and black ocellus in the hindwing anal angle, brown or grey marbled patterns on the forewing, which is dentate on outer margin, and a dark brown dorsal band on the thorax. Retinaculum absent, frenulum reduced. Proboscis nonfunctional. Male antenna distinctly widened above the side groove, with long setae. Abdomen dorsally with narrow, spiniform instead of broad scales. Tibiae without spines; foretibia with apical thorn only in *S. kindermanni, S. minor* Mell and *S. ocellatus* (Van der Sloot, 1967). Hindtibia with distal pair of spurs but proximal pair of spurs absent. Pulvillus and paronychium present.

Genitalia. In male, valva without friction scales. Sacculus simple, rounded or obtusely pointed at its end, and not divided as in *Laothoe*; aedeagus with one or two terminal cornuti directed laterally. In female, lamellae membranous, without a distinct ridge posterior to the ostium bursae.

OVUM: Oval, pale green.

LARVA: Green and rough with fine tubercles; yellow or white oblique lateral stripes present; distinctly sphingiform. Head triangular. Horn erect, straight and granulose. Very similar to *Marumba*.

PUPA (Text fig. 35): Dark and glossy. Very similar to *Marumba*, but lacking the frontal ridges. Antennal sheaths

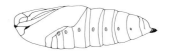

Text Figure 35 Pupa: *Smerinthus o. ocellatus* (Linnaeus)

broad in the male. Cremaster distally upturned, with a pair of small apical spines and some lateral ones.

HOSTPLANT FAMILIES: Shrubs and trees, mainly Salicaceae and Rosaceae.

SMERINTHUS KINDERMANNI Lederer, 1853
Map 12

E – Southern Eyed Hawkmoth

SMERINTHUS KINDERMANNI KINDERMANNI Lederer, 1853

Smerinthus kindermanni Lederer, 1853, *Verh. (Abh.) zool.-bot. Ver. Wien* 2: 92.
Type locality: Argana Maden, Kurdistan [Maden, north of Ergani, Turkey].
(*Taxonomic note.* It is very probable that subsp. *orbatus* Grum-Grshimailo, 1890, is only a form of the nominate subspecies. See note under subsp. *orbatus*.)

ADULT DESCRIPTION AND VARIATION (Pl. 6, fig. 8; Pl. F, fig. 3)

Wingspan: 70–80mm. Sexually dimorphic. Similar to *S. ocellatus* (Linnaeus) (Pl. 6, fig. 11), but forewing elongate, heavily dentate along the outer margin, with the cryptic pattern more distinct and outlined in light brown or buff. Abdominal segments fringed with white. There is very little variation, except between generations, the first being paler and lighter in colour. Populations from hot, arid areas are also paler than those from cooler and moister regions, and resemble subsp. *orbatus* (q.v.).

ADULT BIOLOGY

Frequents open forest, streams, plantation fringes and oases with willows (*Salix* spp.). In Mesopotamia, found almost anywhere where willows grow (Watkins & Buxton, 1923; Wiltshire, 1957); in Iran, restricted to 1500–2000m along the Zagros and Alborz Mountains.

Most adults tend to emerge during the early morning.

FLIGHT-TIME

In northern Iran and northern Turkey, and at higher altitudes, during May and June as one generation. In Lebanon, between early June and mid-September in two overlapping generations. In Mesopotamia and central Iran as three distinct generations – March, May and August/September.

EARLY STAGES

OVUM: Similar to that of *S. ocellatus* and laid in similar locations.

LARVA (Pl. 1, fig. 7; Pl. C, fig. 3): Full-fed 55–60mm. Dimorphic: fern-green or bluish white.

In colour and behaviour, very similar to *S. ocellatus* at all stages; however, of the two colour forms to be found, unlike *S. ocellatus*, the predominant one is primarily fern-green (fig. 7) with yellow oblique side stripes (sometimes edged with red); the other is bluish grey with white markings. When fully grown, shorter and much more slender than *S. ocellatus*, with a stouter but less erect anal horn.

Found from April until October, depending on altitude and latitude.

Hostplants. Principally *Salix* spp.; also *Populus* spp. (Shchetkin *et al.*, 1988).

PUPA: 30–37mm. Deep mahogany brown, but otherwise resembling a short, slender *S. ocellatus* pupa. Pupation occurs at the base of the hostplant or, more often, some distance away under a grass tussock, 2–3cm under the surface of the soil within an earthen cell. The overwintering stage.

PARASITOIDS
None recorded.

BREEDING
All stages as for *Mimas tiliae* and *S. ocellatus* (qqv., pp. 96, 100), but the larva requires more direct sunlight and higher temperatures if reared in northern Europe. Does very well on potted willow (*Salix* spp.) in a greenhouse or sleeved outside in a sunny location, even as far north as Britain. Both larvae and pupae should be kept dry, with good ventilation.

DISTRIBUTION
Not fully known within the western Palaearctic, but has been found as far west as Sinop in northern Turkey and Konya in southern Turkey (de Freina, 1979), on Cyprus, in the Caucasus, Lebanon (Zerny, 1933; Ellison & Wiltshire, 1939) and northern Iraq (Wiltshire, 1957); east to Mashhad (Kalali, 1976) and south to Shiraz (Iran) (Daniel, 1971). One specimen has been recorded from Israel (Eisenstein, 1984) and there is another from Kuwait (Kuwait University Collection) (pers. obs.); both are indistinguishable from typical subsp. *orbatus*.

Extra-limital range. Reported from the Transbaikal area of Russia as subsp. *kindermanni* (Derzhavets, 1984).

SMERINTHUS KINDERMANNI ORBATUS
Grum-Grshimailo, 1890

Smerinthus kindermanni var. *orbata* Grum-Grshimailo, 1890, *in* Romanoff, *Mém. Lépid.* 4: 512–513.
Type locality: Ferghana [Fergana, Uzbekistan].

(*Taxonomic note.* When larvae of subsp. *orbatus* from Turkestan are reared under cool, moist conditions, many of the adults produced are very similar to reared specimens of subsp. *kindermanni* from Turkey (pers. obs.). This would suggest that the former may be a dry-zone ecological form of the latter and may explain the records of subsp. *orbatus* from Israel and Kuwait and subsp. *kindermanni* from the Transbaikal area of Russia (Derzhavets, 1984).)

ADULT DESCRIPTION AND VARIATION (not illustrated)
Wingspan: 70–80mm. Generally like a pale, sun-bleached

version of subsp. *kindermanni*, with the dark bands of the forewing less tawny. The pulvillus is small and pale, and the ridge formed basally by the upper edge of the male valva is smooth.

ADULT BIOLOGY
As subsp. *kindermanni*.

FLIGHT-TIME
Trivoltine; March/April, May/June and August/September, but very much depending on altitude.

EARLY STAGES
As subsp. *kindermanni*.

PARASITOIDS
None recorded.

BREEDING
As for subsp. *kindermanni*.

DISTRIBUTION
Turkmenistan, Uzbekistan to Tajikistan (Shchetkin *et al.*, 1988) and north and east Afghanistan (Rothschild & Jordan, 1903; Ebert, 1969).
Extra-limital range. Xinjiang Province, China (Chu & Wang, 1980b), with adults occurring in July.

OTHER SUBSPECIES
Northern Pakistan (Bell & Scott, 1937) and south-east Afghanistan (Ebert, 1969) as subsp. *obsoletus* Staudinger, 1901. The southern Himalaya (India) as subsp. *meridionalis* Gehlen, 1931.

SMERINTHUS CAECUS Ménétriés, 1857   Map 13
E – Northern Eyed Hawkmoth
Smerinthus caecus Ménétriés, 1857, *Enum. corp. Anim. Mus. Petrop. Lepid.* 2: 135.
Type locality: Amurland, Russia.

ADULT DESCRIPTION AND VARIATION (Pl. 6, figs 9, 10)
Wingspan: 50–65mm. Sexually dimorphic. Very like a small *S. ocellatus* (Pl. 6, fig. 11), but with incomplete small black ocelli containing two small blue streaks. The foretibia lacks an apical thorn. Variation even less than in *S. ocellatus*, mainly consisting of differences in the amount of blue in the ocelli, and in the intensity of forewing markings.

ADULT BIOLOGY
Found in boreal forests, especially in clearings and along streams and lake fringes with a good growth of *Salix* and *Betula*. Like *Marumba quercus* (q.v., p. 95), it occurs at low population densities and may remain undetected for long periods. Males come freely to light.

FLIGHT-TIME
Univoltine; June and early July.

EARLY STAGES
OVUM: Oval, greenish white.

LARVA (Pl. 4, fig. 8): Full-fed 70mm. Probably dimorphic.
Resembles that of *S. ocellatus*. The only known illustrated specimen (from Hokkaido, Japan, photographed by K. Nakatomi), is bluish white with distinct black markings

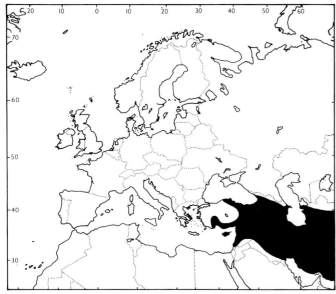

Map 12 – *Smerinthus kindermanni*

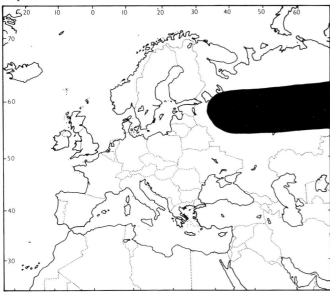

Map 13 – *Smerinthus caecus*

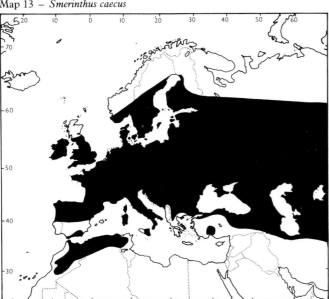

Map 14 – *Smerinthus ocellatus*

on the horn, claspers, prolegs and ventral surface of the head. No information is available for European examples.
Found throughout July and August.

Hostplants. Principally *Salix*, but also on *Populus*.

PUPA: Almost identical to that of *S. ocellatus*. The overwintering stage.

PARASITOIDS
None recorded.

BREEDING
Nothing known.

DISTRIBUTION
Found from the western outskirts of Moscow and Komiland (Russia) (Viidalepp, 1979) to the Urals and western Siberia (Derzhavets, 1984). Its exact European distribution is not known as it is often confused with *S. ocellatus*.

Extra-limital range. Eastwards across Siberia, Amurland (Russia), Korea, northern Japan to north-eastern China (Heilongjiang, Ji Lin and Liaoning Provinces).

SMERINTHUS OCELLATUS (Linnaeus, 1758)
Map 14

E – Eyed Hawkmoth, F – le Sphinx Demi-paon,
G – Abendpfauenauge, Sp – Esfinge ocelada,
NL – Avondpauwoog; Pauwoogpijlstaart,
Sw – Videsvärmare, R – Glazchatyï Brazhnik,
H – esti pávaszem, C – Zubokřídlec paví

SMERINTHUS OCELLATUS OCELLATUS
(Linnaeus, 1758)
Sphinx ocellata Linnaeus, 1758, *Syst. Nat.* (Edn 10) 1: 489. Type locality: Unspecified [Europe].
Sphinx semipavo Retzius, 1783, *Genera et Species Insect.*: 35.
Sphinx salicis Hübner, [1805], *Samml. eur. Schmett.*, Sphingidae: pl.15.
Smerinthus cinerescens Staudinger, 1879, *Stettin. ent. Ztg* 40: 316.

ADULT DESCRIPTION AND VARIATION (Pl. 6, fig. 11; Pl. F, fig. 4)
Wingspan: 70–95mm. Very similar in appearance to the other two *Smerinthus* species found in the western Palaearctic but distinguished by the presence of an apical thorn on the foretibia, and a large, circular hindwing ocellus. In addition to variations in forewing pattern intensity, differences in colour are also to be found on the hindwing, where pink may be replaced by yellow (f. *flavescens* Neumann); grey (f. *pallida* Tutt); or white (f. *albescens* Tutt). In f. *rosea* Bartel, the pink is abnormally deep and often coupled with yellowish brown forewings; in f. *ollivryi* Oberthür, the ocellus may be replaced by a buff brown patch.

ADULT BIOLOGY
A local species occurring year after year in a given area, while nearby localities, which may appear very suitable, never support a population. Even where present, numbers may fluctuate markedly from year to year due to the attentions of parasitic wasps, especially *Microplitis ocellatae*

99

Bouché. Found in many different localities, especially disused railway cuttings overgrown with sallow (*Salix* spp.), riverine shingle bars, wet river valleys, country lanes bordered by willows (*Salix* spp.), apple orchards, coastal sandhills, and suburban gardens. Up to 2000m in the Alps.

At rest (Pl. F, fig. 4), both sexes hang suspended amongst foliage, the cryptic forewing markings mimicking patterns on dead leaves, as with *Mimas tiliae* (q.v., p. 96). Its form is further broken up by the slight projection of the hindwings above the forewing costa. Also, as in *M. tiliae*, pairing takes place in such situations, usually after midnight, the male often hanging freely beneath the female until dusk the following day. After separation, females commence egg-laying almost immediately, continuing over four or five nights, with maximum activity between dusk and midnight. When alarmed, resting adults exhibit a very effective defence mechanism: the body is hunched as the forewings are flicked upwards to reveal a pair of glaring 'eyes' each side of a narrow, beak-like abdomen. Following more disturbance, this startling 'face' is further enhanced: the forewings are repeatedly lowered and raised, causing a blinking effect to the eyes. If this does not succeed, then refuge is sought on the ground. As a non-feeding species, flowers hold no attraction; light, however, can often lure large numbers of males.

FLIGHT-TIME

Univoltine in northern latitudes, from the end of May to early July; farther south bivoltine, during April/May and again in August.

EARLY STAGES

OVUM: Oval, very shiny, pale green (1.6 × 1.4mm), becoming greyish white prior to hatching. Laid singly or in pairs beneath leaves not more than 2m above the ground.

LARVA (Pl. 1, fig. 5): Full-fed: 70–80mm. Dimorphic: bluish green or apple-green.

On hatching, the young larva is about 5mm long, whitish green, with a pale pink horn. It clings tightly with all legs to a vein on the underside of a leaf. In the second instar, pale lateral stripes appear, the head becomes dorsally pointed and, when not feeding, it rests in a typical sphinx-like posture, holding on with the last three pairs of prolegs only. When so positioned beneath a leaf, the larva is very difficult to detect due to its excellent counter-shading. With growth, a number of changes take place: the horn gradually changes from pink, through purple, to blue in the final instar; the counter-shading becomes more noticeable; the dorsal surface becomes much paler than the ventral, with more of the prominent white or yellow tubercles; and the head eventually loses its pointed apex. Two colour forms are found: the commonest is bluish white with white lateral stripes and tubercles; the other (fig. 5), yellowish green with yellow markings and, occasionally, red blotches and stripe edging. Prior to pupation, all become suffused with brownish yellow before the larva wanders off to find a pupation site, which is usually some distance from the hostplant.

In northern regions found from early June until September, but in southern regions from early May until early October in two broods which often overlap.

Hostplants. Principally *Salix* and *Malus* spp.; also *Populus, Betula, Alnus, Prunus, Tilia, Ligustrum* and *Viburnum* spp. Amongst the *Prunus* complex, *P. spinosa* is frequently taken, whereas cherry, almond and peach are less favoured. In London, commonly found on *Prunus laurocerasus* (cherry laurel).

PUPA (Pl. E, fig. 6): 35–41mm. Rich, glossy blackish brown. Thickset, anteriorly blunted and tapering posteriorly, in shape resembling that of *Laothoe populi* (Pl. E, fig. 7). Most are formed in an earthen cell 2–3 cm under the surface of the soil away from the hostplant. The overwintering stage.

PARASITOIDS

Up to 80 per cent of larvae in a colony may perish in any one year due to parasitoids, such as *Microplitis ocellatae* Bouché.
See also Table 4, p. 52.

BREEDING

As for *Mimas tiliae* (q.v., p. 96), except that the larvae must be provided with up to 10cm of loamy soil in which to pupate. Larvae also take longer to mature than those of *M. tiliae*, and consume more food.

DISTRIBUTION

Western and southern Europe (including Sicily (Mariani, 1939)) eastward to the southern Urals (Eversmann, 1844; Derzhavets, 1984) and south to Transcaucasia. Also Turkey (de Freina, 1979), northern Iran (Sutton, 1963), southern Turkmenistan and Uzbekistan. Absent from northern Scandinavia and Arctic Russia.

Extra-limital range. The Urals eastwards to the Altai.

SMERINTHUS OCELLATUS ATLANTICUS
Austaut, 1890 **Stat. rev.**

Smerinthus atlanticus Austaut, 1890, *Naturaliste* 12: 190.
Type locality: Meridje [Meridja, near Kenadsa], Algeria.

 Smerinthus ocellata protai Speidel & Kaltenbach, 1981, *Atalanta*, Würzburg 12: 112–116. **Syn. nov.**

(*Taxonomic notes.* (i) Although Austaut (1890) described this subspecies from specimens obtained *ex larva* from southern Morocco and from adults obtained from the province of Oudja on the Algerian frontier, the three syntypes deposited in the BM(NH) bear the locality Meridje [Meridja], which is a small village near Kenadsa, Algeria. Meridja is very close to the border with Morocco and may once have been 'included' in that country. To avoid confusion, the specimen illustrated on Plate 6 (fig. 12) is hereby designated the **LECTOTYPE** of *S. ocellatus atlanticus*.

 LECTOTYPE ♂ labelled [ALGERIA:] Meridje [Meridja, near Kenadsa], Exampl. type; Type Austaut; *Smer. atlanticus* Austaut *Le Naturaliste* No. 83–15 Août 1890; Rothschild Bequest B.M. 1939-1; [BM(NH)].

(ii) As the male genitalia are almost identical, and there is a great overlap in adult forms, subsp. *protai* Speidel & Kaltenbach (1981), from Sardinia (Pl. 6, fig. 14), is synonymized with subsp. *atlanticus* and relegated to the status of form. The single specimen from Corsica (Pl. 6, fig. 13) is indistinguishable from *S. o. atlanticus* f. *protai*. However, it should be noted that some specimens appear morphologically to be intermediate between subsp. *ocellatus* and subsp. *atlanticus*, although the male genitalia are referable to subsp. *atlanticus*. Individuals superficially similar to *S. o. atlanticus* f. *protai* are also found in southern Spain and may represent a hybrid population.

(iii) Eitschberger *et al.* (1989) reverted *atlanticus* to specific rank but gave no valid reasons for doing so. This placement is rejected, especially as hybrids between subsp. *atlanticus* and subsp. *ocellatus* interbreed quite freely and vigorously over several generations.)

ADULT DESCRIPTION AND VARIATION (Pl. 6, figs 12–14)
Wingspan: 85–110mm. Larger and paler than subsp. *ocellatus*, but still subject to the same degree of variation. Individuals of the spring generation are smaller and darker than those of the summer generation. In the male genitalia, the sacculus is longer and distally narrower than in subsp. *ocellatus*; posterior half of aedeagus heavier and more strongly curved, almost elbowed.

ADULT BIOLOGY
Frequents plantations and river banks overgrown with *Salix* and *Populus*.

FLIGHT-TIME
Bivoltine; mainly April/May and July/August; individuals may be found from March until September.

EARLY STAGES
OVUM: As for subsp. *ocellatus*.

LARVA (Pl. C, fig. 4): Full-fed 70–90mm.
Very similar to that of subsp. *ocellatus*, but larger and with a stouter, uniformly blue-grey horn.
 Found from early May until October.

Hostplants. Populus and *Salix* spp.

PUPA: 38–45mm. Similar in all respects to that of subsp. *ocellatus*.

PARASITOIDS
None recorded.

BREEDING
As for subsp. *ocellatus*. Very easy, even under British conditions. Virgin females of this subspecies will attract males of subsp. *ocellatus* quite freely in Britain and produce very vigorous, fertile offspring over several generations.

DISTRIBUTION
In North Africa, confined to the Atlas Mountains and their surrounding lowlands, from Morocco (Rungs, 1981) to Tunisia; also on Sardinia (Speidel & Kaltenbach, 1981) and Corsica. (This link between North Africa and the two islands is also found in the Saharan *Papilio saharae* Oberthür and the endemic *P. hospiton* Guenée (Pierron, 1990).)

Extra-limital range. None.

OTHER SUBSPECIES
None, but replaced in the eastern Palaearctic by the closely related *S. planus* Walker, 1856.

LAOTHOE Fabricius, 1807

Laothoe Fabricius, 1807, *Magazin Insektenk. (Illiger)* **6**: 287.
Type species: *Sphinx populi* Linnaeus, 1758.
 ‡*Amorpha* Hübner, [1806], *Tentamen determinationis digestionis . . .* [1].

A Palaearctic genus containing five species, of which four occur in the western part of the region.

IMAGO: Differs from *Smerinthus* in having a broad hindwing, sinuate between veins R_1 and $Sc+R_1$ and produced into a lobe at veins Rs and M_1. Ocellus absent and anal angle strongly rounded. Male without, female with reduced frenulum; retinaculum absent. Proboscis non-functional. Male antenna less dentate than in *S. ocellatus*, but with long setae; terminal segment short. Abdominal tergites spiny all over, the spines weak, very dense at and near the apical edges, covered with long scales. Tibiae not spinose. Foretibia without apical thorn. Hindtibia without proximal pair of spurs. Pulvillus and paronychium present.

Genitalia. In male, sacculus bipartite at apex. Aedeagus with spines at the edge of the sheath as well as upon the membrane of the duct. In female, lamellae without processes; the proximal edge of the ostium bursae somewhat raised, wrinkled and more strongly chitinized than rest of vaginal area.

OVUM: Almost spherical, pale green and large for the size of moth.

LARVA: Typically sphingiform; very similar to that of *Smerinthus*, but body stouter and horn smaller.

PUPA (Text fig. 36): Matt black or dark brown, rough (not glossy), unlike that of *Smerinthus*. Cremaster short, dorsoventrally flattened, broad at base and terminating in a sharp point.

HOSTPLANT FAMILIES: Shrubs and trees, mainly Salicaceae.

Text Figure 36 Pupa: *Laothoe p. populi* (Linnaeus)

LAOTHOE POPULI (Linnaeus, 1758)   Map 15

E – Poplar Hawkmoth, **F** – le Sphinx du peuplier,
G – Pappelschwärmer,
Sp – Esfinge del chopo; esfinge del cuerno verde,
NL – Populierpijlstaart, **Sw** – Poppelsvärmare,
R – Topolevyï Brazhnik, **H** – nyárfaszender,
C – Zubokřídlec topolový

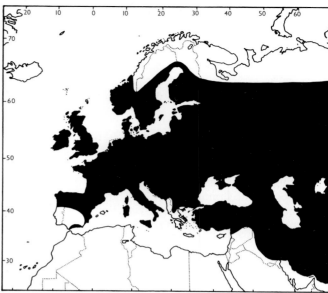

Map 15 – *Laothoe populi*

LAOTHOE POPULI POPULI (Linnaeus, 1758)

Sphinx populi Linnaeus, 1758, *Syst. Nat.* (Edn 10) **1**: 489.
Type locality: Sweden.

> *Sphinx tremulae* Borkhausen, 1793, *Rhein. Mag. Naturk.* **1**: 649.
>
> *Amorpha palustris* Holle, 1865, *Schmett. Deutschl.*: 95.
>
> *Amorpha populi intermedia* Gehlen, 1934, *Ent. Z., Frankf. a. M.* **48**: 60. **Syn. nov.**
>
> *Amorpha populi lappona* Rangnow, 1935, *Ent. Rdsch.* **52**: 189. **Syn. nov.**
>
> *Laothoe populi iberica* Eitschberger, Danner & Surholt, 1989, *Atalanta, Würzburg* **20**: 262-263. **Syn. nov.**

(*Taxonomic notes.* (i) In Turkey, subsp. *populi* and subsp. *populeti* intergrade with each other to produce a number of intermediate forms, such as f. *syriaca* Gehlen, 1932, and f. *intermedia* Gehlen, 1934. The latter is synonymized with subsp. *populi*.

(ii) Under cold climatic conditions, such as in the Arctic Circle and at high altitudes in the Alps, small, dark forms sometimes occur, similar to *L. amurensis* f. *baltica* Viidalepp. These are best referred to as *L. p. populi* f. *lappona* Rangnow.

(iii) *L. philerema* (Djakonov, 1923), from Afghanistan and Tajikistan, is now recognized as being a valid species (Daniel, 1963), as also is *L. austauti* (Staudinger, 1877). Previously, both were considered subspecies of *L. populi*, although Eitschberger *et al.* (1989) did consider *L. austauti* to be a distinct species.

(iv) *L. populi iberica* is not tenable. Superficially, individuals from southern Spain and southern France cannot be distinguished. The creation of a distinct subspecies on the strength of a few less cornuti in the male aedeagus is not warranted, especially as the genitalia within the Smerinthini are naturally somewhat variable.)

ADULT DESCRIPTION AND VARIATION (Pl. 7, fig. 1; Pl. F, fig. 5)

Wingspan: 70–100mm. A very distinct species with rust-red hindwing patches, confusable only with *L. austauti* (Staudinger), *L. philerema* (Djakonov) and *L. amurensis* (Staudinger) (Pl. 7, figs 4–8). Ground colour of the wings varies considerably from pale buff, through pale browns, reds (mainly in females), and greys to an almost black suffusion. Likewise, the transverse lines can be either very prominent, or totally absent, while the hindwing patches may be any shade of red or brown, and any size. Gynandromorphs are common (see Text fig. 11, p. 32), and if each half is a different colour, the effect can be striking.

ADULT BIOLOGY

The region's commonest hawkmoth, and one of the most widespread. Frequents almost any damp, low-lying area, such as country lanes, open woodland rides, railway cut-tings or town parks, particularly where *Populus* spp. but also *Salix* spp. are present; commoner where the former occurs. Up to 1600m in the Alps. Most emerge late at night or early in the morning, clambering up the tree trunk at the base of which the larva had pupated. Not until the following evening does the moth take flight, females quickly selecting a resting position amongst foliage from which the males are attracted at around midnight. Once paired, they remain coupled until the following evening when, after separation, the females start laying eggs almost immediately.

At rest (Pl. F, fig. 5), the hindwings protrude considerably beyond the forewings, so that amongst foliage this attitude, combined with their colour and cryptic pattern, creates a very good resemblance to a cluster of dead poplar leaves. However, individuals are attracted to light in large numbers and are then very conspicuous.

FLIGHT-TIME

Generally univoltine over its northern range, from mid-May to mid-July, with a peak towards mid-June in most years. However, due to the very warm weather experienced in southern England during 1990, this species peaked a month earlier and produced a second generation in August. Over southern Europe and Asia Minor, usually bivoltine in May and early June, and again in August, although in warm years and certain hot localities, adults may be found in April/May, July and early September.

EARLY STAGES

OVUM: Large (1.7 × 1.5mm), almost spherical, pale green, glossy. Up to 200 per female are attached, singly or in pairs, to the undersurface of leaves less than 2m from the ground. However, on the large hybrid black poplar (*Populus × euramericana*) and tree willows lacking side shoots or suckers, eggs are laid high in the crown.

LARVA (Pl. 2, fig. 3; Pl. 4, fig. 2): 65–85mm. Trimorphic: green, whitish grey and bluish white.

The newly-emerged 5mm-long larva, which partially devours its eggshell, is pale green, very rough, with small, yellow tubercles, and has a rounded head and a cream-coloured horn. With growth, yellow lateral stripes appear and the legs and spiracles become pink; the body colour, however, usually remains yellowish green, with yellow tubercles (Pl. 4, fig. 2). Occasionally, some individuals become heavily spotted with red (Pl. 2, fig 3), especially those feeding on *Salix* as opposed to *Populus* spp. Others may be bluish white with cream stripes and tubercles instead of the basic green and yellow coloration (see Grayson & Edmunds, 1989b, for an explanation of this). When full-grown, all individuals are very stocky, whatever the colour form, and develop a brownish tint prior to pupation.

Initially, the young larva rests along a vein on the underside of a leaf but, after the second moult, it adopts a more characteristic sphinx-like posture, hanging beneath a leaf or a stalk by its last two pairs of prolegs. Not very active, it tends to remain in the same feeding area throughout its life, noticeably stripping numbers of shoots.

In northern Europe, occurs from June to early September; farther south, individuals at all stages of development can be present at any time between May and late September.

Hostplants. Principally *Populus* and *Salix* spp., although *Fraxinus* and *Quercus* form a significant part of its diet in central Europe; also on *Betula, Alnus, Malus, Viburnum, Laurus* and, in Spain, *Ulmus*. It should be noted, however, that the ability of populations to feed on these secondary hostplants is a localized phenomenon. Larvae from Oxfordshire, England, refused *Betula, Fraxinus* or *Ulmus* during 1990, resulting in 100 per cent mortality; all larvae of the same brood which were offered *Populus* completed their development.

PUPA (Pl. E, fig. 7): 30–43mm. Matt black, rough and stout, blunt anteriorly. Cremaster triangular, terminating in a sharp, glossy spine beneath which two rough projections are present. Initially, completely covered with a fine dark brown 'skin' which peels off in strips. Formed in an earthen cell 2–3cm below soil level (or under damp leaves) at the base of its hostplant. The overwintering stage.

PARASITOIDS
See Table 4, p. 52.

BREEDING
Very easy. All stages as for *Mimas tiliae* (q.v., p. 96), but with the following differences:
(i) emergence is usually in the early morning;
(ii) pupation is actually in the soil, so up to 10cm of damp, light, gritty loam should be provided;
(iii) with higher temperatures it is possible to obtain three generations per year.

DISTRIBUTION
From Ireland (Lavery, 1991) to just east of the Urals (Eversmann, 1844; Pittaway, 1983b; Derzhavets, 1984) and south to southern Turkey (Daniel, 1939; de Freina, 1979). Contrary to the records of Hariri (1971), this species has never been recorded from Syria. The semi-isolated population in southern and eastern Spain, described as *L. populi iberica* Eitschberger, Danner & Surholt, 1989, (q.v., p. 102) occurs in the provinces of Granada, Almeria, Cuenca and Teruel.

Extra-limital range. None.

LAOTHOE POPULI POPULETI (Bienert, 1870)

Smerinthus populeti Bienert, 1870, *Lepid. Erg. Reise Persiens*: 33.
Type locality: Meschhet [Mashhad, N.E. Iran].

Smerinthus populi var. *populetorum* Staudinger, 1887, *Stettin. ent. Ztg* **48**: 65. **Syn. nov.**
Amorpha populi syriaca Gehlen, 1932, *Ent. Rdsch.* **49**: 184. **Syn. nov.**

(*Taxonomic notes*. (i) Subsp. *populetorum* (Staudinger, 1887) is not tenable as it differs little from subsp. *populeti*, the characters of each being found within the range of the other. It is therefore synonymized with subsp. *populeti*. The same applies to subsp. *syriaca* (Gehlen, 1932a).
(ii) Eitschberger *et al.* (1989) unnecessarily reverted *populeti* to specific rank without giving any reason, other than that the cornuti in the male aedeagus are fewer in number and less robust. As the genitalia are naturally variable in most Smerinthini, minor variations tend to have little taxonomic significance except, possibly, to identify populations.)

ADULT DESCRIPTION AND VARIATION (Pl. 7, figs 2, 3)
Wingspan: 70–120mm. Very like subsp. *populi*, but with no violet tint to the grey pigmentation, which is itself paler and faintly pinkish orange. Some may have a reddish tone (f. *vera* Staudinger), or grey replacing the pinkish tint (f. *populetorum* Staudinger (fig. 2)). In the genitalia of most males, the lobes of the sacculus are of equal length but more slender than those in subsp. *populi*, especially the upper one; the aedeagus has fewer and thinner cornuti. The uncus is obviously broad and the gnathos longer than in subsp. *populi* (cf. Text figs 37a,b, p. 104).

ADULT BIOLOGY
As subsp. *populi*.

FLIGHT-TIME
Trivoltine; April/May, June/July and September.

EARLY STAGES
OVUM: As subsp. *populi*.

LARVA: (not illustrated)
Very similar to that of *L. austauti* (Staudinger) (q.v., p. 104; Pl. 4, fig. 9) in all its stages. The caudal horn is noticeably longer, more curved and stouter than in larvae of subsp. *populi* from western Europe, but never orange as in *L. austauti*. The forehead may be pale blue in a few individuals.

Found from April until October.

Hostplants. *Populus* and *Salix* spp.

PUPA: As subsp. *populi*.

BREEDING
As for subsp. *populi*.

PARASITOIDS
None recorded.

DISTRIBUTION
Transcaucasia, eastern Turkey, north-east Iraq (Wiltshire, 1957), the mountainous Iranian plateau (Brandt, 1938), and Turkmenistan, Uzbekistan, Tajikistan, Kyrgyzstan

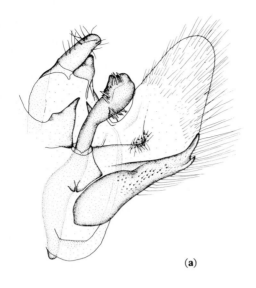

(a)

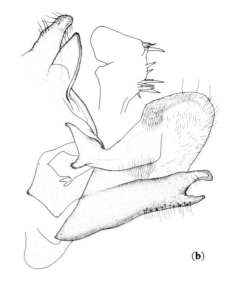

(b)

Text Figure 37 Male genitalia
(a) *Laothoe p. populi* (Linnaeus) (England)
(b) *Laothoe p. populeti* (Bienert) (Fergana, Uzbekistan)

and Kazakhstan (Alphéraky, 1882; Myartzeva & Tokeaew, 1972; Tashliev, 1973; Derzhavets, 1984), usually above 1000m.

Extra-limital range. The Tian Shan and limited areas of western Xinjiang Province, China.

LAOTHOE AUSTAUTI (Staudinger, 1877) Map 16

E – Maghreb Poplar Hawkmoth,
G – Riesenpappelschwärmer

Smerinthus austauti Staudinger, 1877, *Petites Nouv. ent.* **2**: 190.
Type locality: Nemours [Ghazaouet], Algeria.

  Smerinthus poupillieri Bellier, 1878, *Petites Nouv. ent.* **2**: 193.

(*Taxonomic note.* L. *austauti* is almost certainly a biospecies closely related to *L. populi*; this is in agreement with the findings of Eitschberger *et al.* (1989). F$_1$ hybrids between *L. austauti* and *L. p. populi* (either way) are intermediate between both species. These hybrids, when back-crossed with either parent species, produce F$_2$ hybrids in which the genitalia are slightly to severely atrophied – a phenomenon which was noted earlier by Denso (1913b) – and of which the pupae rarely overwinter and tend to produce a preponderance of males. Furthermore the adult male genitalia and early stages are distinct from those of *L. populi*.)

ADULT DESCRIPTION AND VARIATION (Pl. 7, fig. 4)

Wingspan: 95–120mm. Considerably larger and generally paler than *L. p. populi*, although manifesting all the colour forms of that species, with a dark brown form (f. *brunnea* Huard) being quite common. In the male, the antenna more triangular in cross-section and compressed than in *L. p. populi*; apical lobes of the sacculus more triangular (pointed); uncus obviously broader.

ADULT BIOLOGY

Frequents poplar- and willow-lined streams and rivers, even in desert areas; also fields, plantations and oases.

FLIGHT-TIME

Usually bivoltine; April/May and July/August. In some years there is a third brood.

EARLY STAGES

OVUM: Identical to that of *L. p. populi*, but slightly larger.

LARVA (Pl. 4, fig. 9): Full-fed 90mm. Dimorphic: green or bluish white.

The newly-emerged larva is about 6mm long and pale green. The rough body is covered with fine yellow tubercles and yellow spots, upon which there are seven yellow, oblique lateral stripes and a noticeable dorso-lateral line of the same colour. The head is rounded and the long horn pale orange.

In most individuals, this colour scheme and pattern persists into the fourth instar, with the horn becoming stouter and the head more pointed. The first and last oblique side-stripes become much more prominent than the rest. The yellow cheek-stripes become more pronounced and the dorso-lateral line disappears in the second instar. In the bluish white form these markings are paler, almost white.

When full-grown, much larger than *L. p. populi*, and very stout. The area within the now orange, inverted V-mark on the forehead is glossy and blue in colour. Compared with *L. p. populi*, the horn is mainly orange but dorsally blue-green, more glossy, longer and cylindrical in cross-section.

Found from April until October, usually high in a tree, rarely near ground level.

Hostplants. Populus and *Salix* spp., including most of the introduced cultivars.

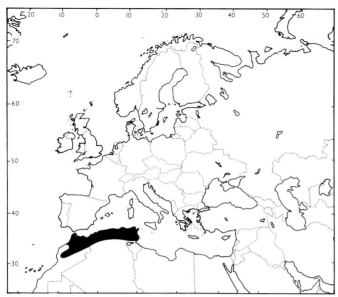

Map 16 – *Laothoe austauti*

PUPA: 45–55mm. More reddish than *L. p. populi* but in shape similar to *Smerinthus ocellatus* (Pl. E, fig. 6); in colour and texture very similar to *Mimas tiliae* (see Text fig. 5b, p. 23). Pupation takes place up to 10cm deep in the soil. The overwintering stage.

PARASITOIDS
None recorded.

BREEDING
As for *Mimas tiliae* (q.v., p. 96), but dislikes high humidity and should not be reared in closed containers. At least 15cm of soil must be provided for pupation.

DISTRIBUTION
The Atlas Mountains and coastal plains of Morocco, northern Algeria and Tunisia, and the neighbouring desert areas (Rungs, 1981; Pittaway, 1983b).
Extra-limital range. None.

LAOTHOE PHILEREMA (Djakonov, 1923)
Map 17

E – Pamir Poplar Hawkmoth
Amorpha populi philerema Djakonov, 1923, *Ezheg. zool. Muz.* **24**: 104–115, pl. 4.
Type locality: Termez, near Bukhara [Amu-Dar'ya River, southern Uzbekistan].

(*Taxonomic note*. Gehlen (1932b) stated that this species is allied to or is even a subspecies of *L. amurensis* (Staudinger), basing his conclusions on the lack of a rust-red hindwing patch in both species; however, examination of the male genitalia shows them to be very different from those of *L. amurensis* and that there is a closer relationship between *L. philerema* and *L. populi*.)

ADULT DESCRIPTION AND VARIATION (Pl. 7, figs 5, 6)
Wingspan: 80–120mm. Resembles a very pale, poorly marked *L. p. populi*, but with the rust-red hindwing patch

of that species vestigial or absent. The male antennae are unusually long, being half the length of the forewing, which itself bears a very distinct antemedial band. In the male genitalia, the two apical processes of the sacculus are even more reduced than in *L. populi populeti*, although similar in outline; the aedeagus is more slender, less elbowed and with fewer and smaller cornuti at the edge of the vesica.

ADULT BIOLOGY
Very little is known except that it is an inhabitant of river valleys in the Hindu Kush, Pamirs and Tian Shan.

FLIGHT-TIME
Bivoltine, or possibly trivoltine. Adults have been captured in late April and late June.

EARLY STAGES
OVUM: As *L. p. populi*.

LARVA (not illustrated): Undescribed. Full-fed 65–75mm.
Hostplants. *Populus* spp. of the subgenus *Turanga* (Derzhavets, 1984).

PUPA: Undescribed. Presumably the overwintering stage.

PARASITOIDS
None recorded.

BREEDING
Unknown, but presumably as for *L. p. populi* except that drier conditions would be required.

DISTRIBUTION
Known only from south-eastern Turkmenistan (one specimen in BM(NH) coll.), southern Uzbekistan (Djakonov, 1923), Tajikistan (Derzhavets, 1984) and eastern Afghanistan (Ebert, 1969; Daniel, 1971), i.e. the Hindu Kush, Pamirs and western Tian Shan.
Extra-limital range. Unknown, but it may occur in north-western China, northern Pakistan (Hindu Kush) and most of the rest of the Tian Shan range.

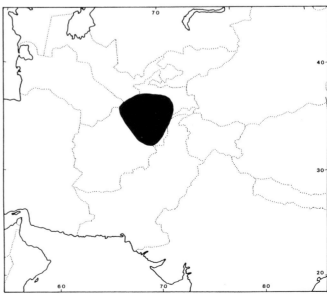

Map 17 – *Laothoe philerema*

LAOTHOE AMURENSIS (Staudinger, 1892)
Map 18

E – Aspen Hawkmoth, **F** – le Sphinx du tremble,
G – Zitterpappelschwärmer, **Sw** – Aspsvärmare,
R – Osinovyĭ Brazhnik

Smerinthus tremulae var. *amurensis* Staudinger, 1892, *in* Romanoff, *Mém. Lépid.* **6**: 155.
Type locality: Amurland [Far Eastern Russia].

> *Smerinthus tremulae* Boisduval, 1828, *Eur. Lepid. Index meth.*: 34 [homonym]. [This name is often incorrectly attributed to G. Fischer de Waldheim, 1830, *Oryct. Gouv. Moscou* (Edn 1): [25], pl. 10, figs 1, 2.]
>
> *Laothoe tremulae baltica* Viĭdalepp, 1979, *Uchen. Zap. tartu gos. Univ.* **12**: 32. **Syn. nov.**

(*Taxonomic notes.* (i) This species is often referred to as *L. tremulae*. However, as *Smerinthus tremulae* Boisduval is a junior primary homonym of *Sphinx tremulae* Borkhausen, 1793 (see *L. p. populi*, p. 102), the next oldest available name must be used.

(ii) Subsp. *baltica* Viĭdalepp, 1979, is not tenable as it is clearly only a form that appears occasionally within the European range of the nominate subspecies.

(iii) *Laothoe sinica* (Rothschild & Jordan, 1903) **stat. nov.**, formerly referred to as a subspecies of *L. amurensis*, and found in the cold mountains of central China from the Mongolian border to the foothills of the Tibetan plateau and Sichuan Province, appears to be a distinct biospecies allied to *L. amurensis*. Notable localities are the Daba (Tapa) Shan and Qinling (Tsinling) Shan ranges south of Xi'an. Chu & Wang (1980b) also record *L. sinica* from western China and there is a specimen in BM(NH) coll. from the 'eastern Sayan Mountains' west of Irkutsk, Russia, but this record requires confirmation.)

ADULT DESCRIPTION AND VARIATION (Pl. 7, figs 7, 8)

Wingspan: 75–95mm. Closely resembles *L. p. populi*, but immediately distinguishable by the absence of the rust-red patch at base of hindwing, although there may be a tint of brown present in some individuals. Ground colour equally variable, ranging from buff through pink to various shades of grey. The small, dark-grey form is referable to f. *baltica* Viĭdalepp (fig. 8). According to Rothschild & Jordan (1903), in the male genitalia, the sacculus is much broader than in *L. p. populi*, the two lobes (processes) being very broad; the aedeagus is multispinose around almost all the edge, with the spines being small on one side; uncus shorter than in *L. p. populi*, gnathos more evenly convex below, its sides straighter, the apex more obtuse.

ADULT BIOLOGY

This species is found in woodland clearings and on lake margins abounding in *Populus tremula* (aspen); rarely elsewhere.

The behaviour of *L. amurensis* is very similar to that of *L. populi*. It rarely flies before midnight and has often been seen to fly to and fro over still water, dipping into it repeatedly. Certain tropical sphingids such as species of *Pachylia* Walker, *Hemeroplanes* Hübner and *Perigonia* Herrich-Schäffer are known to drink and this species may do the same. Individuals in any one population tend to emerge within 5–7 days of one another and have a very short lifespan of only 6–10 days. This suggests that this

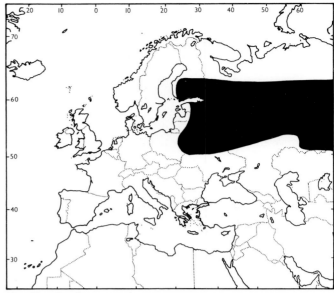

Map 18 – *Laothoe amurensis*

species is a true denizen of boreal climatic conditions, i.e. short, warm summers and long cold winters.

FLIGHT-TIME

Univoltine. Both sexes are strongly attracted to light between late May and early July, depending on locality, with the main flight period around the first 10 days of June (Skvortzov & Thomson, 1974).

EARLY STAGES

OVUM: Glossy, greenish yellow, rather large for size of moth (2.3 × 2.0 × 1.9mm) and dorso-ventrally flattened. Each female deposits fewer than 100 eggs, singly, on the underside or petiole of a leaf, hatching 10–12 days later.

LARVA (Pl. 4, fig. 5): Full-fed 65–80mm. Dimorphic: bluish green or dark green.

The newly-emerged larva is pale green with faint yellow markings appearing after feeding. In the second instar, it assumes its final coloration and form, which is bluish green (sometimes dark green) with white oblique lateral stripes on abdominal segments 1–8 and covered in white tubercles. Resembles that of *L. p. populi*, except that the spiracles are larger and a pair of enlarged tubercles are present dorsally on thoracic segment 2; these tubercles become more noticeable as the larva grows. The spiracles are white, ringed with brown, and the short horn is bluish, dotted laterally with white. As in *L. p. populi*, it assumes a light plum colour prior to pupation.

On hatching, the larva commences feeding immediately on mature leaves. At first, these are skeletonized but, with increasing size, whole leaves are consumed except for the midrib (Skvortzov & Thomson, 1974). However, like the larva of *L. p. populi*, it is a wasteful feeder, with large segments of leaf being chewed off and falling to the ground. It feeds at night and, at first, rests by day stretched out under a leaf. In the second and later instars it assumes a sphinx-like attitude with its head hunched under the thorax; only the last 2–3 pairs of prolegs cling to the leaf. Fully-grown larvae tend to rest singly along a petiole, high up in mature trees where their colour enables them to blend superbly with the surrounding blue-green aspen leaves. When feeding on *Populus*, most individuals complete only four instars before pupation; some, however, pass through five instars, as do those feeding on *Salix*. This stage takes about 40–50 days, the resulting mature larva being larger than that of *L. p. populi*.

Occurs from late June to early August.

Hostplants. Principally *Populus tremula* and *P. lancifolia* (Seppänen, 1954), but also *Salix* and other *Populus* spp.

PUPA: 30–43mm. In colour and shape, almost identical to that of *L. p. populi*, except for the presence of a two-tier cremaster. Pupation normally takes place after only about four days in an unlined chamber, 2–3cm deep in the soil at the base of a tree or in a grass tussock. The overwintering stage.

PARASITOIDS

See Table 4, p. 52.

BREEDING

Treat as for *L. populi*, but note that the eggs hatch in seven days (Pelzer, 1989).

DISTRIBUTION

Southern Finland (Seppänen, 1954; Nordström *et al.*, 1961), the European part of Russia, Belarus (Merzheevskaya *et al.*, 1976), Estonia (Petersen, 1924; Thomson, 1967; Skvortzov & Thomson, 1974) and Lithuania (Kazlauskas, 1984)), and eastern Poland (Derzhavets, 1984). Although quite widely distributed in northern Europe, its occurrence may go undetected for long periods due to low population densities and a preference for remote, inaccessible regions. However, new localities are regularly being discovered as moth-traps are increasingly deployed in wilder regions.

Extra-limital range. The boreal regions of Siberia to Sakhalin Island (Russia) (Derzhavets, 1984), northern Japan, Korea and north-eastern China (Chu & Wang, 1980b) as far south as Beijing.

SUBSPECIES

None (see Taxonomic note (iii) above).

SUBFAMILY **Macroglossinae** Harris, 1839

Macroglossiadae Harris, 1839, *Am. J. Sci. Arts* **36**: 287. Type genus: *Macroglossum* Scopoli, 1777.

Distributed worldwide and comprising more than 600 species.

IMAGO: First segment of labial palpus on mesal side with short sensory hairs on the naked area; proboscis long; frenulum and retinaculum always well developed; spines on caudal margins of abdominal terga either strong or weak, sometimes diffuse; midtibia with a single pair of spurs, hindtibia with two pairs.

Genitalia. In male, symmetrical or slightly to moderately asymmetrical with gnathos and uncus divided or undivided; saccus sometimes developed; sacculus with a single apical process; eighth sternum usually modified by being sclerotized medially, laterally and basally. In female, lamella postvaginalis developed, lamella antevaginalis absent.

LARVA: Variable. Head large or small, but always rounded. Horn may be absent and replaced by a 'button' in the final instar. Tapering anteriorly from thoracic segment 3 or abdominal segment 1 to head. Eye-spots may be present on these segments.

PUPA: Proboscis fused with body, but may project into a ridge (carinate).

TRIBE DILOPHONOTINI Burmeister, 1856

Type genus: *Dilophonota* Burmeister, 1856 (=*Erinnyis* Hübner, [1819]).

A tribe of nearly 150 species distributed throughout the tropical and temperate regions of the Old and New Worlds, but with its stronghold in the Neotropical Region.

IMAGO: Pattern of forewing variable.

Genitalia. Very diverse and often asymmetrical in both sexes. In male, uncus with beak-like extension, entire or divided, but with a marked furrow on the medial line; gnathos spatulate (entire or split), but sometimes absent; sacculus with complex apical process armed with spines; saccus well developed. In female, lamella postvaginalis broad and developed; sclerotized part of ductus bursae short but the two slender bands on the posterior margin always well defined.

LARVA: Diverse; the horn may be very slender and movable. Many have tubercles or excrescences on the thoracic segments and/or anal segments.

PUPA: Proboscis fused with body. Labrum displaced ventrally.

SUBTRIBE Aellopodina Carcasson, 1968

Type genus: *Aellopos* Hübner, [1819].

(*Taxonomic note*. Derzhavets (1984) established a new subtribe, Aellopina (Type genus: *Aellopos* Hübner, [1819]), without realising that Carcasson had already done so.)

IMAGO: Pattern of forewing with a few crossbands, or both fore- and hindwing with transparent fenestrae.

LARVA: With a pronounced dorso-lateral line and no oblique lateral stripes in final instar. Body granulose.

PUPA: Within a loosely spun cocoon.

HEMARIS Dalman, 1816

Hemaris Dalman, 1816, *K. svenska VetenskAkad. Handl.*
1816(2): 207.
Type species: *Sphinx fuciformis* Linnaeus, 1758.

Hemeria Billberg, 1820, *Enumeratio Insect. Mus. C.J.*
Billberg: 82.
Haemorrhagia Grote & Robinson, 1865, *Proc. ent. Soc.*
Philad. **5**: 149, 173.
Aege Felder, 1874, *in* Felder & Rogenhofer, *Reise öst.*
Fregatte Novara (Zool.) **2** (Abschnitt 2): pl. 75, fig. 6.
Chamaesesia Grote, 1873, *Bull. Buffalo Soc. nat. Sci.* **1**:
18.
Cochrania Tutt, 1902, *Nat. Hist. Br. Lepid.* **3**: 503.

(*Taxonomic note.* Given the variability of the genus *Hemaris*
and the paucity of consistent adult characters separating it
from the Old World tropical genus *Cephonodes* Hübner,
[1819], the latter should perhaps be synonymized with the
former.)

A Holarctic genus consisting of about 17 species.

IMAGO: Small, diurnal and resembling a bumblebee in
shape. Forewing with hyaline areas or fully scaled, although in those species which do have hyaline areas, these
may be covered initially by deciduous scales which are lost
during the first flight. Tips of veins R_2+R_3 and R_4 of
forewing united. Rs and M_1 of hindwing on short stalk, or
from common point on the cell. Cell short, but twice as long
as broad; M_3 and Cu_1 always separate. Antenna strongly
clubbed in both sexes, thin basally and abruptly narrowed
before apex and forming a slender, recurved hook. Lateral
hair-scales of frons hanging down to eye. Abdomen with
large 'fan-tail'. Segment-margin scales spinose and strongly
developed, flat and arranged in several rows, those of the
first row rounded, and broader than long. (These spinose
scales are on plates which can be removed leaving a semi-
membranous unit below.) Seventh abdominal sternite of
female with entire apical margin spinose. Midcoxa triangularly widened posteriorly; tip of dilated part sharply
pointed and directed away from body. Hindcoxa similar
but obtusely broadened. Foretibia with few spines at apex.
Pulvillus and paronychium variable.

Genitalia. Asymmetrical in male. Uncus divided, the two
lobes subequal, heavily sclerotized, with rounded apex.
Gnathos with either two long processes or with the right
process absent. Right and left sacculi and valvae differ; left
sacculus always vestigial; right often clavate and enlarged.
Aedeagus slender with distal part developed as a slender
lobe. Ostium bursae of female angled to left.

OVUM: Small, spherical, pale glossy green.

LARVA: Small, cylindrical, distinctly granulose, the gran-
ules often with small bristles. Many different colour forms
exist, the most common being green and brown; all have a

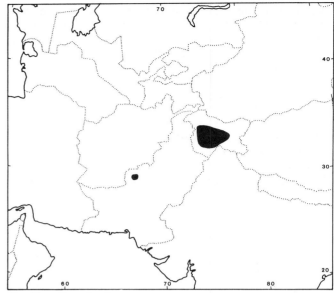

Map 19 – *Hemaris rubra*

distinctive, pale, dorso-lateral, longitudinal stripe from
head to horn. Ventral surface darkly coloured.

PUPA (Text fig. 38): Rugose and glossy in most species.
Proboscis keeled. Alongside each eye a prominent tubercle,
or hook, present. Cremaster large, flattened, triangular,
with a pair of apical spines and a few lateral ones. In a
loosely spun cocoon.

HOSTPLANT FAMILIES: Mainly herbs and shrubs of the
Dipsacaceae and Caprifoliaceae.

HEMARIS RUBRA Hampson, [1893]　　Map 19
E – Kashmir Bee Hawkmoth

Hemaris rubra Hampson, [1893], *Fauna Br. India (Moths)*
1: 120.
Type locality: Sind and Gurais Valleys, Kashmir.

ADULT DESCRIPTION AND VARIATION (Pl. 8, fig. 1)
Wingspan: 44–58mm. Wings almost entirely covered with
reddish brown scales. Male genitalia similar to those of *H.*
croatica and *H. tityus*, but process of right sacculus longer.

ADULT BIOLOGY
Diurnal. A species of flower-rich meadows at around
2500m.

FLIGHT-TIME
Univoltine; from mid-June to early August.

EARLY STAGES
Unknown. Presumably overwinters in the pupal stage.

PARASITOIDS
Unknown.

BREEDING
Nothing known.

DISTRIBUTION:
Currently known only from a number of valleys in Kashmir
(Bell & Scott, 1937), including Ladakh (in BM(NH) coll.).

Text Figure 38 Pupa: *Hemaris f. fuciformis* (Linnaeus)

Extra-limital range. Ziarat, near Quetta, Pakistan (one specimen in BM(NH) coll.). It is likely, therefore, that *H. rubra* also occurs at suitable altitudes in an arc between these two localities, i.e. in western and northern Pakistan, eastern Afghanistan and northern India.

HEMARIS DENTATA (Staudinger, 1887)   Map 20
E – Anatolian Bee Hawkmoth

Macroglossa ducalis var. *dentata* Staudinger, 1887, *Stettin. ent. Ztg* **48**: 66.

Type locality: Aïntab [Gaziantep], near Antiochia, [Turkey].

(*Taxonomic note*. Evidence supplied by de Freina (1988) indicates that this taxon may be a natural hybrid between *H. fuciformis* and *H. croatica*.)

ADULT DESCRIPTION AND VARIATION (Pl. 8, fig. 2)

Wingspan: 36–45mm. Intermediate in colour between *H. fuciformis* and *H. croatica*, but with male genitalia more closely resembling those of the former. Forewing cell undivided. A variable species, with the hyaline areas absent (i.e. scaled over) in some individuals; the abdominal pattern may approach that of *H. fuciformis* (de Freina, 1988).

ADULT BIOLOGY

Diurnal. A very local species, with one or two specimens turning up here and there in flower-rich mountain meadows at about 1500m. Very similar in behaviour to *H. croatica*.

FLIGHT-TIME

Univoltine; mid- to late July. The last two weeks of July at 1400–1700m (de Freina, 1988).

EARLY STAGES

Unknown. Presumably overwinters in the pupal stage.
Hostplants. Probably *Cephalaria* (as for *H. fuciformis* and *H. croatica* occurring in the same habitat).

PARASITOIDS

Unknown.

BREEDING

Nothing known.

DISTRIBUTION

Southern Turkey as far west as the Toros Mountains. Confirmed localities in Turkey are Gaziantep; Ala Dağları (near Niğde) and Anamas Dağları (near Egridir) (de Freina, 1988); and Sihli (near Tekir), Palaz Dağı (near Antalya), Ilica (near Süleymanli), Nemrut Dağı (near Adiyaman) and Harput (near Elâziğ) (W. Hogenes, pers. comm.). Records from 'Syria' (Rothschild & Jordan, 1903) refer to what is now southern Turkey.

Extra-limital range. None.

HEMARIS DUCALIS (Staudinger, 1887)   Map 21
E – Pamir Bee Hawkmoth

Macroglossa ducalis Staudinger 1887, *Stettin. ent. Ztg* **48**: 66.

Type locality: Namangan, Transalai [Fergana Valley, western Tian Shan, Uzbekistan].

Macroglossa temiri Grum-Grschimailo, 1887, *in* Romanoff, *Mém. Lépid.* **3**: 401.

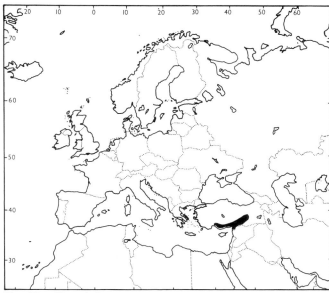

Map 20 – *Hemaris dentata*

ADULT DESCRIPTION AND VARIATION (Pl. 8, fig. 3)

Wingspan: 40–50mm. Differs from *H. dentata* in having the forewing almost fully scaled, the third abdominal segment yellowish white with any brown being confined to lateral tufts, and pulvillus absent. According to Rothschild & Jordan (1903), the male genitalia are rather distinctive; left valva more slender than in *H. tityus*, approaching that of *H. fuciformis*; process of left sacculus represented by a conspicuously spinose hump; right valva ventrally slightly emarginate beyond middle; process of right sacculus long, slender, slightly club-shaped, upper surface in apical half armed with long spines. Juxta rather densely covered with long hairs apically; aedeagus rather long, not very sharply pointed. A high-altitude montane form from Tajikistan, f. *efenestralis* Derzhavets (1984), has the hyaline windows of the forewing partially obscured by dark scales on the side towards the outer wing margin; the windows are completely obscured on the hindwing.

ADULT BIOLOGY

Diurnal. Found in woodland and scrub, rarely below 2300m and occurring up to 3500m.

FLIGHT-TIME

Univoltine; late June to early August.

EARLY STAGES

OVUM: Unknown.

LARVA (not illustrated): Full-fed 33–42mm. Early instars undescribed.

According to Degtyareva & Shchetkin (1982), the fully-grown larva resembles that of *H. fuciformis* (q.v., p. 111): dorsally and laterally pale green with a prominent yellowish white dorso-lateral line running from the first thoracic segment up to a slightly curved caudal horn; however, unlike *H. fuciformis*, *H. ducalis* bears a line of pale yellow V-shaped markings dorsally, one per segment but less pronounced on the thoracic segments, all pointing caudad. Ventrally, reddish brown with a fine, light yellow ventro-lateral line separating the dark lower surface from the green

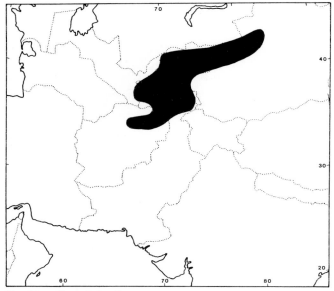

Map 21 – *Hemaris ducalis*

sides. Where this line crosses the prolegs it thickens and becomes darker; true legs brownish cream. Horn 5mm long, violet-black at base and whitish green or cream towards the tapering tip. As in *H. fuciformis*, the rounded head is bluish grey and offset by a cadmium-yellow collar to the first thoracic segment. A line of the same colour also edges the anal flap. The entire head, body and horn are covered with pale tubercles, light yellow dorsally, mostly becoming cream below the dorso-lateral line. Oval spiracles reddish brown and cream.

Found throughout July, August and September.

Hostplants. Lonicera spp.

PUPA: 37mm. Blackish brown, rugose, very similar to that of *H. fuciformis*. The overwintering stage.

PARASITOIDS
None recorded.

BREEDING
Nothing known.

DISTRIBUTION
The mountains of southern Uzbekistan, Tajikistan (Grum-Grshimailo, 1890), Kyrgyzstan, the Chinese portion of the Tian Shan, the Pamirs and northern Afghanistan (Ebert, 1969).

Extra-limital range. The mountains of south-western Xinjiang Province, China (several specimens in BM(NH) coll.).

HEMARIS FUCIFORMIS (Linnaeus, 1758)
Map 22

E – Broad-bordered Bee Hawkmoth,
F – le Sphinx bourdon; le Sphinx fuciforme,
G – Hummelschwärmer, **Sp** – Cristalina borde ancho,
NL – Glasvleugelpijlstaart, **Sw** – Humlelik dagsvärmare,
R – Brazhnik Shmelevidnyi zhimolostnyi,
H – dongószender, **C** – Dlouhozobka chrastavcová

HEMARIS FUCIFORMIS FUCIFORMIS
(Linnaeus, 1758)

Sphinx fuciformis Linnaeus, 1758, *Syst. Nat.* (Edn 10) **1**: 493.
Type locality: Europe.

  Sphinx variegata Allioni, 1766, *Mélang. Soc. Turin*: 193.
  Macroglossa milesiformis Treitschke, 1834, *Schmetterlinge Europa* **10**(1): 125.
  Macroglossa lonicerae Zeller, 1869, *Stettin. ent. Ztg* **30**: 387.
  Macroglossa caprifolii Zeller, 1869, *Stettin. ent. Ztg* **30**: 387.
  Macroglossa robusta Alphéraky, 1882, *Horae Soc. ent. ross.* **17**: 17.
  Hemaris simillima Moore, 1888, *Proc. zool. Soc. Lond.* **1888**: 391.
  Haemorrhagia fuciformis syra Daniel, 1939, *Mitt. münch. ent. Ges.* **29**: 94. **Syn. nov.**

(*Taxonomic note.* The descriptions and illustrations provided by Daniel (1939) of *H. fuciformis* subsp. *syra* Daniel, 1939, indicate that it may be *Hemaris dentata* (Staudinger, 1887). This matter requires clarification.)

ADULT DESCRIPTION AND VARIATION (Pl. 8, fig. 4; Pl. G, fig. 1)

Wingspan: 38–45mm. Distinguished from *H. tityus* (Pl. 8, fig. 7) by a much broader marginal band to the wings; forewing discal cell divided longitudinally by a fold. More variable than *H. tityus*: in f. *musculus* Wallengren, all olive green coloration is replaced by reddish grey; in f. *heynei* Bartel, black is the predominant colour in the central abdominal belt; in f. *milesiformis* Trimen, inner border of submarginal band of forewing is toothed.

ADULT BIOLOGY

Diurnal. In northern Europe, frequents the *Lonicera*-entwined edges of sunny glades in open woodland. Farther south, displays a marked preference for sand- and chalk hills lightly overgrown with conifers and shrubby honeysuckles, such as *L. xylosteum* and *L. tatarica*, where it is often extremely abundant. It was formerly also abundant in many marginal areas (Tatchell, [1926]), but modern practices of woodland management have severely reduced its numbers. Generally avoids open meadows, preferring to fly along woodland margins or rides.

Behaviourally, this species closely resembles *H. tityus*: active by day, usually from about 10.00 hours, it avidly searches along rides and forest edges for the flowers of *Rhododendron*, *Silene*, *Ajuga*, *Lychnis* and *Pulmonaria*. Some may even venture into gardens to be seen at *Syringa* and *Phlox*. It is in such situations that courtship and pairing take place; the female then takes flight in search of *Lonicera* growing in partial shade on which to lay her eggs. If disturbed at any time, this species flies off at great speed.

FLIGHT-TIME

Bivoltine throughout its southern range. In northern Europe, late May to mid-June, with a partial second brood in August; one generation in June normal for the Urals (Eversmann, 1844) and Tian Shan (Grum-Grshimailo, 1890).

EARLY STAGES

OVUM: Small (1.1 × 1.0mm), pale glossy green, spherical.

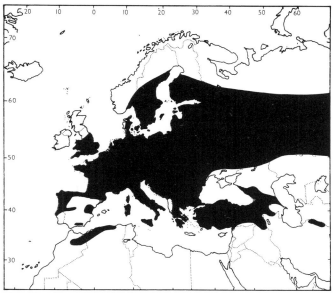

Map 22 – *Hemaris fuciformis*

Laid singly on the underside of the leaves of its hostplant; plants growing in either deep shade or in the open are avoided.

LARVA (Pl. 2, fig. 13): Full-fed 40–45mm.

On hatching, the 3mm-long, whitish yellow larva rests along the midrib on the underside of a leaf, occasionally nibbling oval holes in the latter. Most feed at night, resting during daylight hours in a typical sphinx-like attitude. Fully grown, some have reddish spots and blotches bordering the dorso-lateral line and spiracles which, combined with a very rough, tubercled skin, camouflage the larva well. If disturbed or alarmed, it will drop to the ground, although not as readily as *H. tityus*.

Occurs from mid-June until early August over central and northern Europe, but from May until September farther south.

Hostplants. Principally *Lonicera* spp. in northern Europe, particularly *L. periclymenum*; in central and southern parts of its range, the shrubby *L. nigra*, *L. xylosteum* and *L. tatarica* are favoured. Also on *Symphoricarpos*, *Galium*, *Deutzia* and *Knautia*; in Turkey mainly *Cephalaria procera* (de Freina, 1979).

PUPA (Pl. E, fig. 8): 23–25mm. Similar to that of *H. tityus*: blackish brown with glossy highlights and brown, intersegmental cuticle. Formed on the ground in a loosely spun silken cocoon interwoven with debris. The overwintering stage.

PARASITOIDS

See Table 4, p. 52.

BREEDING

Pairing. As for *H. tityus* (see p. 114) but the flight-cage should be situated amongst shrubs and trees.

Oviposition. A large flight-cage situated amongst shrubs is required.

Larvae. As for *H. tityus*; can be readily fed on cut *Lonicera* and *Symphoricarpos* placed in water, but preventative measures must be taken against drowning.

Pupation. As for *H. tityus*.

DISTRIBUTION

As for *H. tityus*, but absent from most of the Iberian Peninsula (Gómez Bustillo & Fernández-Rubio, 1976), although it has recently (1986) been found at 1250m in the Sierra de la Yedra mountains, Granada Province, southern Spain (Pérez-López, 1989). Otherwise, from Spain through Europe to the Urals and western Siberia (Eversmann, 1844). Also Turkey (Daniel, 1932, 1939; de Freina, 1979), the Caucasus, the Kopet Dağ Mountains of Turkmenistan (Derzhavets, 1984), northern Afghanistan (Ebert, 1969) and the Tian Shan of Tajikistan (Alphéraky, 1882; Grum-Grshimailo, 1890; Derzhavets, 1984). Absent from Ireland (Lavery, 1991) and Scotland (Gilchrist, 1979), northern Scandinavia and Arctic Russia.

Extra-limital range. The south-eastern Altai, Pamirs, Hindu Kush and north-west Pakistan (Bell & Scott, 1937).

HEMARIS FUCIFORMIS JORDANI
(Clark, 1927)

Haemorrhagia fuciformis jordani Clark, 1927, *Proc. New Engl. zool. Club* **9**: 105.
Type locality: Ras-el-Ma, Morocco.

ADULT DESCRIPTION AND VARIATION (Pl. 8, fig. 5)

Wingspan: 45–51mm. Larger than subsp. *fuciformis*; many specimens are also darker in colour with small dentate areas on the wing margins.

ADULT BIOLOGY

Diurnal. Frequents damp mountain forest and scrub, especially areas of deciduous oak (*Quercus* spp.).

FLIGHT-TIMES

Bivoltine; April/May and June/July.

EARLY STAGES

OVUM: Large; laid singly beneath an older leaf.

LARVA (Pl. C, fig. 6): Full-fed 42–48mm.
Very similar to subsp. *fuciformis*, but a higher percentage of individuals have red patches and a red border to the yellow dorso-lateral line.

Commonly found during May, June, August and early September.

Hostplants. Lonicera spp., particularly small clumps growing on the ground under trees.

PUPA: As subsp. *fuciformis*.

PARASITOIDS

None recorded.

BREEDING

Pairing. As for *H. tityus* (see p. 114), but the flight-cage should be situated amongst shrubs and trees.

Oviposition. A large flight-cage situated amongst shrubs is required.

Larvae. As for *H. tityus*; can be readily fed on cut *Lonicera* and *Symphoricarpos* placed in water, but preventative measures must be taken against drowning.

Pupation. As for *H. tityus*.

DISTRIBUTION

Confined to the Atlas Mountains of Morocco and Algeria (Rougeot & Viette, 1978; Rungs, 1981).

Extra-limital range. None.

OTHER SUBSPECIES

Xizang Province/Tibet as subsp. *ganssuensis* Grum-Grshimailo, 1891. (*Hemaris affinis* Bremer, 1861, is a distinct species occurring in Japan, Ussuri, Amurland (far eastern Russia), Korea (Chistyakov & Belyaev, 1984) and northern China (Chu & Wang, 1980b).)

HEMARIS CROATICA (Esper, 1800)   Map 23

E – Olive Bee Hawkmoth, **F** – Macroglosse de Croatie,
G – Olivegrüner Hummelschwärmer

Sphinx croatica Esper, 1800, *Die Schmett.* (Suppl.)(Abschnitt 2): 33, pl. 45, fig. 2.
Type locality: Karlstadt [Karlovac, Croatia].

> *Sphinx sesia* Hübner, [1805], *Samml. eur. Schmett.*, Sphingidae: 94.

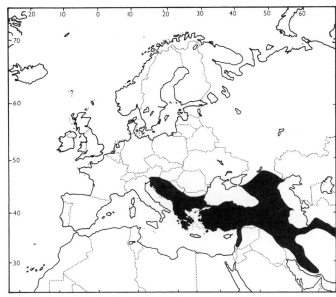

Map 23 – *Hemaris croatica*

ADULT DESCRIPTION AND VARIATION (Pl. 8, fig. 6)

Wingspan: 36–48mm. Easily distinguished from other western Palaearctic bee hawkmoths in that the wings remain fully scaled. Not a very variable moth. In some, the green is replaced by grey. In f. *obscurata* Closs the forewing is dark, blackish green as are the two anterior segments of the abdomen, which has two very dark red belts. In f. *rangowi* Closs, the reddish brown submarginal bands of the forewing are very narrow, the hindwing small and pale with no dark margin, and the posterior abdominal segments are sulphur-yellow tipped with a velvet-black anal tuft with no red central mark. This latter form is normally found only in the steppe areas of Russia to the north-west of the Caspian Sea.

ADULT BIOLOGY

Diurnal. Tends to follow hill and mountain chains, where it prefers open, hot, dry meadows interspersed with patches of woodland on mountain- and hillsides.

FLIGHT-TIME

Usually univoltine across its northern range and at altitude, during July; bivoltine in its southern range, May/June and again in August.

EARLY STAGES

OVUM: Almost spherical (1.1 × 1.0mm), pale glossy green. Laid singly on the underside of leaves of wild *Scabiosa* spp., with rarely more than one to a plant.

LARVA (Pl. 3, fig. 2): Full-fed 45–50mm. Polymorphic.

On hatching, the 3mm-long larva is pale yellow with a dark horn.

Most full-grown larvae are primarily bright green and densely speckled with white spots. However, larvae are highly variable, with other colour forms ranging from yellowish white, through blue-green (fig. 2) to pale red. All have the same pattern: dorsally a darker 'heart line' runs from head to horn, often bordered on either side by yellow, with a much more substantial white dorso-lateral line always present and blackish patches between the prolegs. Spiracles red, ringed with pale yellow; horn orange to reddish yellow to blue.

Throughout all instars, the larva exhibits the same feeding behaviour, resting along the midrib on the undersurface of a leaf, when not engaged in nibbling oval holes in it. Like *H. tityus*, it will usually fall to the ground if disturbed.

Double-brooded throughout most of its range, being found in June and July, and again during August and September; where only a single generation occurs, between July and September.

Hostplants. Principally *Scabiosa* spp., but also on *Asperula* and *Cephalaria* spp.

PUPA: 24–27mm. Reddish brown, tapering sharply towards the head. Proboscis prominent, almost keeled; head tubercles, or hooks, much more reduced than in the previous species (cf. Pl. E, fig. 8). Formed in a loosely-spun, light brown cocoon amongst ground debris. The overwintering stage.

PARASITOIDS

None recorded.

BREEDING

Treat as for *H. tityus* (q.v., p. 114), with one important exception: larvae of this species rarely survive on *Knautia* and should always be offered *Scabiosa* or *Cephalaria*.

DISTRIBUTION

From Italy (Trieste), Austria (Heinemann, 1859; Hoffmann & Klos, 1914) and Hungary (Károlyváros) (Abafi-Aigner *et al.*, 1896) south-eastward through Yugoslavia, Greece, southern Bulgaria (Ganev, 1984), Turkey (Daniel, 1932; de Freina, 1979) to the Lebanon (Zerny, 1933; Ellison & Wiltshire, 1939) and Caucasus, then southwards along the Iranian plateau to Shiraz (pers. obs.). Also north to Sarepta on the lower Volga (Russia) and possibly as far north as Kazan (Eversmann, 1844).

Freyer (1836) related that he received a larva of this species from Kindermann, who was then living in Switzerland but there is no mention of where it was found. This specimen may possibly have given rise to the rumour that *H. croatica* was resident in Switzerland. In the same work, Freyer recorded that this species was to be found in small numbers in the neighbourhood of Salzburg, Austria, and Heinemann (1859) also recorded this species from the province of Kärnten, Austria, where it is now extinct. In view of the extinction in Switzerland this century of the equally thermophilic papilionid *Zerynthia polyxena* ([Denis & Schiffermüller], 1775) (Schweizerischer Bund für Naturschutz, 1987) and its virtual disappearance from Austria, *H. croatica* may perhaps have occurred at one time in a few favourable locations farther north than the present-day limit of its range, which is the south-east corner of Austria (pers. obs.). Kovács (1953) does not list this species for Hungary and it may also have died out there since its discovery in 1896 (Abafi-Aigner *et al.*, 1896).

Extra-limital range. None.

HEMARIS TITYUS (Linnaeus, 1758)   Map 24

E – Narrow-bordered Bee Hawkmoth,
F – le Sphinx bombyliforme; le Sphinx gazé,
G – Skabiosenschwärmer, **Sp** – Cristalina borde estrecho,
NL – Hommelpijlstaart, **Sw** – Svärfluglik dagsvärmare,
R – Shmelevidka skabiozovaya,
H – pöszörszender, **C** – Dlouhozobka zimolezová

HEMARIS TITYUS TITYUS (Linnaeus, 1758)

Sphinx tityus Linnaeus, 1758, *Syst. Nat.* (Edn 10) **1**: 493.
Type locality: Unspecified [Europe].

> *Sphinx bombyliformis* Linnaeus, 1758, *Syst. Nat.* (Edn 10) **1**: 493.
> *Sphinx musca* Retzius, 1783, *Genera et Species Insect.*: 33.
> *Macroglossa scabiosae* Zeller, 1869, *Stettin. ent. Ztg* **30**: 387.
> *Macroglossa knautiae* Zeller, 1869, *Stettin. ent. Ztg* **30**: 387.

ADULT DESCRIPTION AND VARIATION (Pl. 8, fig. 7)

Wingspan: 40–50mm. Distinguished from *H. fuciformis* by the narrow band of scaling along the outer margin of each wing and the presence of an undivided forewing cell. Although some confusion may arise with freshly-emerged examples which have their entire wings covered with loose, grey scales, these are lost after the first flight, after which identification is made easier by a near absence of variation.

ADULT BIOLOGY

Diurnal. An insect of meadows, roadside verges and woodland glades rich in flowers, especially on chalk and sand where *Ajuga* (bugle) thrives. In such localities, vast numbers can be found in some years but very few in others, though it is never totally absent. Where this species occurs, *Scabiosa* (scabious) is nearly always present; however, many areas rich in this plant are devoid of *H. tityus*. Up to 2000m in the Alps (Vorbrodt & Müller-Rutz, 1911; Forster & Wohlfahrt, 1960).

Unlike most other sphingids, this species is active during the day, generally between 10.00 and 15.00 hours, when it can be seen flitting rapidly from flower to flower of its favourite plants – *Ajuga, Lychnis, Viscaria, Salvia* and *Knautia* (Bergmann, 1953). Although mimicking a

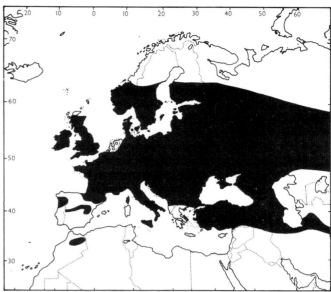

Map 24 – *Hemaris tityus*

bumblebee in coloration, its flight is much more rapid and agile. Its skill is superbly demonstrated during courtship when pairs chase each other low over the ground as well as spiralling upwards like a whirlwind. Copulation generally follows, with pairs remaining *in copula* for up to two hours.

FLIGHT-TIME

Mainly univoltine in northern Europe, mid-May to mid-June with a partial second brood in August; around Saratov, Russia, this species flies in June and July (Kumakov, 1977). Bivoltine farther south, during late April or early May, and again in August.

EARLY STAGES

OVUM: Small (1.1 × 1.0mm), almost spherical, with a small depression on top; pale glossy green. Laid singly on the underside of a leaf of the hostplant, with up to six per plant or group of plants not uncommon.

LARVA (Pl. 2, fig. 7): Full-fed 50mm.

The newly-hatched, white larva is about 3mm long and covered with small black tubercles bearing forked, black hairs. The caudal horn is small with two black hairs at its apex. With each moult and feeding the body and tubercles gradually become green, the horn redder, and pale dorso-lateral and ventro-lateral lines appear along the side. Fully grown, most are whitish green with a dull red horn, purple-red ventral surface and with purple-red surrounds to the spiracles. The body is very rough to the touch. Many are heavily spotted with additional reddish blotches, especially those feeding on purple-spotted scabious leaves. When fully grown, such blotches are masked as the larva turns a rich plum colour prior to pupation.

Throughout its life, the larva generally remains beneath a leaf, nibbling holes on either side of the midrib on which it rests. Most active at night; if disturbed during the day, it will drop to the ground, many even taking shelter there when not feeding. A noteworthy feature of this species is the variation in growth-rate of larvae of the same brood; however, all tend to mature much more rapidly than those of other sphingids.

In Finland, usually found in July/August; farther south, in June/July; and, in the warmest parts of its range, from May to September.

Hostplants. In northern and central Europe, principally *Scabiosa succisa* and *Knautia arvensis*. Also other *Scabiosa* spp., *Galium, Lonicera, Symphoricarpos, Succisa, Dipsacus* and *Lychnis*.

PUPA: 24–27mm. Typically *Hemaris* in shape but blackish brown, tinted with reddish brown at the juncture of each segment. Body tapers at either end, with two sharp head tubercles and a flat triangular cremaster. Formed in a rather strong but coarse cocoon among grass tussocks, or slightly into the soil. The overwintering stage.

PARASITOIDS
See Table 4, p. 52.

BREEDING
Pairing. Difficult; most successful in a large flight-cage placed outside in full sun with an ample supply of flowers and hostplant.

Oviposition. As above. The female can be tethered to the hostplant.

Larvae. Does very well on potted *Scabiosa* in partial sunlight. Cut food in water should be avoided as this usually leads to great losses. As many larvae will leave the plant by day to hide, precautions should be taken to prevent their becoming lost. For pupation, larvae should be provided with damp moss on damp peat.

Pupation. Easy to keep in an airtight tin placed in a sheltered location, but must not be allowed to dry out. Also very easy to force out of season and, if so desired, should be placed on damp earth under damp moss and brought into warmth during January.

DISTRIBUTION
From Ireland (Lavery, 1991) across temperate Europe (except the Iberian Peninsula, where it occurs only in a few mountain refugia, and the Netherlands, where it is almost extinct (Meerman,1987)) to the Urals (Eversmann, 1844; Kumakov, 1977), and from Turkey (de Freina, 1979) to northern Iran (Brandt, 1938; Sutton, 1963) and the Tian Shan (Alphéraky, 1882). Absent from northern Scandinavia and Arctic Russia. (See also p. 36; Text figs 15a,b, p. 37.)
Extra-limital range. From the Urals eastward to northeastern China (Derzhavets, 1984) and southwards to Xizang Province/Tibet, including Qinghai Province (Lake Qinghai), China, and the Altai.

HEMARIS TITYUS AKSANA (Le Cerf, 1923)
Stat. rev.

Haemorrhagia tityus aksana Le Cerf, 1923, *Bull. Soc. ent. Fr.* **1923**: 199.
Type locality: Azrou, Middle Atlas, Morocco.

(*Taxonomic note.* Eitschberger *et al.* (1989) raised subsp. *aksana* to specific rank but gave no valid reasons for doing so. This action is not accepted here.)

ADULT DESCRIPTION AND VARIATION (Pl. 8, fig. 8; Pl. G, fig. 2)
Wingspan: 44–51mm. Similar to subsp. *tityus*, but the green of the thorax is less yellowish. Grey basal area of the forewing extended and marginal band blackish and broader than in subsp. *tityus*.

ADULT BIOLOGY
Diurnal. Found in habitats similar to those of subsp. *tityus*, but between 1300 and 2000m. A species of flower-rich meadows.

FLIGHT-TIME
In its warmest locations, on the wing in March, June and August; at higher altitude, the latter half of May as a single brood, sometimes with a partial second brood in August (Rungs, 1981).

EARLY STAGES
OVUM: As subsp. *tityus*.

LARVA: As subsp. *tityus*.
Hostplants. Scabiosa spp.

PUPA: As subsp. *tityus*.

PARASITOIDS
None recorded.

BREEDING
As for subsp. *tityus*.

DISTRIBUTION
Known only from the Middle Atlas Mountains of Morocco (Rungs, 1981), but may be more widespread.
Extra-limital range. None.

OTHER SUBSPECIES
The western Tian Shan (Tajikistan) as subsp. *alaiana* (Rothschild & Jordan, 1903), although Derzhavets (1984) considers this merely a form of subsp. *tityus*.

TRIBE MACROGLOSSINI Harris, 1839
Type genus: *Macroglossum* Scopoli, 1777.

A cosmopolitan tribe of more than 400 species.

IMAGO: Pattern of forewing consisting of transverse (Macroglossina) or oblique (Choerocampina) bands; basal tone brown-grey, olive-green or brown-purple; hindwing with contrasting median area and dark submarginal crossbands.

Genitalia. In male, uncus and gnathos with rostrate extensions directed caudad so as to resemble a crab's claw; the apex of each heavily sclerotized and tapering to a sharp or blunt point. Tegumen extended caudad into a broad tube; sacculus with apical process, accessory armature weakly developed or absent; saccus in the form of a narrow fold of vinculum. In female, lamella postvaginalis short.

LARVA: Head small or large, rounded; horn developed or replaced by a tubercle in the final instar; cuticle smooth in all instars; body with longitudinal stripes or faint oblique lateral stripes; several ocellate spots may be present along the body, that on the third thoracic or first abdominal segment being well developed in some species.

PUPA: Proboscis fused with body; may be raised into a protruding ridge; in *Pergesa*, proboscis free of body.

SUBTRIBE Macroglossina Harris, 1839

Type genus: *Macroglossum* Scopoli, 1777.

IMAGO: Pattern of forewing with transverse undulate or straight crossbands; outer margin entire or serrate.

LARVA: Head rounded; cuticle smooth in final instar; thoracic segments often with ocellate spots.

PUPA: Labrum displaced ventrally (except in some *Acosmeryx*).

DAPHNIS Hübner, [1819]

Daphnis Hübner, [1819], *Verz. bekannter Schmett.*: 134.
Type species: *Sphinx nerii* Linnaeus, 1758.

Histriosphinx Varis, 1976, *Notul. ent.* **56**: 127.

A genus of the Old World tropics and subtropics, containing seven to nine species.

IMAGO: Wing entire in outline, with forewing apex pointed. Usually some shade of green or brown with curved bands and triangles of different colours. Eye large, without pendant lashes. Labial palpus large, obtuse, smoothly scaled. Antenna slightly clubbed in female; setiform in male, with a long thin terminal segment. Spines of abdomen in several rows, elongate, weak; first tergite large. Tibiae without spines. Midtarsus with basal comb. Hindtibia with two pairs of spurs, the proximal being considerably longer than the distal. M_2 of hindwing before centre of cell.

Genitalia. In male, valva with fewer than ten large, erect, modified friction scales on outer surface. Sacculus with two processes, one proximal, the other distal, both dorsal. Aedeagus with one or two left processes, and a longer right process. In the female, lamella postvaginalis suddenly narrowed at end where it is concave; apical margin raised and projecting ventrad.

OVUM: Ovoid, pale green, shiny.

LARVA: Head small, body tapering sharply forward from abdominal segment 1, rest of body almost cylindrical. Usually bears a prominent eye-spot on its third thoracic segment and a pale dorso-lateral line from abdominal segment 1 to the horn. Horn stout and curved downward in final instar. Body smooth in final instar.

PUPA (Text fig. 39): Proboscis fused with body, a thin, black line along the former; cremaster long, slender, with a pair of apical spines; black surrounds to the spiracles.

HOSTPLANT FAMILIES: Trees and shrubs of the Rubiaceae and Apocynaceae.

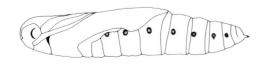

Text Figure 39 Pupa: *Daphnis nerii* (Linnaeus)

DAPHNIS NERII (Linnaeus, 1758)   Map 25

E – Oleander Hawkmoth, **F** – le Sphinx du laurier-rose,
G – Oleanderschwärmer, **Sp** – Esfinge de la adelfa,
NL – Oleanderpijlstaart, **Sw** – Oleandersvärmare,
R – Oleandrovyĭ Brazhnik, **H** – oleanderszender,
C – Lyšaj bobkovnicový

Sphinx nerii Linnaeus, 1758, *Syst. Nat.* (Edn 10) **1**: 490.
Type locality: Unspecified [Europe].

> *Deilephila nerii bipartita* Gehlen, 1934, *Bull. Mus. r. Hist. nat. Belg.* **10**(3): 2.

(*Taxonomic note*. Subsp. *bipartita* cannot be justified as the original description was based on an abnormal specimen of *nerii*.)

ADULT DESCRIPTION AND VARIATION (Pl. 8, fig. 9; Pl. G. fig. 3)

Wingspan: 90–110mm. Most easily confused with *Daphnis hypothous* (Cramer, 1780), a rare vagrant from India (q.v., p. 117). Variation confined to the intensity of wing coloration, especially the greens and pinks; most strikingly in f. *nigra* Schmidt, where green is replaced by black.

ADULT BIOLOGY

A species of dry river-beds, oases and warm hillsides with scattered oleander bushes; however, localities overgrown with this shrub tend to be shunned.

Rests by day, either on a solid surface or suspended among foliage with which it blends; the head is tucked in, with the thorax and abdomen raised off the underlying substrate. Most emerge late in the evening but do not take flight until just before dawn, to feed avidly from such flowers as *Nicotiana*, *Petunia*, *Lonicera*, *Saponaria* and *Mirabilis*. Thereafter, flight periods are mainly just after dusk and before dawn. Under warm conditions, adults are extremely wary and, if disturbed, will take flight even during daylight hours. Pairing is a short affair usually lasting, at most, four hours but, occasionally, a couple will remain *in copula* until morning. Rarely found at light, unlike *D. hypothous* (I. J. Kitching, pers. comm.).

FLIGHT-TIME

Migrant and multivoltine in its resident range, from May to September in four or five generations, often overlapping, unlike *Acherontia atropos* or *Agrius convolvuli* (qq.v.). In southern Europe, only two migrant-induced generations are evident – during June and August, with individuals from the latter migrating into central and northern Europe.

EARLY STAGES

OVUM (Pl. C, fig. 1): Almost spherical, small (1.50 × 1.25mm), light green. Laid singly on both the upper and lower surfaces of young leaves of isolated, preferably sheltered, bushes, especially those at the foot of cliffs or near houses, or in clearings amongst trees. Females often fly around a plant several times before approaching with a pendulous flight. Most take up to twelve days to hatch but, during hot weather, some hatch in only five.

LARVA (Pl. 2, fig. 1; Pl. C, fig. 5): Full-fed 100–130mm. Dimorphic: green or brown.

Newly-hatched larvae (3–4mm), which consume their eggshells, are bright yellow with an unusually long, very thin, blackish horn. However, with feeding the yellow quickly assumes a greenish hue and, after the first moult, the primary colour becomes apple-green with a white dorso-lateral line from abdominal segment 1 to the horn, a small, white eye-spot on the third thoracic segment, black spiracles and pink legs. With growth, the eye-spots become blue with white centres, ringed in black. The horn has an unusual bulbous 'cap' until the penultimate instar (Text fig. 40). Fully-grown larvae show little difference from younger ones, except for the change in eye-spots, and the horn losing its bulbous cap and becoming orange with a black tip, finely warted, and downward curved. In some individuals the dorsal surface is rosy, while in most the dorso-lateral line becomes edged in blue. In their final instar, some assume a bronze colour with rosy red anterior segments, which tends to mask the pre-pupation plum coloration; in all, however, the newly acquired blue-black dorsal pigmentation, the now black eye-spots and unchanged white spots on either side of the dorso-lateral line remain prominent.

When young, larvae feed fully exposed on the topmost leaves and flowers; when larger, they tend to conceal themselves further down the branches, or even, when not feeding during daylight hours, on the ground under stones or leaf-litter. Those choosing to remain on the hostplant rest along the lower surface or stem of a leaf, with the first four segments of the body slightly hunched. When first disturbed the caterpillar stretches out to resemble an oleander leaf. With further disturbance, the anterior segments are arched up, suddenly revealing the startling eye-spots; at this point the noxious gut contents may also be regurgitated.

Found throughout its resident range between May and October; across southern Europe, sometimes in July but usually in August/September. Rarely found north of the Alps.

Hostplants. Principally the flowers and young leaves of *Nerium oleander*; also *Vinca, Vitis, Gardenia, Asclepias, Jasminum, Trachelospermum, Amsonia, Carissa, Tabernaemontana, ?Mangifera, Rhazya, Adenium, Catharanthus, Ipomoea* and *Thevetia*.

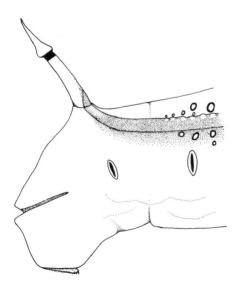

Text Figure 40 Terminal abdominal segments of third-instar larva of *Daphnis nerii* (Linnaeus) showing bulbous cap on horn

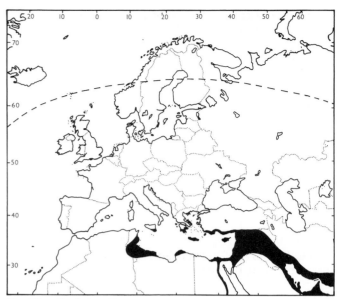

Map 25 – *Daphnis nerii*

PUPA (Pl. E, fig. 9): 60–75mm. Light brown, sprinkled with brown dots, particularly on wings and abdominal segments. A black line runs up the proboscis, over the head and thorax, before fading away on darker coloured abdominal segments. Cremaster short and straight. Spiracles set in a very dark brown, almost black spot. A crescent-shaped mark partly encircles each eye. Formed in a loosely spun yellow cocoon among dry debris on the ground. Rarely survives European winters, but the main overwintering stage throughout its resident range.

PARASITOIDS
See Table 4, p. 52.

BREEDING
Pairing. Difficult, but possible in a 160 × 160cm cylinder cage containing flowers and oleander twigs. A 10:1 diluted honey solution is recommended for force-feeding but the moths need not be fed except under the driest conditions. Better results seem to be achieved by suspending a weak orange light near the cage, rather than keeping it in total darkness. Pairing usually takes place three to four nights after emergence at a temperature of 15–20°C.

Oviposition. Relatively easy under the above conditions. As most eggs are laid some time after pairing, feeding is essential if quantities of eggs are required (Heinig, 1976).

Larvae. Not difficult to rear given plenty of fresh hostplant and dry, uncrowded conditions. Daily cut food, with plenty of flowers, placed in an indoor, net-covered, cardboard box, will suffice. Rearing in glass jars beyond the second instar or on cut hostplant in water both usually prove fatal. Sunlight or an artificial heat-source are beneficial, as most larvae like basking; however, larvae reared under the long hours of daylight of northern Europe often produce infertile adults. The day-length should, therefore, be shortened artificially to a 12:12 hour cycle, or less. With this short day and rearing temperatures of 21°–26°C, most resulting adults become fertile only after several days, during which time they must be repeatedly fed. From central Europe southwards, sleeving outside is best. Those fed on *Vinca* tend to be smaller than normal. Alternatively, young larvae can be readily induced to feed on *Ligustrum ovalifolium* if it has previously been sprayed with an extract from oleander or *Vinca* leaves; however, any extract sprayed on a host-plant must be allowed to dry prior to use. Larger larvae will accept this plant without treatment. This species can also be changed from one hostplant to another after a moult.

Pupation. Easy if the larvae are provided with loose leaves on dry, sandy soil. If kept warm and absolutely dry, emergence will take place after three to four weeks. Excessive moisture, frost or prolonged temperatures below 10°C are all fatal.

NOTE: Several generations are possible throughout the year with ample supplies of hostplant and warmth (Koch & Heinig, 1976).

DISTRIBUTION

The southern Mediterranean region, North Africa and the Middle East to Afghanistan (Ebert, 1969). Along the Mediterranean, there is no clear distinction between resident and migrant populations. Permanent populations exist in suitable locations in Sicily, Crete and Cyprus; however, over a number of favourable years further colonies may be established in those islands and also in southern Italy and southern Greece, all of which die out during a hard winter.

Extra-limital range. From Afghanistan eastward to southeast Asia and the Philippines; as a migrant, it penetrates northwards into central Europe and central southern Asia. In 1974, this species was recorded as having established itself in Hawaii (Beardsley, 1979).

DAPHNIS HYPOTHOUS (Cramer, 1780)   Map 26
E – Jade Hawkmoth

DAPHNIS HYPOTHOUS HYPOTHOUS
(Cramer, 1780)

Sphinx hypothous Cramer, 1780, *Uitlandsche Kapellen (Papillons exot.)* **3**: 165.
Type locality: Amboina [Ambon Island, Seram, Indonesia].

ADULT DESCRIPTION AND VARIATION (Pl. 8, fig. 10)

Wingspan: 86–120mm. Very like a dark *D. nerii*, but distinguished by a white spot at the forewing apex; head and collar uniformly dark purple-brown, thorax and first two abdominal segments dark green.

ADULT BIOLOGY

D. hypothous is a very fast flyer, attracted to both sweet-smelling flowers and light, but little else is known of its behaviour.

FLIGHT-TIME

Migrant. Multi-brooded in the Oriental region.

EARLY STAGES

OVUM: Very similar to that of *D. nerii*.

LARVA (not illustrated): Full-fed 60–100mm. Dimorphic: green or pale red.

Has never been found within the western Palaearctic region, but fully-grown examples from India (Bell & Scott, 1937) are usually pale apple-green, the thoracic segments

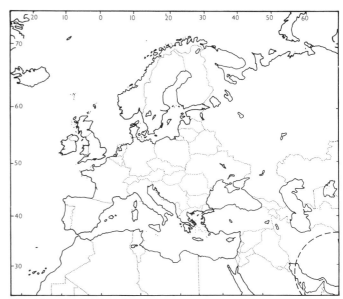

Map 26 – *Daphnis hypothous*

being darker than the abdominal segments. A white dorso-lateral line, tinged with orange at each end, extends from thoracic segment 2 to the base of the horn, and, on the abdominal segments, is edged dark green above and sky-blue below. Laterally these segments bear dark green, oblique stripes and yellow dots, with a large, blue eye-spot on thoracic segment 3. The brown, tubercled horn, unlike that of *D. nerii*, is fully formed, stout, erect and slightly curved, similar in appearance to that of *Agrius convolvuli*. Although most larvae are green, some may be pale red. Before pupation, the whole body assumes a plum colour blotched with dark red.

Occurs on bushes, usually near water.

Hostplants. Principally Rubiaceae, such as *Cinchona*, *Wendlandia* and *Uncaria* in India (Bell & Scott, 1937).

PUPA: 60–80mm. Similar to that of *D. nerii*, but with a thin black line running along the proboscis, over the head and along the whole dorsal surface, terminating, in this species, in a broad, triangular cremaster tipped with two points. The spiracles are located in a dark lateral band which runs parallel to two others along the ventral surface of the abdomen. Formed in a loosely spun cocoon among debris on the ground. Does not overwinter.

PARASITOIDS

None recorded.

DISTRIBUTION

The nominate subspecies sometimes occurs in the western Palaearctic region as a rare vagrant. However, the frequency is not known because of its confusion with *D. nerii*. During the last hundred years a number have been discovered within the Middle East; one was even found in Scotland late last century but was probably imported as a pupa with cargo.

Extra-limital range. India, Sri Lanka, Thailand, Malaysia and Indonesia.

OTHER SUBSPECIES

New Guinea and northern Australia as the very distinct subsp. *pallescens* Butler, 1875.

117

CLARINA Tutt, 1903

Clarina Tutt, 1903 [March 15], *Entomologist's Rec. J. Var.* **15**: 76.

Type species: *Deilephila syriaca* Lederer, 1855 (=*Clarina kotschyi syriaca* (Lederer, 1855)).

> *Berutana* Rothschild & Jordan, 1903 [April 21], *Novit. zool.* **9** (Suppl.): 502 (key), 519.

A western Palaearctic genus containing one species.

IMAGO: Distal margin of wings slightly dentate. Pendant lashes at upper edge of eye. Head with a distinct dorsal crest. Long apical spur of mid- and hindtibiae less than half length of the first tarsal segment, which is shorter than the tibia. No comb on midtarsus. Pulvillus small. Paronychium with only one lobe either side. Veins Rs and M_1 of hindwing on a short stalk, M_2 before centre of cell, lower angle of cell acuminate. D_4 less than half length of D_3. Very similar to the Oriental genus *Ampelophaga* Bremer & Grey and the Nearctic genus *Darapsa* Walker.

Genitalia. In male, almost identical to those of the Oriental *Ampelophaga rubiginosa* Bremer & Grey, 1852, but with the two processes of the aedeagus thinner, not dentate. Otherwise, valva ladle-shaped; outer friction scales large. Sacculus spatulate and dilated, partly dentate on the upperside. In female, lamella postvaginalis suddenly narrowed, as in *Daphnis*. Ostium bursae large, free, edges slightly raised.

OVUM: Oval, dorso-ventrally flattened, pale yellowish green.

LARVA: Typically sphingiform, with the anterior segments retractile into the third thoracic and first abdominal segments. With a series of inverted V-marks down the dorsal surface, one per segment, and a pale, non-ocellate dorso-lateral line. Very similar to that of *Ampelophaga* and *Darapsa*.

PUPA (Text fig. 41): Proboscis fused to body. Brown dotted lines on the wings, legs and antennae. Spiracles set in dark surrounds. Cremaster short and pointed. Similar to *Daphnis*.

HOSTPLANT FAMILY: Vitaceae.

Text Figure 41 Pupa: *Clarina k. kotschyi* (Kollar)

CLARINA KOTSCHYI (Kollar, 1850)   Map 27

E – Grapevine Hawkmoth

CLARINA KOTSCHYI KOTSCHYI
(Kollar, 1850)

Deilephila kotschyi Kollar, 1850, *in* Kollar & Redtenbacher, *Denkschr. Akad. Wiss. Wien* **1**: 53.

Type locality: Schiraz [Shiraz, Iran].

> *Metopsilus mardina* Staudinger, 1901, *in* Staudinger & Rebel, *Cat. Lepid. palaearct. Faunengeb.*: 104.

ADULT DESCRIPTION AND VARIATION (Pl. 8, fig. 11)

Wingspan: 60–80mm. Much larger than subsp. *syriaca* (Pl. 8, fig. 12), with slightly different wing pattern and form, i.e. fewer and fainter transverse bands to the forewing and margin less serrate. Wing coloration very variable, ranging from reddish brown to pale grey with a brown suffusion (Ebert, 1976).

ADULT BIOLOGY

Rests by day amongst the dense foliage of its hostplant, or on rocks on the ground. At dusk, even those individuals that have emerged that afternoon take flight and, for a short period, go in search of nectar-bearing flowers.

Local in central Iran on hillsides up to 2000m and in mountain valleys with shrubs, vineyards and isolated trees. Can be very common in some grape-growing areas.

FLIGHT-TIME

Trivoltine; early May to late August in three overlapping generations.

EARLY STAGES

OVUM: Oval (1.50 × 1.25mm), pale yellowish green, often containing a distinct bubble of air; very similar to that of *Mimas tiliae*. Laid singly on both the upper and lower leaf-surface of its hostplant, with preference shown for *Vitis* shrubs growing on the edge of gorges, in plantation corners or in gardens (Pittaway, 1982b).

LARVA (Pl. 2, fig. 5): Full-fed 55–70mm.

Initially 3–4mm long, pale yellow, cylindrical, with a black, straight, almost upright horn 1.25mm in length. With feeding, it becomes yellowish green. In the second instar, pink replaces black on the horn and the final pattern appears, i.e. a yellow, dorso-lateral line from over the head to the horn, yellow body spots and a dorsal inverted V-mark on each segment. In the final instar, further changes occur: the third thoracic and first abdominal segments become slightly inflated so that, when alarmed, the small head and first two thoracic segments can be retracted into them. A blue tint suffuses the basic green colour below the dorso-lateral line and a narrow, ventro-lateral streak appears above the thoracic legs.

At all stages of development, the larva rests stretched out along the midrib on the lower surface of a leaf, rarely exposing more than its head and first two segments when feeding. It feeds constantly and, by the time it is fully grown, large quantities of food have been consumed (Pittaway, 1982b).

Prior to pupation, the larva becomes reddish brown and anoints itself with 'saliva' before descending from its hostplant under the cover of darkness.

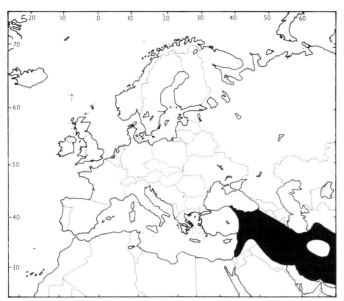

Map 27 – *Clarina kotschyi*

Found from early May to September, often with many stages of development present together.

Hostplants. Vitis vinifera, Parthenocissus spp. and *Ampelopsis* spp.

PUPA (Pl. E, fig. 10): 35–48mm. Similar to that of *Daphnis nerii* (q.v.). The wings are translucent brown, with the veins marked out with lines of dark brown spots. As in *D. nerii*, a crescent-shaped mark partly encircles each eye and the spiracles are set in dark surrounds. Formed in a loosely spun cocoon of light brown silk amongst debris and stones at the base of a rock or a grass tussock. The overwintering stage.

PARASITOIDS
None recorded.

BREEDING
Pairing. Nothing recorded.
Oviposition. Nothing recorded.
Larvae. Very easy to rear at normal room temperatures (14°–23°C), given a continuous supply of food, clean surroundings and no crowding; glass jars or net cages are equally suitable. Great quantities of leaves are eaten, but vine leaves may be stored for many days in a refrigerator and are then just as acceptable as fresh ones. The larval stage normally lasts three weeks.
Pupation. Again easy in an open box with dry soil and leaves. Summer pupae should be placed on dry sand or peat with good ventilation, as they will emerge later that year. Do not spray.

DISTRIBUTION
The Iranian plateau (Brandt, 1938; Barou, [1967]; Ebert, 1976) and Mesopotamia (Wiltshire, 1957). In northern Iran, northern Iraq and Turkey, subsp. *kotschyi* forms a clinal hybrid population with subsp. *syriaca*.
Extra-limital range. None.

CLARINA KOTSCHYI SYRIACA (Lederer, 1855)
Deilephila syriaca Lederer, 1855, *Verh. (Abh.) zool.-bot. Ver. Wien* **5**: 195.
Type locality: Beirut, Syria [Lebanon].

ADULT DESCRIPTION AND VARIATION (Pl. 8, fig. 12)
Wingspan: 50–65mm. Smaller than subsp. *kotschyi*, resembling *Dendrolimus pini* (Linnaeus) (European Pine Lappet). There is considerable variation within the subspecies, especially in ground colour, which ranges from deep reddish brown to pale grey. The transverse lines on the wings are sometimes very faint; distal margin of forewing sometimes without indentations. There is a noticeable sexual dimorphism, with males being lighter in colour and smaller than females.

This subspecies appears to have evolved in isolation in the Levant and spread to Turkey, Iraq and northern Iran since the last ice age, where it has come into contact and interbred with the nominate race, which spread westwards. It is sometimes difficult to assign individuals from these countries to either subsp. *syriaca* or subsp. *kotschyi*.

ADULT BIOLOGY
Favours cultivated valley floors with boulder-strewn streams bordered by vine-covered trees and shrubs, where large numbers are often present. It is also found on hillsides and in mountain valleys with shrubs and isolated trees, and in vineyards where, in Lebanon, it occurs up to 1000m (Zerny, 1933; Ellison & Wiltshire, 1939).

FLIGHT-TIME
Bivoltine; from May to early July, depending on latitude, and again in August/September.

EARLY STAGES
OVUM: As subsp. *kotschyi* (q.v.).

LARVA (not illustrated): Full-fed 55–65mm.
Very similar to that of subsp. *kotschyi* both in colour and behaviour; retains its basic pattern and coloration throughout its development but thoracic segments 1 and 2 are retractile in the penultimate instar.
Found commonly in suitable areas during June and July and again in September and October.

Hostplants. Vitis and *Parthenocissus* spp.

PUPA (Pl. E, fig. 10): 30–36mm. Identical to that of subsp. *kotschyi*.

PARASITOIDS
See Table 4, p. 52.

BREEDING
As for subsp. *kotschyi*.

DISTRIBUTION
From northern, southern and eastern Turkey to north-east Iraq, north-west Iran and north-west Mesopotamia (Wiltshire, 1957); also, northern and western Syria, Lebanon and Israel (Eisenstein, 1984), i.e. south-west of a line across central Mesopotamia and northern Iran.
Extra-limital range. None.

ACOSMERYX Boisduval, [1875]

Acosmeryx Boisduval, [1875], *in* Boisduval & Guenée, *Hist. nat. Insectes* (Spec. gén. Lépid. Hétérocères) **1**: 214.
Type species: *Sphinx anceus* Stoll, 1781.

A genus containing eight to ten species from the Oriental, Australasian and eastern Palaearctic regions. One penetrates the western Palaearctic.

IMAGO: Upperside of wings predominantly brown and grey, the markings forming a tessellated pattern; outer margin of forewing serrate, sinuate between veins R_4 and R_5. Antenna rod-like, with terminal segment very long, filiform and rough-scaled. No eyelashes. Labial palpus large, rounded in side-view. Spines of abdomen numerous, the short ones pale, rather weak, the long ones stronger. Eighth tergite deeply sinuate, separate from sternite. Hindtibia with two pairs of spurs, the proximal spur twice as long as the distal.

Genitalia. In male, uncus simple, long, slender, slightly curved; gnathos shorter, broader, somewhat boat-shaped, with apex always sinuate. Valva large, sole-shaped, with three or four rows of large friction scales. Sacculus dilated at end, the dilated part armed with spine-like teeth which are directed upwards. Aedeagus with a dentate lobe on left side, continuous with a slender, acute process at right side. In the female, lamella postvaginalis suddenly narrowed distally; ostium bursae transverse, postmedian, sometimes covered with a bilobate ridge.

OVUM: Ovoid, smooth, shiny green.

LARVA: Head small, body tapering forward from the first abdominal segment, thoracic segment 3 and abdominal segment 1 with ventro-lateral flanges; horn of medium length, erect but sharply curved. Dorso-lateral longitudinal line present. Thoracic segments dark ventrally.

PUPA (Text fig. 42): Proboscis fused with body; abdominal segment 7 deeply undercut around its posterior margin; cremaster thick and short with a pair of apical spines, or extended into a shaft (*A. naga* Moore).

HOSTPLANT FAMILIES: Climbers and ramblers of the Vitaceae, Actinidiaceae and Dilleniaceae.

Text Figure 42 Pupa: *Acosmeryx n. naga* (Moore)

ACOSMERYX NAGA (Moore, 1857)   Map 28
E – Naga Hawkmoth, **R** – Gissarskiĭ vinogradnyi Brazhnik

ACOSMERYX NAGA HISSARICA
Shchetkin, 1949

Acosmeryx naga hissarica Shchetkin, 1949, *Soobshch. Akad. Nauk tadzhik. SSR* (19): 24–27.
Type locality: Vakhsh Valley, Gissar Mountains, Tajikistan.

ADULT DESCRIPTION AND VARIATION (Pl. 8, figs 13, 14; Pl. G, fig. 5)

Wingspan: 82–103mm. The most obvious feature which separates this from the nominate subspecies is the form of the dark brownish bar which stretches from the forewing costa, along vein M_2, towards the centre of the outer margin. In *A. naga naga* this is straight, in *A. n. hissarica* it is noticeably arched. The primary body colour in subsp. *hissarica* is also greyer and lighter in appearance than in subsp. *naga*.

ADULT BIOLOGY

Occurs in cultivated as well as uncultivated mountain valleys and ravines. Along the southern slopes of the Gissar Mountains of Tajikistan (Shchetkin, 1956) it is confined to damp locations in gorges at 1100–1700m which support *Juglans regia* (walnut) and *Acer turkestanicum* (Turkestan maple) overgrown with wild *Vitis vinifera* (grapevines). In Afghanistan it occurs at 1500–2500m (Daniel, 1971).

FLIGHT-TIME

Bivoltine; late April to June, and again in late July (Ebert, 1969; Daniel, 1971; Derzhavets, 1984).

EARLY STAGES

OVUM: Size unrecorded; almost spherical, smooth, a deep rich green, becoming whitish before hatching (Bell & Scott, 1937).

LARVA (Pl. 4, fig. 3): Full-fed 70-90mm. Dimorphic: green or brown.

According to Bell & Scott (1937), in the first instar, the head and body are green, the horn black, long and straight. In the second instar, thoracic segment 1 is as narrow as the head, with the body tapering sharply anteriorly from the first abdominal segment. The black horn is still long and straight, but reddish at the base and with a white tip. The head and anal segments are yellow, the body green dotted with white, with the spiracle of abdominal segment 1 surrounded by a black spot. A yellow dorso-lateral stripe runs from the head to the horn. This colour pattern remains the same for the next two instars, but a lateral flange develops on the third thoracic and first abdominal segments and a yellow ventro-lateral stripe appears on the thoracic segments and abdominal segment 1. In the final instar of the green form, the head is grass-green, with a narrow, pale yellowish dorso-lateral stripe and a broader stripe of the same colour separating face from cheek; thoracic segments 1 and 2 grass-green with short darker stripes; the rest of the body bluish green, mottled with yellow above the dorso-lateral stripe and pale greyish blue below. The dorso-lateral stripe is narrow and white on thoracic segments 1 and 2, broader and pale yellow on thoracic segment 3 and abdomi-

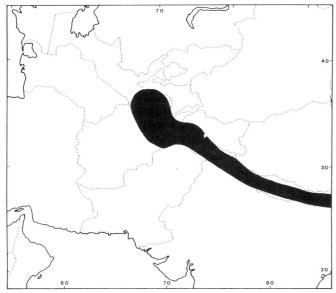

Map 28 – *Acosmeryx naga*

nal segment 1, then white to the base of the horn, and edged narrowly above with orange on segments 3–7. The narrow, white ventro-lateral stripe on thoracic segments 1 and 2 becomes broad and yellow as it outlines the flange before turning upwards on abdominal segment 1 to form an oblique stripe. There are also pale yellow, oblique lateral stripes on segments 1–7. The horn is lilac-grey dotted with purple. There are dark purple patches on the body above the bases of the legs, increasing in size caudally, and extending along the lower edge of the flange on thoracic segment 3 and abdominal segment 1. Upper part of prolegs bluish, the lower part pale yellow; feet brown, anal clasper bluish, anal flap edged broadly with pale yellow. Spiracles deep orange. In the brown form the green pigmentation is replaced by pinkish brown. The whole body is smooth and moderately shiny.

At rest on the underside of a leaf or stem, the larva throws back its head and anterior segments in a sharp curve, the head held so that the face is in the same plane as the dorsum of thoracic segment 1. The thoracic legs are pressed close to the body and the flange is dilated laterally. As in *Clarina kotschyi*, which it closely resembles, the alarmed larva withdraws the head and first two thoracic segments into segment 3; unlike *C. kotschyi*, thoracic segment 3 and abdominal segment 1 are puffed out and the flanges further dilated.

Occurs from June until September.

Hostplants. Principally *Vitis* and *Ampelopsis* (both Vitaceae), but also feeding on *Actinidia* and *Saurauia* (both Actinidiaceae).

PUPA (Pl. E, fig. 11): 44–55mm. Very similar in shape to *Clarina* (Pl. E, fig. 10) but dark brown with the hind margin of abdominal segment 7 deeply undercut. Between the frons and the proboscis, there is a knobbly pentagonal sclerite. Abdomen marked with dark pits and stripes. Cremaster small, basal half bulbous, distal half a cylindrical shaft ending in two small hooks. Formed in a slight cocoon on the ground. Very sensitive to desiccation (Shchetkin,

1956), but highly tolerant of moisture. The overwintering stage.

PARASITOIDS
None recorded.

BREEDING
As for *Clarina k. kotschyi* (q.v., p. 119), but the pupae must not be allowed to dry out.

DISTRIBUTION
As subsp. *hissarica*, only southern Tajikistan (Shchetkin, 1949; Shchetkin, 1956; Derzhavets, 1984) and Afghanistan (Ebert, 1969; Daniel, 1971; C. M. Naumann, pers. comm.). *A. naga hissarica* appears to be an arcto-tertiary relict (along with numerous plants (Shchetkin, 1956)) which developed in the western Tian Shan and which has since spread to Afghanistan, where it has come into contact with the nominate subspecies.

Extra-limital range. None.

OTHER SUBSPECIES
As subsp. *naga*, the Himalayan regions of Pakistan, India, Nepal and China. Also Peninsular Malaysia, Thailand, most of eastern China, Korea and Japan.

RETHERA Rothschild & Jordan, 1903

Rethera Rothschild & Jordan, 1903, *Novit. zool.* **9** (Suppl.): 500 (key), 547.
Type species: *Deilephila komarovi* Christoph, 1885.

    Borshomia Austaut, 1905, *Ent. Z., Frankf. a. M.* **19**: 30.

A western Palaearctic genus containing four species.

IMAGO: Wings entire. Eye with moderate lashes. Labial palpus obtuse, rounded in dorsal and lateral aspect. Antenna club-shaped, abruptly tapering to form a short hook. Terminal segment short, not long and filiform; fasciculate setae in male vestigial. Epiphysis of foretibia extending beyond apex. Mid- and hindtibial spurs unequal, long spur less than half as long as the respective first tarsal segment. First hindtarsal segment longer than first midtibial one, and about as long as hindtibia. Pulvillus and paronychium absent. Veins Rs and M_1 of hindwing stalked.

Genitalia. Friction scales of male valva truncate, rather small. Valva elongate, sole-shaped; sacculus ending in an obtuse process with parallel sides and concave upper surface. Aedeagus with an oblique series of cornuti.

OVUM: Known only for *R. komarovi* (Pittaway, 1979a). Large, ovoid, glossy blue-green with a white encircling line which changes to brown after a few days.

LARVA: Known only for *R. komarovi*. Head small; tapering anteriorly, with the head and thoracic segments 1 and 2 retractile into thoracic segment 3 and abdominal segment 1. Horn absent in final instar; spiracles of abdominal segments 1 and 8 surrounded with eye-spots. Diffuse oblique lateral stripes present on abdominal segments.

PUPA (Text fig. 43): Known only for *R. komarovi*. Dark brown, very glossy. Proboscis fused with body; produced forward as in *Hippotion* but not laterally compressed. Cremaster large with a pair of apical spines.

HOSTPLANT FAMILIES: Principally herbaceous plants of the Rubiaceae.

Text Figure 43 Pupa: *Rethera komarovi manifica* Brandt

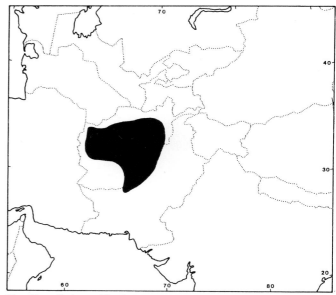

Map 29 – *Rethera afghanistana*

RETHERA AFGHANISTANA Daniel, 1958

Map 29

E – Afghan Madder Hawkmoth
Rethera afghanistana Daniel, 1958. *Beitr. naturk. Forsch. SüdwDtl.* **17**: 83–84.
Type locality: Herat, Afghanistan.

ADULT DESCRIPTION AND VARIATION (Pl. 9, fig. 1)
Wingspan: 46–51mm. Similar to, but larger than, *R. brandti* (q.v., p. 126; Pl. 9, fig. 6).

ADULT BIOLOGY
Nothing known, except that it occurs at 1000–2100m.

FLIGHT-TIME
Probably univoltine; mid-April (Herat) and late May (Kabul).

EARLY STAGES
Unknown.

PARASITOIDS
Unknown.

BREEDING
Nothing known.

DISTRIBUTION
First discovered at Herat in western Afghanistan, this species is now known to occur throughout Afghanistan (Ebert, 1969). It may yet be discovered in central Iran.
Extra-limital range. Western Pakistan (Quetta) (Daniel, 1971).

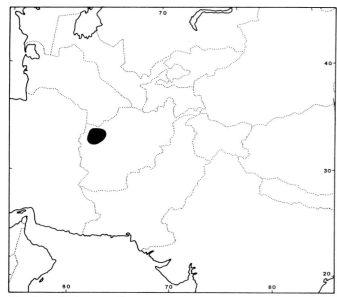

Map 30 – *Rethera amseli*

RETHERA AMSELI Daniel, 1958   Map 30
E – Amsel's Hawkmoth
Rethera amseli Daniel, 1958, *Beitr. naturk. Forsch. Südw-Dtl.* **17**: 83–84.
Type locality: Herat, Afghanistan.

ADULT DESCRIPTION AND VARIATION (Pl. 9, fig. 2)
Wingspan: 34–43mm. Larger, with paler markings, than its close relative, *R. brandti* (Pl. 9, fig. 6), from which it is clearly separated by the distinct genitalia (Kernbach, 1959).

ADULT BIOLOGY
Unknown.

FLIGHT-TIME
Univoltine; mid-April.

EARLY STAGES
Unknown.

PARASITOIDS
Unknown.

BREEDING
Nothing known.

DISTRIBUTION
R. amseli was discovered near Herat (western Afghanistan) at the same time as *R. afghanistana*, i.e. 1958, but has not been seen since.
Extra-limital range. None.

RETHERA KOMAROVI (Christoph, 1885) Map 31
E – Madder Hawkmoth, **R** – Komorova Brazhnik

RETHERA KOMAROVI KOMAROVI
(Christoph, 1885)
Deilephila komarovi Christoph, 1885, *in* Romanoff, *Mém. Lépid.* **2**: 169.
Type locality: Askhabad [Ashkhabad, Turkmenistan].
    Choerocampa stipularis Swinhoe, 1885, *Trans. ent. Soc. Lond.* **1885**: 346.

ADULT DESCRIPTION AND VARIATION (not illustrated)
Wingspan: 55–65mm. Almost indistinguishable from subsp. *drilon* (Rebel & Zerny) (Pl. 9, fig. 3), but paler and larger. The species is confusable only with *R. brandti* (Pl. 9, fig. 6) and the other two *Rethera* species from Afghanistan (Pl. 9, figs 1, 2), but is considerably larger than any of them.

Exhibits very little variation except for the intensity of 'tint' colours, which range, ventrally, from pale orange to deep pink. Newly-emerged examples often have a magnificent rose sheen. However, due to the fact that the dorsal green coloration fades in strong sunlight, those collected from drier regions tend to be paler than examples from more verdant areas. The same applies when killing agents such as diethyl ether and ethyl acetate are used.

ADULT BIOLOGY
Occurs in mountains and hilly areas where it frequents 'vegetation islands' on sparsely vegetated cliffs and rocky slopes which are cold in winter and become parched in summer. Little is known of the behaviour of this species, except that it is attracted to light.

FLIGHT-TIME
Univoltine; mid-May to the second week of June.

EARLY STAGES
OVUM: Unknown; probably identical to that of subsp. *manifica* (q.v., p. 125).

LARVA (not illustrated): Full-fed, 70–90mm.
    Descriptions of early stages not available; when fully grown, identical to that of subsp. *manifica* (Pl. 2, fig. 14).
    Occurs during June and July.
Hostplants. Principally *Rubia* and *Galium* spp. (Pittaway, 1979a); also, reputedly, *Euphorbia* (Buresch & Tuleschkow, 1931).

PUPA: Unknown, but probably as subsp. *manifica*. Overwintering stage.

PARASITOIDS
None recorded.

BREEDING
Nothing known, but presumably as for subsp. *manifica*.

DISTRIBUTION
Transcaucasia, eastern Turkey (Daniel, 1979), southern Turkey (Hariri, 1971), northern Iraq (Wiltshire, 1957) to southern Turkmenistan (Tashliev, 1973) and Tajikistan (Derzhavets, 1984). A very sparsely distributed species, though large colonies sometimes occur in certain localities

in Kurdistan, between 1000-2000m altitude (Wiltshire, 1957). It may have a wider distribution than indicated due to the remote, inaccessible nature of its habitat.

Extra-limital range. The Pamirs.

RETHERA KOMAROVI DRILON
(Rebel & Zerny, 1932)

Deilephila komarovi drilon Rebel & Zerny, 1932, *Denkschr. Akad. Wiss. Wien* **103**: 85.
Type locality: the environs of Brutti, Albania.

ADULT DESCRIPTION AND VARIATION (Pl. 9, fig. 3)
Wingspan: 45–52mm. Very similar to the nominate race, but smaller with better-defined, darker markings.

ADULT BIOLOGY
A very local subspecies occurring in dry, hilly regions. Especially partial to dry, herb-rich hillsides and vegetation-fringed, dry riverbeds. Generally at 600–1200m.

Many individuals are attracted to most forms of light and, by day, seem to have a preference for resting on boulders (Rebel & Zerny, 1932).

FLIGHT-TIME
Univoltine; early April/May until early June, depending on altitude.

EARLY STAGES
OVUM: As subsp. *manifica* (q.v., p. 125).

LARVA: As subsp. *manifica* (Pl. 2, fig. 14). A full-grown larva is illustrated by Pelzer (1991).
Occurs from May to July.

PUPA: As subsp. *manifica*.

PARASITOIDS
None recorded.

BREEDING
As for subsp. *manifica* (q.v., p. 125).

DISTRIBUTION
The mountains of eastern Albania (Rebel & Zerny, 1932), southern Yugoslavia, northern Greece, southern Bulgaria (Ganev, 1984; S. V. Beschkow, pers. comm.) and western and central Turkey (Daniel, 1932; de Freina & Witt, 1987).
Extra-limital range. None.

RETHERA KOMAROVI RJABOVI
O. Bang-Haas, 1935 **Stat. rev.**

Rethera komarovi rjabovi O. Bang-Haas, 1935, *Ent. Z., Frankf. a. M.* **48**: 184.
Type locality: Ordubad, Armenia [Nakhichevan, Azerbaijan].

(*Taxonomic note*. Kernbach (1959) doubted the validity of this subspecies on the strength of one specimen from Nakhichevan which was superficially and in genitalia identical to subsp. *komarovi*. Such a conclusion, based on just one individual, is not acceptable as virtually all specimens from the same area can be referred to subsp. *rjabovi*. Moreover, as there are no differences in genitalia between

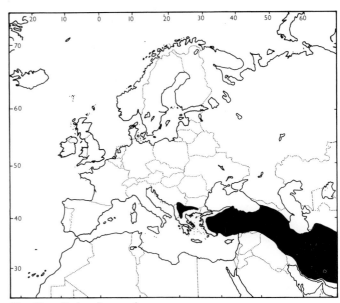

Map 31 – *Rethera komarovi*

any of the various subspecies of *R. komarovi*, his use of this character is invalid.)

ADULT DESCRIPTION AND VARIATION (Pl. 9, fig. 4)
Wingspan: 60–72mm. Male usually smaller than female. Unlike subsp. *komarovi*, coloration very bright and contrasting: ventral surface a rich rose, which extends up over the head. In f. *roseocingulata* O. Bang-Haas (1937), this roseate coloration also replaces the white on the dorsal surface of abdominal segment 4.

ADULT BIOLOGY
Occurs in similar habitats to those of subsp. *komarovi*, but normally only between 600–1500m. Readily attracted to light.

FLIGHT-TIME
Univoltine; mid-May to mid-June.

EARLY STAGES
Unknown, but presumably as subsp. *manifica* (q.v., p. 125).

PARASITOIDS
Unknown.

BREEDING
Nothing known, but presumably as for subsp. *manifica* (q.v., p. 125).

DISTRIBUTION
Principally the area between the Caucasus and the Anatolian highlands to the south-west (O. Bang-Haas, 1936; 1937). Kalali (1976) reports subsp. *rjabovi* also from Sarakhs on the Iran/Turkmenistan border east of Mashhad, but this may be erroneous.
Extra-limital range. None.

RETHERA KOMAROVI MANIFICA
Brandt, 1938 **Stat. rev.**

Rethera komarovi manifica Brandt, 1938, *Ent. Rdsch.* 55: 698.

Type locality: Mian-Kotal & Sine-Sefid [southern Zagros Mountains], Iran.

(*Taxonomic note.* Kernbach (1959) doubted the validity of this subspecies on the strength of examining one paratype which was superficially and in genitalia identical to subsp. *komarovi*. Such a conclusion, based on just one individual, is not acceptable as nearly all specimens from the same area can be referred to subsp. *manifica*. There are no differences in genitalia between the various subspecies of *R. komarovi* and therefore his use of this character is invalid.)

ADULT DESCRIPTION AND VARIATION (Pl. 9, fig. 5)

Wingspan: 65–81mm. Larger and much paler than the nominate subspecies.

ADULT BIOLOGY

A local subspecies, occurring in well-defined colonies amongst richer than normal vegetation. Forms a cline with subsp. *komarovi* where they 'overlap' in range in Iraqi Kurdistan. Occurs in mountainous areas, usually between 1500–2000m on steep, herb- and grass-covered hillsides, especially those strewn with boulders and grazed by live-stock; also on the sides of rocky gullies (Pittaway, 1979a).

During daylight hours, hides between boulders, amongst low vegetation at their base, or even on the ground in some shady location. Light is very attractive, with large numbers often arriving over a short period.

FLIGHT-TIME

Usually the last two weeks of May, with a partial second brood in mid-August.

EARLY STAGES

OVUM: Large (1.6 × 1.4mm), slightly oval. Initially, glossy blue-green with a faint, incomplete, whitish, encircling line which changes to brown after a few days. Prior to hatching, the whole ovum assumes a greyish colour.

Laid singly, with up to five per plant, on the upper- and undersides of leaves, young stems, unopened flower-heads and even dead shoots. Preference is shown for clumps of basally-sprouting, little-grazed *Rubia* growing on the ground or between rocks in a dense mass, rather than those climbing up shrubs (Pittaway, 1979a).

LARVA (Pl. 2, fig. 14; Pl. D, fig. 1): Full-fed, 80–100mm.

Newly hatched, the 4mm-long larva is bluish yellow with a small, black, upright horn, black prolegs and claspers. The light brown head is disproportionately large, as are the anterior segments which, together with the rest of the body, bear longitudinal rows of dark, bristle-bearing tubercles. Ignoring the eggshell, the larva commences to feed on plant material, flower-buds being preferred. This causes the body colour to darken to bluish green, on which a white dorso-lateral line appears. In the second and third instars, the now cylindrical larva is frosty blue-green; this frostiness is imparted by a dense covering of very fine, white spots, each bearing a short, white bristle, and which even extend up the horn. The now brilliant white, dorso-lateral line extends from the green head to the short, straight, reddish horn. At this stage the larva feeds quite openly, day and

night, amongst the flowerbuds and leaves, relying on matching coloration for camouflage. However, if in immediate danger from predators or grazing animals, it drops down into the tangled mass of its hostplant.

After the next moult a sudden transformation in appearance takes place: the green coloration and the horn are replaced by dark and light brown mottling and a small tubercle, respectively. The head is brown, with three fine, dark longitudinal stripes, whilst the spiracles of abdominal segments 1 and 8 are ringed in violet and yellow, set in a black blotch so as to resemble eyes. Feeding now takes place only at night, the larva hiding during daylight hours at the base of the hostplant or amongst debris and stones.

In its final instar, the appearance of the larva is striking. When alarmed, the head and first two thoracic segments are tucked under the third thoracic and the first abdominal segments which have become slightly swollen so that, dorsally, the whole animal resembles a small viper (Pl. D, fig. 1). This illusion is further enhanced by the larva swinging its false 'head' from side to side. If further disturbed, the larva will drop down amongst its hostplant, remaining in this viper-like position for up to an hour. Such a defence must be very useful as, in order to satisfy its large appetite, a fully-grown larva must now also ascend the hostplant two or three times during daylight hours to feed, thus exposing itself to predatory birds.

Feeds from late May to mid-June and again in September, as two generations.

Hostplants. Principally *Rubia* spp.; also *Galium*.

PUPA (Pl. E, fig. 12): 58–65mm. Slender, dark brown, almost black, with intersegmental cuticle reddish brown; very glossy and hard but also very mobile. Remarkable for its projecting, bulbous proboscis and elongated head region. Cremaster conical with a glossy tip. Formed between rocks or in a clump of plant debris inside a large, gravel- and soil-impregnated silken cell. The overwintering stage.

PARASITOIDS

None recorded.

BREEDING

Pairing. Nothing known.

Oviposition. Nothing known.

Larvae. Difficult, even with good ventilation and potted hostplant or cut sprigs in water. Normal room temperatures are adequate, but moisture or prolonged high humidity are usually fatal, so glass containers are not recommended. If *Rubia* is unobtainable, *Galium aparine* (goose-grass) is the best substitute. Unfortunately, most other European species of *Galium* have too high a water-content, which invariably causes digestive disorders and death. Good results have been achieved with under-watered potted hostplant placed in a small netting cage in a sunny window. As large larvae leave the plant when not feeding, a cage is essential. If cut goose-grass is to be used for feeding, to keep losses to a minimum, plants growing in a dry situation should be selected as these have a lower water-content, but those growing under trees or bushes should be avoided as they can be contaminated by harmful bacteria and viruses from above.

Pupation. Larvae shrink considerably for some days and wander a great deal before settling down; at this stage they also void drops of black fluid from time to time, due to excessive moisture in the food. They should, therefore, be confined to individual, open-topped glasses, one-third full of dry, light, gritty soil with a top layer of dry leaves, absorbent paper and other debris. Overwintering pupae (still in their individual glasses), should be placed in a warmish, dry location and left until early November, then transferred to a cold, dry location until March. Once again placed in warm surroundings, provided that the soil they are on is slightly dampened, the moths will emerge in May. At all stages, good ventilation, low humidity and lack of disturbance are essential; even then, many die.

DISTRIBUTION
Iraqi Kurdistan (Wiltshire, 1957) south to the Zagros Mountains of Iran (Barou, [1967]) and east to Afghanistan (Ebert, 1969).
Extra-limital range. None.

RETHERA BRANDTI O. Bang-Haas, 1937
Map 32

E – Lesser Madder Hawkmoth

RETHERA BRANDTI BRANDTI
O. Bang-Haas, 1937

Rethera brandti brandti O. Bang-Haas, 1937, *Ent. Z., Frankf. a. M.* **50**: 562.
Type locality: Elburz [Alborz] Mountains, northern Iran.

ADULT DESCRIPTION AND VARIATION (Pl. 9, fig. 6)
Wingspan: 42–46mm. Quite distinct, though very similar to *R. afghanistana* (Pl. 9, fig. 1); it can be confused only with a small *R. komarovi* (see Pl. 9, fig. 3) but may be distinguished by its abdominal segments being edged with grey and the entirely green fourth segment. There is practically no variation.

ADULT BIOLOGY
Found on sparsely vegetated slopes up to 2000m. Like *R. komarovi*, this species is widely distributed as local populations containing large numbers of individuals. Comes to light, sometimes in great numbers.

FLIGHT-TIME
Univoltine; April to mid-May.

EARLY STAGES
Unknown.
Hostplants. Not known, but probably *Galium*.

PARASITOIDS
Unknown.

BREEDING
Nothing known.

DISTRIBUTION
Restricted to the Alborz and Kopet Dağ Mountains of northern Iran.
Extra-limital range. None.

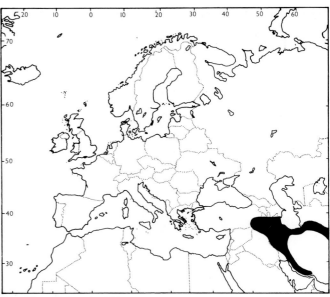

Map 32 – *Rethera brandti*

RETHERA BRANDTI EUTELES Jordan, 1937

Rethera brandti euteles Jordan, 1937, *Novit. zool.* **40**: 325.
Type locality: C'urum [between Shiraz and Bushehr], Iran.
(*Taxonomic note.* Kernbach (1959) doubted the validity of this subspecies on the strength of two specimens from southern Iran, which were superficially similar to subsp. *brandti*. Examination of an extensive series in the collection of E. P. Wiltshire (deposited at the Natural History Museum, London), show that his doubts were unjustified.)

ADULT DESCRIPTION AND VARIATION (Pl. 9, fig. 7)
Wingspan: 40–45mm. Paler and greyer than subsp. *brandti* (Pl. 9, fig. 6) and lacking any pink coloration in most individuals; however, some specimens are similar to subsp. *brandti* in coloration. As in that subspecies, there is little variation.

ADULT BIOLOGY
Does not overlap with subsp. *brandti*. Found in hilly steppe and desert-edge vegetation as local populations.

FLIGHT-TIME
Univoltine; late March to mid-May, depending on altitude.

EARLY STAGES
Unknown.
Hostplants. Unknown, but probably *Galium*.

PARASITOIDS
Unknown.

BREEDING
Nothing known.

DISTRIBUTION
From south-east Turkey and north-east Iraq (Wiltshire, 1957) to southern Iran along the Zagros Mountains (Brandt, 1938). May be more widespread than indicated due to the remoteness of its habitat.
Extra-limital range. None.

SPHINGONAEPIOPSIS Wallengren, 1858

Sphingonaepiopsis Wallengren, 1858, *Öfvers. K. Vetensk-Akad. Förh. Stockh.* **15**: 138.

Type species: *Sphingonaepiopsis gracilipes* Wallengren, 1858.

Pterodonta Austaut, 1905, *Ent. Z., Frankf. a. M.* **19**: 25.

A genus containing seven species and occurring in the southern Palaearctic, Oriental and Afrotropical regions.

IMAGO: Distal margin of forewing dentate or lobate. Labial palpus rough-scaled, first segment with lateral apical fan. Eye lashes and a high head-tuft present. Spines of abdomen very weak. Antenna dentate or pectinate in male, simple and club-shaped in female; terminal segment short. Tibia spinose; midtarsus with a basal comb; hindtarsus with a few basal spines. Hindtibia with two spurs of equal length. Paronychium with lateral lobes very small, the ventral ones absent; tarsi long. Veins M_3 and Cu_1 of forewing originating close together; Cu_2 at apical third of cell. Costal margin of hindwing almost straight, convex near base; Cu_1 and Cu_2 close together, some distance from angle of cell.

Genitalia. In male, uncus elongate-triangular, apex more or less rounded-truncate; gnathos either strongly chitinized, with the upperside transversely ribbed distally, or short, broad, membranous. Valva without friction scales; sacculus variable; aedeagus with or without apical processes. In female, lamella postvaginalis triangular, apical edge projecting.

OVUM: Small, spherical, green.

LARVA: Typically sphingiform, with a noticeably thin body and small horn, and pale, longitudinal stripes. Head large; thorax weakly constricted anteriorly.

PUPA: Proboscis fused with body. Small, very glossy; similar in shape to that of *Hyles* spp. (Pl. E, fig. 15) but boldly marked with black. Cremaster with rounded tip.

HOSTPLANT FAMILIES: Herbaceous plants of the Rubiaceae.

SPHINGONAEPIOPSIS GORGONIADES
(Hübner, [1819]) Map 33

E – Gorgon Hawkmoth, **F** – Sphinx Gorgon, **G** – Chalcedongrauer Schwärmer, **R** – Gorgona Brazhnik

SPHINGONAEPIOPSIS GORGONIADES GORGONIADES (Hübner, [1819])

Proserpinus gorgoniades Hübner, [1819], *Verz. bekannter Schmett.*: 132.

Type locality: Southern Volga, Russia.

‡*Sphinx legitima gorgon* Esper, [1806], *Die Schmett.* (Suppl.) (Abschnitt 2): 49 [homonym].

ADULT DESCRIPTION AND VARIATION (Pl. 9, fig. 8)

Wingspan: 25–30mm. Cannot be confused with any other hawkmoth from the region, except *S. kuldjaensis* (Pl. 9, fig. 11). Forewing whitish grey. Exhibits very little variation except in intensity of coloration and extent of buff or pale orange speckling on the hindwings. In drier locations, this speckling occurs on the forewing of some paler, brownish individuals.

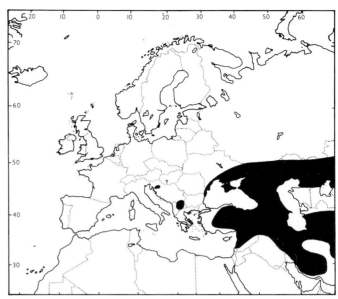

Map 33 – *Sphingonaepiopsis gorgoniades*

ADULT BIOLOGY

Crepuscular, occurring in open scrub and hilly steppe in small, local colonies which are easily overlooked due to its small size. Generally at 2000–2500m in the Caucasus. Little is known of its behaviour, though it is strongly attracted to flowers at dusk.

FLIGHT-TIME

Bivoltine; late May and early June, with a second brood in August.

EARLY STAGES

OVUM: Very small, spherical, pale green.

LARVA (not illustrated): Full-fed, 30–40mm. Polymorphic: green, red or whitish. Similar to that of subsp. *chloroptera* (Pl. 2, fig. 6).

Early instars undescribed. In fully-grown larvae, the cylindrical body may be green, red or whitish, with two dorsal and two lateral, brilliant white, longitudinal stripes, sometimes bordered with red. White-striped head rounded, with short bristles; horn pink or orange, narrow, short, straight or slightly curved upwards; spiracles white, set in a thin, white line, often bordered with pink.

Occurs mainly in June, July and September.

Hostplants. Principally *Galium* spp., also other Rubiaceae.

PUPA: No description available. The overwintering stage.

PARASITOIDS

None recorded.

BREEDING

No information available.

DISTRIBUTION

From eastern Bulgaria (Beschkow, 1990) and Romania across the southern Ukraine and Crimea (Zerny, 1933), southern Russia as far north as Kazan, Kazakhstan (Eversmann, 1844)), the Caucasus (Derzhavets, 1984) and Turkey to the Tian Shan. A very disjunct and, as yet, poorly-understood distribution. May occur in a great many areas other than those indicated.

Extra-limital range. Across southern Siberia, including the Altai and Amurland (Russia). This subspecies may also occur in Mongolia and northern China.

SPHINGONAEPIOPSIS GORGONIADES PFEIFFERI Zerny, 1933 **Stat. rev.**

Sphingonaepiopsis gorgon pfeifferi Zerny, 1933, *Dt. ent. Z. Iris* 47: 60.

Type locality: Northern Lebanon.

(*Taxonomic note.* Mentzer (1974) separated *S. pfeifferi* from *S. gorgoniades* on minor morphological grounds, which cannot be justified. In Turkey, the range of subsp. *pfeifferi* is contiguous with that of subsp. *gorgoniades*, yielding intermediate forms; individuals from more humid areas are darker and tend towards subsp. *gorgoniades*.)

ADULT DESCRIPTION AND VARIATION (Pl. 9. fig. 9)

Wingspan: 25–28mm. Distinguished from subsp. *gorgoniades* by its much lighter, brownish forewings and very extensive buff or orange speckling on both the fore- and hindwing.

ADULT BIOLOGY

Crepuscular. Occurs in similar habitat to those of subsp. *gorgoniades*, but in drier conditions above 1800m, though in Lebanon it occurs at around 1700m.

FLIGHT-TIME

Bivoltine; May and July/August.

EARLY STAGES

OVUM: Not known.

LARVA: As subsp. *gorgoniades* (q.v., p. 127).

Hostplants. Galium spp.

PUPA: No description available. The overwintering stage.

PARASITOIDS

None recorded.

BREEDING

No information available.

DISTRIBUTION

From southern Turkey (Daniel, 1932; Hariri, 1971) and Lebanon (Zerny, 1933; Ellison & Wiltshire, 1939) eastward across northern Iraq (Wiltshire, 1957), Iran to Afghanistan (Ebert, 1969; Daniel, 1971). Little is known of its distribution in Iran and Afghanistan.

Extra-limital range. None.

SPHINGONAEPIOPSIS GORGONIADES CHLOROPTERA Mentzer, 1974 **Stat. rev.**

Sphingonaepiopsis pfeifferi chloroptera Mentzer, 1974, *Acta ent. Jugosl.* 10: 147–153.

Type locality: Matka, near Skopje, southern Yugoslavia [Macedonia].

(*Taxonomic note.* Mentzer (1974) described this taxon as a subspecies of *S. pfeifferi*, having separated the latter from *S. gorgoniades* on minor morphological grounds. This is not warranted, hence subsp. *chloroptera* is here regarded as a subspecies of *S. gorgoniades*. Supporting this view, some individuals of subsp. *gorgoniades* from the Crimea are almost indistinguishable from subsp. *chloroptera*, and the early stages of all three subspecies are identical.)

ADULT DESCRIPTION AND VARIATION (Pl. 9, fig. 10)

Wingspan: 30–32mm. Larger than the two subspecies described previously, with grey coloration of the forewing darker and less marked than in subsp. *gorgoniades* and with a faint violet tint. Additionally, apex and anal angles of forewing, together with basal area of hindwing, heavily marked with pale orange spots.

ADULT BIOLOGY

Crepuscular, occurring in hilly steppe with scattered trees. Attracted to light in very small numbers; few individuals are caught, possibly because the species is crepuscular.

FLIGHT-TIME

Bivoltine; the second half of May and late July.

EARLY STAGES

OVUM: Small (size unrecorded), spherical, green.

LARVA (Pl. 2, fig. 6): Polymorphic, as subsp. *gorgoniades* (q.v., p. 127), green, red or whitish.

Occurs during May, June, August and September.

Hostplants. Galium spp.

PUPA: No description available. The overwintering stage.

PARASITOIDS

None recorded.

BREEDING

No information available.

DISTRIBUTION

So far, this subspecies has been found only in the Balkans in Croatia (near Senj) (series in BM(NH) coll. and one male in the Vienna Natural History Museum (Zerny, 1933)), and in Macedonia (Matka) (Mentzer, 1974). There are, however, old as well as recent unconfirmed reports that this subspecies may occur in Hungary (Rougeot & Viette, 1978).

Extra-limital range. None.

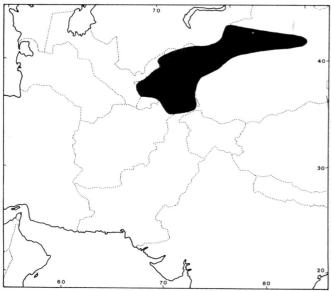

Map 34 – *Sphingonaepiopsis kuldjaensis*

SPHINGONAEPIOPSIS KULDJAENSIS
(Graeser, 1892) Map 34

E – Kuldja Hawkmoth

Pterogon kuldjaensis Graeser, 1892, *Berl. ent. Z.* **37**: 299.

Type locality: Kuldja [Yining/Gulja, Tian Shan, China].

ADULT DESCRIPTION AND VARIATION (Pl. 9, fig. 11)

Wingspan: 30–34mm. Resembles *S. gorgoniades chloroptera* (Pl. 2, fig. 6), but hindwing bright orange with a brown border.

ADULT BIOLOGY

Nothing known.

FLIGHT-TIME

Bivoltine; early May and early July, possibly later.

EARLY STAGES

Nothing known.

PARASITOIDS

Unknown.

BREEDING

Nothing known.

DISTRIBUTION

Known, at present, only from the western Tian Shan and Pamirs (Derzhavets, 1984).

Extra-limital range. The eastern Tian Shan.

SPHINGONAEPIOPSIS NANA (Walker, 1856)
Map 35

E – Savanna Hawkmoth

Lophura nana Walker, 1856, *List Specimens lepid. Insects Colln. Br. Mus.* **8**: 107.

Type locality: Natal, South Africa.

‡*Pterogon nanum* Boisduval, 1847, *in* Delegorgue, *Voyage dans L'Afrique Australe* **2**: 594.

Sphingonaepiopsis gracilipes Wallengren, 1858, *Öfvers. K. VetenskAkad. Förh. Stockh.* **15**: 138.

Sphingonaepiopsis nanum, Boisduval [1875], *in* Boisduval & Guenée, *Hist. nat. Insectes* (Spec. gén. Lépid. Hétérocères) **1**: 314.

ADULT DESCRIPTION AND VARIATION (Pl. 9, fig. 12)

Wingspan: 25–30mm. A very small, brown, blunt-winged moth which is very easily overlooked.

ADULT BIOLOGY

Crepuscular. Frequents savanna, steppe and semi-desert at about 1000m. In Africa, often seen hovering at flowers well before nightfall.

FLIGHT-TIME

March and April, but may fly later as a second generation.

EARLY STAGES

OVUM: Small (size unrecorded), spherical, pale green.

LARVA (not illustrated): Full-fed, 40mm. Polymorphic: green, reddish or whitish.

According to Carcasson (1968), in the penultimate instar of the dark form, the small head is dull orange-brown. The body is blackish brown with a black dorsal stripe, and with a dorso-lateral line which is orange-brown on the first thoracic segment and anterior part of the second and composed of minute white dots thereafter. On the second and third thoracic segments there is a transverse line of six white dots. The true legs are orange, the venter and prolegs blackish brown, the claspers orange. The anal flap is outlined in orange. Horn very long, straight and erect, black with an orange basal line. This tends to be 'waved' up and down along the axis of the body. The final instar is very similar, but with dorso-lateral stripe extending to the whole of the second segment. There is also a trace of a dull orange-brown ventro-lateral line speckled with white. Spiracles white and horn red, laterally compressed, apically blunt, very long (8mm).

In the green form the ventro-lateral line is brilliant white and edged with red. The fainter dorso-lateral line is of the same colour.

Instead of being voided as individual pellets, the frass forms 'sticks' 1–4cm in length.

Occurs mainly during April and May.

Hostplants. Various Rubiaceae, including *Kohautia*, *Galium*, *Rubia* and *Jaubertia*.

PUPA: Similar in shape to that of *Hyles* (cf. Pl. E, fig. 15). Yellowish olive with intersegmental sutures black on thorax and brown on the abdomen, the latter with a dark green dorsal stripe and transverse lines of small, black dots. Wings with black lines from base to just before termen, interrupted in discoidal cell. Eyes, proboscis, antennae and legs outlined in black. Apex of abdomen black, cremaster a

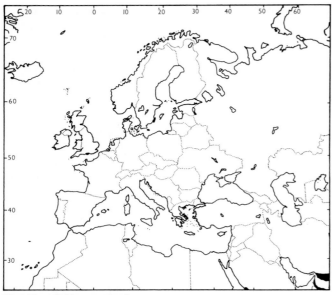

Map 35 – *Sphingonaepiopsis nana*

short black spike. Formed in a loosely-spun cocoon of brown silk among debris on the ground (Carcasson, 1968). The overwintering stage.

PARASITOIDS

None recorded.

BREEDING

No information available.

DISTRIBUTION

Baluchistan (southern Iran) and western Saudi Arabia (Jeddah) (Wiltshire, 1990). This is essentially a moth of the Afrotropical zoogeographical region which just enters the western Palaearctic region.

Extra-limital range. Southern Arabian Peninsula (including Oman) and eastern Africa to Natal.

PROSERPINUS Hübner, [1819]

Proserpinus Hübner, [1819], *Verz. bekannter Schmett.*: 132. Type species: *Sphinx oenotherae* [Denis & Schiffermüller], 1775.

> *Pterogon* Boisduval, 1828, *Eur. Lepid. Index meth.*: 32.
> *Pteropogon* Meigen, 1829, *Syst. Beschreibung eur. Schmett.* **1**(3): 129.
> *Lepisesia* Grote, 1865, *Proc. ent. Soc. Philad.* **5**: 38.
> *Pogocolon* Boisduval, [1875], *in* Boisduval & Guenée, *Hist. nat. Insectes* (Spec. gén. Lépid. Hétérocères) **1**: 314.
> *Dieneces* Butler, 1881, *Ann. Mag. nat. Hist.* (5) **8**: 308.

A Holarctic genus containing six species, with only one occurring in the Palaearctic region.

IMAGO: Forewing with dark transverse bands; distally dentate (more so in some species than others). Genal process large, triangular, nearly reaching tip of pilifer. Eye with long lashes dorsally. Antenna club-shaped, abruptly narrowing to an apical hook; terminal segment at least three times as long as broad. Abdominal spines weakly chitinized; abdomen with short tufts of scales laterally; anal tuft truncate. Tibiae spinose; foretibia with apical thorn and a lateral row of long spines, the basal one the shortest. Spurs of mid- and hindtibiae unequal, longer ones equalling or surpassing in length second tarsal segment. Mid- and hindtarsus without basal comb. Ventral lobes of paronychium very small; pulvillus present. Vein M_2 of hindwing central; M_3 and Cu_1 rather close together; D_2 transverse, slightly concave, D_3 oblique, lower angle of cell little produced.

Genitalia. In male, uncus and gnathos simple, similar to those in *Macroglossum*. Valva without friction scales; sacculus vestigial. Saccus relatively narrow and short. Aedeagus with a horizontal, apical, pointed process, directed dextro-laterad.

OVUM: Small, almost spherical, green, glossy.

LARVA: Bears dark, oblique, lateral stripes sloping up towards the head. Head small, not granulose, but hairy. Thorax constricted anteriorly. Young larvae striped longitudinally. In final instar, horn small or replaced by a button.

PUPA (Text fig. 44): Proboscis fused with body, not cariniform. Slender and glossy with two frontal tubercles. Mesonotum with an interrupted, transverse carina. Cremaster long and thin, terminating in two spines.

HOSTPLANT FAMILIES: Herbaceous plants, mainly of the Onagraceae and Lythraceae.

Text Figure 44 Pupa: *Proserpinus p. proserpina* (Pallas)

PROSERPINUS PROSERPINA (Pallas, 1772)

Map 36

E – Willowherb Hawkmoth, **F** – le Sphinx de l'épilobe,
G – Nachtkerzenschwärmer, **Sp** – Esfinge proserpina,
NL – Teunisbloempijlstaart, **R** – Zubokrylyĭ Brazhnik,
H – törpeszender, **C** – Chobotnatka pupalková

PROSERPINUS PROSERPINA PROSERPINA (Pallas, 1772)

Sphinx proserpina Pallas, 1772, *Spicilegia Zool. quibus novae . . . et obscurae anim. species . . . illustrantur* 1: 26.
Type locality: Vienna district [Wien, Austria].

Sphinx oenotherae [Denis & Schiffermüller], 1775, *Ankündung syst. Werkes Schmett. Wienergegend*: 43, 239.
Sphinx schiffermilleri Fuessly, 1779, *Magazin Liebh. Ent.* 2: 69.
Proserpinus aenotheroides Butler, 1875, *Proc. zool. Soc. Lond.* **1875**: 621.

ADULT DESCRIPTION AND VARIATION (Pl. 9, figs 13, 14)

Wingspan: 36–60mm.; most about 46mm. Difficult to confuse with any other sphingid from the region. Although highly variable in size, colour variation tends to be minimal, being confined to the primary ground coloration, which is normally shades of green: f. *schmidti* Schmidt has yellow-grey forewings and grey hindwings; f. *brunnea* Geest has a pale 'leather' coloration with a reddish median band; f. *grisea* Rebel has the green coloration entirely replaced by grey.

ADULT BIOLOGY

A local species, often disappearing from a locality for a number of years, but appearing in another. Mainly found in damp, woodland clearings and edges of woods, especially in valleys. Also sandy, waste ground in and around towns where *Epilobium* and *Oenothera* are common. In the Alps occurs up to 1500m, and in Spain to 2000m (Gómez Bustillo & Fernández-Rubio, 1976).

Few adults are found due to their predominantly green coloration combined with their preference for resting among low vegetation at the base of the plants. Flies for only a short period each day, usually at dawn and dusk (Rondou, 1903), although individuals have been seen in sunlight at midday (pers. obs.). Attracted to strong-smelling flowers such as *Jasminum* and *Echium*.

FLIGHT-TIME

Univoltine. Normally during late May and early June. Over its southern range, including the Lebanon, in mid-May. At altitude in the Pyrenees, during June and July (Rondou, 1903). However, Lhonoré (1988) has evidence that this species may be partially bivoltine in France, with larvae occurring in June and September.

EARLY STAGES

OVUM: Small (1.1 × 1.0mm), green and glossy. Laid singly on the under-surface of leaves, close to the flower-heads, often more than one to a plant.

LARVA (Pl. 2, fig. 11): Full-fed, 60–70mm. Dimorphic: green or brown.

Matt green when young with a yellow dorso-lateral line, a yellow head and a matt yellow spot supplanting the horn. Spiracles black, ringed with yellow. At this stage it feeds quite openly, either on the upperside of a leaf or, when on *Oenothera*, on the flowers. Growth is slow at first, but speeds up considerably on reaching the fifth instar, at which stage most individuals leave the hostplant to hide during the day, either at its base or somewhere nearby. The majority will now have acquired their final coloration: dorsal surface brown with black dots; sides and ventral surface buff, criss-crossed by a network of fine, dark lines. Spiracles brown, each set in a brown spot elongated into an oblique streak. Head brown; posterior button (in place of horn) yellow or orange, brown-centred and black-ringed. A few, however, remain primarily green throughout the larval stage but still bear the standard pattern. Similar to the larva of *Rethera komorovi* (Pl. 2, fig. 14) (Pittaway, 1979a; Pelzer, 1991).

Always found singly but often locally common. When on the move, the larva creeps along with a rather hesitant, jerky motion.

Occurs most commonly during June, July and August.
Hostplants. Principally *Epilobium* and *Oenothera* spp.; also *Lythrum*.

PUPA (Pl. E, fig. 13): 25–30mm. Reddish brown with darker head and abdomen. Cremaster broad basally, flattened dorso-ventrally. With two prominent head tubercles; proboscis only slightly keeled. Very similar in shape to that of *Daphnis* but much smaller. Formed in a loosely spun cocoon among debris on the ground. The overwintering stage.

PARASITOIDS

See Table 4, p. 52.

BREEDING

Pairing. Fairly easy in a medium-sized flight-cage placed outside on the ground and stocked with flowers and potted hostplant, although it may take place several nights after the emergence of the moths. It is advisable to force feed the adults for optimum results.

Oviposition. On the second or third night after pairing, and under the same conditions as above. Even with unlimited food, this species rarely lays well in captivity.

Larvae. Somewhat difficult. Do best on potted food outside; *Fuchsia* can be used as an alternative. If reared in glass containers, larvae should be kept singly, not crowded, and as dry as possible. Dead leaves or moss should be provided at the base of the hostplant as a hiding place during daylight hours. Prior to pupation, some larvae will wander for up to three days or more.

Pupation. Good results have been achieved by providing a gauze- or netting-covered flowerpot, three-quarters filled with damp sand, on which a layer of dead leaves and moss has been placed. This should initially be placed in sunlight, otherwise larvae seem to have some difficulty in going down. After pupation, which takes place at the sand/moss interface, the undisturbed container should be left outside in a dry, protected location until spring. Never allow the pupae to dry out or get too wet. If pupae have to be kept in tins, never spray directly. Place under damp peat in the spring. Unfortunately, under such conditions losses can be great.

DISTRIBUTION

Central and southern Europe (Newman, 1965) eastward to Kazan and the lower Volga (Eversmann, 1844); south-

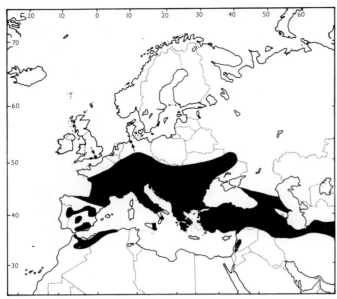

Map 36 – *Proserpinus proserpina*

eastwards across Bulgaria (Ganev, 1984), Turkey to Trans-caucasia and Kazakhstan, Lebanon (Zerny, 1933; Ellison & Wiltshire, 1939) to north-eastern Iran (Kalali, 1976). There is a single record from Israel (Eisenstein, 1984).

This species appears to be expanding its range in Europe. On 26 May 1985, the first British specimen, a male, was captured at Denton, near Newhaven, East Sussex (Pratt, 1985). He also records it as having colonized the Ardennes, Belgium. Individuals have been found in northern Germany, the Netherlands (Meerman, 1987) and around Paris and, on 4 May 1989, a single male was captured in Corsica (Guyot, 1990).

Extra-limital range. None.

PROSERPINUS PROSERPINA GIGAS Oberthür, 1922

Proserpinus proserpina gigas Oberthür, 1922, *Etud. Lépid. comp.* **19**: 128.
Type locality: Timahdit, Middle Atlas, Morocco.

ADULT DESCRIPTION AND VARIATION (Pl. 9, fig. 15; Pl. G, fig. 6)

Wingspan: 50–60mm. Very similar in size and coloration to many individuals of subsp. *proserpina* from Asia Minor. This suggests that subsp. *gigas* is insufficiently distinct from subsp. *proserpina* to warrant subspecific status. However the majority of North African individuals are larger, paler and greener than most European ones.

ADULT BIOLOGY

Occurs along the sides of streams or in open woodland clearings, usually near deciduous trees, where *Epilobium* is present.

FLIGHT-TIME

Bivoltine; March and June/July (Rungs, 1981).

EARLY STAGES

OVUM: As subsp. *proserpina*.

LARVA: As subsp. *proserpina*.

Hostplants. Epilobium spp., especially *E. hirsutum* (Rungs, 1981).

PUPA: As subsp. *proserpina*.

PARASITOIDS

Heavily parasitized by Tachinidae (Rungs, 1981), but the species involved are not recorded.

BREEDING

As for subsp. *proserpina*.

DISTRIBUTION

The Atlas Mountains of Morocco and Algeria (Rungs, 1981), where it is confined to high elevations. Also the Rif and north-western lowlands of Morocco (Rungs, 1981).

Extra-limital range. None.

PROSERPINUS PROSERPINA JAPETUS (Grum-Grshimailo, 1890)

Pterogon proserpina var. *japetus* Grum-Grshimailo, 1890, *in* Romanoff, *Mém. Lépid.* **4**: 513.
Type locality: Kabadian [Shaartuz, south-western Tajikistan].

ADULT DESCRIPTION AND VARIATION (Pl. 9, fig. 16)

Wingspan: 50–60mm. Paler than subsp. *proserpina*, with the marginal band of the hindwing narrower below.

ADULT BIOLOGY

Occurs at 1900m in Afghanistan (Daniel, 1971).

FLIGHT-TIME

Univoltine; June and July in Afghanistan (Ebert, 1969). The holotype, a female (Pl. 9, fig. 16), was found in mid-May.

EARLY STAGES

As subsp. *proserpina*.

Larvae have been found in July in Afghanistan (Daniel, 1971).

Hostplants. Species of *Epilobium*.

PARASITOIDS

None recorded.

BREEDING

Presumably as for subsp. *proserpina*.

DISTRIBUTION

The western Pamirs (Grum-Grshimailo, 1890), the Tian Shan (Derzhavets, 1984) and eastern Afghanistan (Ebert, 1969; Daniel, 1971).

Extra-limital range. Western Xizang Province/Tibet (Chu & Wang, 1980b) and western Xinjiang Province, China.

MACROGLOSSUM Scopoli, 1777

Macroglossum Scopoli, 1777, *Introd. Hist. nat.*: 414.

Type species: *Sphinx stellatarum* Linnaeus, 1758.

‡*Bombylia* Hübner, [1806], *Tentamen determinationis digestionis . . .* [1].

Psithyros Hübner, [1819], *Verz. bekannter Schmett.*: 132.

Bombylia Hübner, 1822, *Syst.-alphab. Verz.*: 10–13.

Macroglossa Boisduval, 1833, *Nouv. Annls Mus. Hist. nat. Paris* 2(2): 226.

Rhamphoschisma Wallengren, 1858, *Öfvers. K. Vetensk-Akad. Förh. Stockh.* **15**: 139.

An Old World genus containing 80 tropical and subtropical species.

IMAGO: Genal process very large, triangular. Distal margin of wings never dentate or lobate; forewing apex always pointed, but never falcate. Head broad, without distinct tuft. Antenna club-shaped in both sexes, tapering abruptly to a short hook. Terminal segment slender, long, almost filiform in some species, shorter in others. Eye with distinct pendant lashes. Labial palpus broad with porrect, pointed tip and triangular terminal surface. Thorax and abdomen broad, flattened; latter bears lateral and fan-tail tufts. Abdominal spines heavily chitinized as in *Hemaris* (q.v., p. 108), arranged in several rows. First row rounder and broader than long, with exception of proximal segment. Seventh abdominal sternite triangular with a membranous, non-spinose apex in female (unlike *Hemaris*). Mid- and hindtibiae different in length, the midtibial spur with a comb of stiff bristles or spines. Midtarsus with basal comb. Spurs of hindtibia very unequal. Pulvillus present, paronychium with two pairs of lobes. Veins R_2+R_3 and R_4 of forewing not joined distally as in *Hemaris*. Veins Rs and M_1 of hindwing originating separately from upper angle of cell; M_3 and Cu_1 likewise never stalked.

OVUM: Ovoid, small, pale yellow, or greenish yellow to green.

LARVA: Head small, rounded, semi-oval or subquadrate. Thorax slightly tapering forward. Longitudinally striped and granulose; horn straight and erect. No eye-spots present.

PUPA (Text fig. 45): Proboscis fused with body, but produced anteriorly, carinate. Dorsum of abdominal segments 1–5 flattened. Yellowish or brownish with black markings. Abdominal spiracles set in dark patches. Cremaster variable. In a loosely spun, silk cocoon on the ground.

HOSTPLANT FAMILIES: Herbaceous plants, principally of the Rubiaceae, Valerianaceae and Caryophyllaceae.

Text Figure 45 Pupa: *Macroglossum stellatarum* (Linnaeus)

MACROGLOSSUM STELLATARUM
(Linnaeus, 1758) Map 37

E – Hummingbird Hawkmoth, **F** – le Moro-Sphinx,
G – Taubenschwanz, **Sp** – Esfinge cola de paloma,
NL – Kolibrievlinder; Onrust; Meekrapvlinder,
Sw – Stor dagsvärmare, **R** – Brazhnik-yazykan,
H – kacsafarkú lepke, **C** – Dlouhozobka svízelová

Sphinx stellatarum Linnaeus, 1758, *Syst. Nat.* (Edn 10) **1**: 493.

Type locality: Unspecified [Europe].

Sphinx flavida Retzius, 1783, *Genera et Species Insect.*: 33.

Macroglossa nigra Cosmovici, 1892, *Naturaliste* **14**: 280.

ADULT DESCRIPTION AND VARIATION (Pl. 9, fig. 17)

Wingspan: 40–45mm. A very distinct species confusable only with some freshly emerged *Hemaris* (q.v.) species which have not yet lost their wing scales. Not prone to much variation though sometimes very pale, albinistic specimens occur, as do individuals with blackish brown abdomens and hindwings (f. *subnubila* Schultz).

ADULT BIOLOGY

Diurnal. In behaviour, this moth is exceptional amongst European Sphingidae: whilst preferring to fly in bright sunlight, it will also take wing at dawn, at dusk or at night; in rain, or on cool, dull days. Very hot weather tends to induce a state of torpidity in many, with activity then confined to the relative cool of the morning and late afternoon. Whatever the flight-time, this species is very strongly attracted to flowers yielding plentiful supplies of nectar, such as *Jasminum*, *Buddleja*, *Nicotiana*, *Primula*, *Viola*, *Syringa*, *Verbena*, *Echium*, *Phlox* and *Stachys*, hovering in front of and repeatedly probing each bloom before darting rapidly to the next. A great wanderer, being present right across Europe from the alpine tree-line to city centres, wherever nectar flowers may be found. Its powers of flight are amazing, and have been studied in detail by Heinig (1987). Apparently, this species also has a fine memory, as individuals return to the same flower-beds every day at about the same time.

When not feeding, pairs in courtship can be seen dashing up and down around steep cliffs, buildings, or over selected stretches of open ground. Pairs *in copula* can occasionally be found in such locations, although they seldom stay together for more than an hour. Whilst *in copula*, unlike other hawkmoths, this species is still capable of flight in a manner similar to butterflies. After a further period of feeding, gravid females search for patches of *Galium* growing in sunny locations. While hovering, each patch is carefully examined, sprig by sprig, before a single ovum is placed amongst the flower-buds. Up to 200 ova may be deposited by each female, therefore egg-laying can take a considerable time.

M. stellatarum is unique among sphingids of the region in overwintering as an adult, although north of the Alps very few survive. With the onset of cooler weather, individuals can be seen examining caves, rocky crags, holes in trees or sheds, before selecting a suitable place for hibernation. This state of torpidity is not absolute, however, for warm days in December and January may bring some out to feed.

FLIGHT-TIME

Migrant and multivoltine. Over its southern range, most of

the year in three or four broods, retiring into a state of semi-hibernation during winter. Farther north, migrants appear around about April or May, giving rise to two main generations, one in June, the other in August/September, with a number of smaller broods in between.

EARLY STAGES

OVUM: Almost spherical, approximately 1mm in diameter, glossy, pale green, resembling unopened flower-buds of *Galium*. This stage lasts from six to eight days.

LARVA (Pl. 2, fig. 8): Full-fed, 45mm. Polymorphic.

Newly-hatched larvae are about 2–3mm in length, cylindrical and clear yellow. They eat copiously from the outset and so grow rapidly. In the second instar they assume their green coloration, covered with tiny yellow dots. A dark grey, cream-bordered, dorso-lateral line runs from head to horn, matched by an identical ventro-lateral band. The dorsal heart line is deep green; the tail horn purplish red with an orange tip. This pattern remains to the final instar, with only the predominant colour of the horn changing from reddish to blue with an orange tip. In some, however, the green body colour is entirely replaced by reddish brown, which effectively masks the final, pre-pupation, rich plum coloration.

Feeding takes place fully-exposed at the top of the plant, with no time preference being shown. When resting, or moulting, most larvae retire to within the bed of tangled stems. Often, several fully-grown specimens may be found in the company of a number of *Deilephila porcellus* larvae; both share the same habitat preferences. With ample food and sunshine, this stage may be as short as 20 days.

Can be found throughout the summer months, with two peaks of abundance in the north: one during May and June, the other during August. Occurs almost anywhere where *Galium* is present in open situations, i.e. commons, field edges, roadside verges, sand-dunes, railway embankments, and even high alpine meadows over its southern range.

Hostplants. Principally *Galium* and, from central Europe south, *Rubia*; also various other Rubiaceae (*Jaubertia* in Oman), and *Centranthus*, *Stellaria* and *Epilobium*.

PUPA (Pl. E, fig. 14) 30–35mm. Pale brownish cream, splashed with darker brown, especially on wings and spiracles. Proboscis prominent, keeled, laterally compressed. Cremaster terminating in two sharp spines. Enclosed in a silken cocoon spun low down among the hostplant, or among debris on the ground. Does not overwinter.

PARASITOIDS

See Table 4, p. 52.

BREEDING

Pairing. Very difficult, as an extensive courtship flight is usually necessary; males have been known to mate with females at communal roosts (Friedrich, 1986).

Oviposition. Quite easy to obtain as long as the flight-cage is placed outside in sunlight with the hostplant touching the top of the cage. Ample food should be provided for the adults, which may have to be force-fed. It should be remembered, however, that a proportion of the autumn generation of moths hibernate without laying eggs; this has been studied in detail by Heinig (1981a, 1984).

Larvae. Not difficult if uncrowded and placed on growing

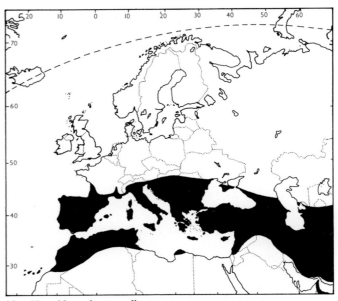

Map 37 – *Macroglossum stellatarum*

hostplants. Best results are achieved on potted hostplant placed in sunlight; fully-grown larvae then pupate at the base of the plant. Also does well in individual glass jars, provided they are sterilized at every food change.

Pupation. Easy to achieve in a cardboard box, the floor of which should be littered with leaves, peat and moss. This stage rarely lasts longer than three weeks.

DISTRIBUTION

Southern Europe, North Africa, across Central Asia to Japan. A noted summer migrant to the north. In the southernmost part of its resident range, confined to mountains, as in Iran and Oman.

Extra-limital range. In winter, a migrant as far south as southern India and the Gambia in Africa.

SUBTRIBE **Choerocampina** Grote & Robinson, 1865

Type genus: *Choerocampa* Duponchel, 1835 (=*Deilephila* [Laspeyres], 1809).

IMAGO: Pattern of forewing generally with apically convergent crossbands; outer margin entire, except in some species of *Deilephila* and two Neotropical genera, *Xylophanes* Hübner [1819] and *Phanoxyla* Rothschild & Jordan, 1903. Pilifer consisting of an apical part bearing short (or vestigial) bristles and a proximal part bearing long ones, modified to act as an auditory mechanism in conjunction with the maxillary palps (Roeder, 1975). Genal process short, not much projecting. Inner surface of second segment of labial palpus more or less naked. Terminal segment of antenna elongate, but not filiform, with six or more very long bristles. Antenna never dentate or pectinate. Abdomen conical in all forms, generally long and ending in a simple pointed tuft with a rudimentary tuft on each side; spines multiseriate; seventh sternite without spines, obtusely triangular, membranous at end. Scent-organ of fore coxa more or less distinct. Tibiae never spinose. Midtibial spurs unequal in length (except in *Cechenena* Rothschild & Jordan, 1903, an Oriental genus), the outer one the shorter. Hindtibia always with two pairs of spurs. Paronychium always with two lobes, but pulvillus without a pad in some species (e.g. *Hyles euphorbiae*).

Genitalia. Armature simple. In male, friction scales always present on the valva.

LARVA: Head small, round or semi-elliptical. Body tapering anteriorly from abdominal segment 1. Horn variable. Body rarely granulose in final instar. Eye-spots nearly always present in or as a dorso-lateral line.

PUPA: Proboscis fused with body (except in the Oriental genus *Pergesa* Walker, 1856); usually projecting forward. Surface usually dull and smooth with no carina on mesonotum; ante-spiracular ridges poorly developed. Cremaster and coloration variable.

HYLES Hübner, [1819]

Hyles Hübner, [1819], *Verz. bekannter Schmett.*: 137.
Type species: *Sphinx euphorbiae* Linnaeus, 1758.
‡*Celerio* Oken, 1815, *Okens Lehrbuch Naturgesch.* 3(1): 761.
   Celerio Agassiz, [1846], *Nomencl. zool.* (Nom. syst. Generum Anim.) Lepid.: 14.
   Turneria Tutt, 1903, *Entomologist's Rec. J. Var.* 15: 76.
   Hawaiina Tutt, 1903, *Entomologist's Rec. J. Var.* 15: 76.

A cosmopolitan genus containing more than fourteen species.

(*Taxonomic notes.* (i) Within the *Hyles* genus there is a complex of species, subspecies and forms, all closely related to *H. euphorbiae* and all of which are highly polymorphic with an amazing variety of colour forms, some geographic in nature, others environmental. The *Hyles euphorbiae* complex is rather difficult to classify for it would seem to be in the process of diverging into a number of species. Many of the following species and subspecies are regarded by some authors simply as subspecies and forms respectively of *H. euphorbiae*. Electrophoresis has not yet been undertaken on this complex; it may solve some of the problems of relationship as it has for the *Papilio* (*Pterourus*) *glaucus* (Linnaeus) complex in North America (Hagen *et al.*, 1991; Hagen & Scriber, 1991). Examination of the genitalia is of little use as they show practically no difference between taxa.

Larval coloration and pattern do, however, provide a very good guide to the relationships between the many 'species' and 'subspecies'; this same conclusion was also reached by Shchetkin (1956). By reference to these, in combination with other features, it is possible to group together certain species, subspecies and forms.

    Hyles euphorbiae
        H. e. euphorbiae
        H. e. conspicua
        H. e. robertsi
    Hyles tithymali
        H. t. tithymali
        H. t. mauretanica
        H. t. deserticola
        H. t. gecki
        H. t. himyarensis
    Hyles dahlii
    Hyles nervosa
    Hyles salangensis

An intriguing explanation for the variability of *H. euphorbiae* s. str. has been put forward by Harbich (1976a). Primary hybrids between *H. euphorbiae* s. str. and other *Hyles* species were able to produce viable offspring when back-crossed with *H. euphorbiae* s. str., but not, however, with the other parent. Thus the variability of *H. euphorbiae* s. str. could be the result of a genetic influx from other members of the genus, such as *H. hippophaes*, *H. vespertilio* and *H. gallii*, which themselves would be protected from the influence of *H. euphorbiae*. Of course, the degree of influence the other species have on *H. euphorbiae* would depend on how often interspecific crosses occur, which, in turn, depends on how closely related the species are. Additionally, gene flow between *H. tithymali*, *H. dahlii* and *H. euphorbiae* may be decreased by genetic linkage between ecologically important traits in sex chromosomes, such as an adaptation, or lack of adaptation, to desert conditions. These species appear to be primarily maintained by ecological selection and not genetic barriers.

(ii) *Hyles costata* (Nordmann, 1851), found from the Altai through Mongolia to Amurland and northern China, is a valid species with a very distinct larva. (*Celerio euphorbiae sinensis* Closs, 1917, is a subspecies of *H. costata* (Nordmann). **Stat. nov.**))

IMAGO: Forewing entire with apex pointed. Forewing pattern very distinctive, consisting of a pale cream or white, oblique median stripe in most species. Antenna distally incrassate in both sexes and, in female, strongly clavate. Pilifer with long setae medially, short setae laterally. Eye lashed. Labial palpus smoothly scaled with no erect hairs, concealing base of proboscis; scaling at apex of first segment not arranged in a regular border on inner surface; second segment without apical tuft on inner side. Abdomen conically pointed, with strong dorsal spines usually arranged in three rows. External spines of foretarsus more or less prolonged, always longer than the respective spines on the inner side of tarsus; comb of mid- and hindtarsus vestigial, the spines not much prolonged. First segment of hindtarsus shorter than tibia. Inner distal spur of hindtibia more than twice as long as outer and about half tibial length. Pulvillus absent or vestigial.

Genitalia. Very simple and similar in all species, thus of little use in distinguishing species. In male, valva broadly sole-shaped; friction scales numerous and small in most species. Sacculus ending in a thin, more or less curved, simple, tapering process. Aedeagus with dorsal apical edge incrassate, dentate, produced at left side into a short process; length of brim-like incrassation, as well as dentation, slightly different in various species.

Behaviour. An interesting characteristic of all *Hyles* species is their reaction to disturbance. Dropping to the ground, the abdomen is hunched, the wings are swung upwards and the antennae laid forward until they touch. The insect then hops about in short jumps.

OVUM: Small, almost spherical, pale green or blue-green.

LARVA: Typically sphingiform; thorax slightly tapering anteriorly. Usually with a conspicuous dorso-lateral line of eye-spots. Many are polymorphic, the forms themselves varying in background colour, spots and the colour of these spots. Horn present or absent.

PUPA (Text fig. 46): Proboscis fused with body, not projecting anteriorly, not obviously carinate. Light tan, smooth, covered with short, fine dark lines. Cremaster large, sometimes bent ventrally, with an apical pair of spines. In a fairly substantial cocoon on the ground.

Text Figure 46 Pupa: *Hyles vespertilio* (Esper)

HOSTPLANT FAMILIES: Principally herbaceous plants of the Euphorbiaceae, Onagraceae, Rubiaceae, Zygophyllaceae, Elaeagnaceae and Liliaceae.

HYLES EUPHORBIAE (Linnaeus, 1758)   Map 38

E – Spurge Hawkmoth, **F** – le Sphinx de l'euphorbe,
G – Wolfsmilchschwärmer, **Sp** – Esfinge de la lechetrezna,
NL – Wolfsmelkpijlstaart,
Sw – Vitsprötad skymningssvärmare,
R – Molochaïnyĭ Brazhnik, **H** – kutyatejszender
C – Lyšaj pryšcový

HYLES EUPHORBIAE EUPHORBIAE
(Linnaeus, 1758)

Sphinx euphorbiae Linnaeus, 1758, *Syst. Nat.* (Edn 10) **1**: 492.
Type locality: Unspecified [Europe].

> *Sphinx esulae* Hufnagel, 1766, *Berlin. Mag.* **2**: 180.
> *Deilephila esulae* Boisduval, 1834, *Icones hist. Lépid.* **2**: 26 [homonym].
> *Deilephila paralias* Nickerl, 1837, *Böhm. Tag.*: 22.
> *Deilephila euphorbiae grentzenbergi* Staudinger, 1885, *Ent. Nachr. Berlin* **11**: 10.
> *Celerio euphorbiae vandalusica* Rebel, 1910, *Dt. ent. Z. Iris* **23**: 212.
> *Celerio euphorbiae etrusca* Verity, 1911, *Boll. Soc. ent. ital.* **42**: 278. **Syn. nov.**
> *Celerio euphorbiae strasillai* Stauder, 1921, *Dt. ent. Z. Iris* **35**: 31. **Syn. nov.**
> *Celerio euphorbiae rothschildi* Stauder, 1928, *Lepid. Rdsch.* **2**: 113. **Syn. nov.**
> *Celerio euphorbiae subiacensis* Dannehl, 1929, *Mitt. münch. ent. Ges.* **19**: 99. **Syn. nov.**
> *Celerio euphorbiae dolomiticola* Stauder, 1930, *Ent. Z., Frankf. a. M.* **43**: 268. **Syn. nov.**
> *Celerio euphorbiae filapjewi* O. Bang-Haas, 1936, *Ent. Z., Frankf. a. M.* **50**: 255. **Syn. nov.**
> *Hyles euphorbiae lucida* Derzhavets, 1980, *Ent. Obozr.* **59**: 349. **Syn. nov.**

(*Taxonomic note*. 'Subspecies' *strasillai*, *rothschildi*, *subiacensis*, *dolomiticola* and *filapjewi* are all forms which were incorrectly given subspecific status; they can be found intermittently throughout the range of *H. e. euphorbiae*. 'Subsp.' *lucida* (Pl. 10, fig. 1) does not warrant subspecific status as it falls within the normal variation of *H. e. euphorbiae*.)

ADULT DESCRIPTION AND VARIATION (Pl. 10, figs 1, 2; Pl. H, fig. 1)

Wingspan: 70–85mm. Very similar to the other subspecies and to *H. tithymali*, *H. centralasiae* and *H. nicaea*. Normal specimens of each are illustrated (Pl. 10, figs 7, 8; Pl. 11, figs 4, 5; Pl. 11, fig. 8) but, due to the great wing-pattern variation exhibited by *H. e. euphorbiae*, some confusion may arise. Such variation includes wings with no trace of red; wings deeply flushed with red (f. *grentzenbergi* Staudinger); hindwings rust-brown (f. *brunnescens* Schultz), or yellow (f. *lafitolei* Thierry-Mieg) in place of the normal red. All other coloration can vary widely, while in the most extreme form the entire forewing is brown with a series of small, cream, dagger-like markings replacing the oblique median stripe (f. *restricta* Jordan).

ADULT BIOLOGY

Can be very common in warm, dry years but very scarce in cold, wet ones. Found where perennial herbaceous *Euphorbia* species are present, with open, sunny locations being preferred, such as along the edges of fields and woodland, on coastal sand-dunes and on mountain-sides up to 1900m in the Alps.

By day, most individuals rest on stones, low walls, amongst low vegetation, or even on the ground, taking flight just after dusk for a short period and again later, in three or four bouts of activity. It is during one of these that pairing takes place, usually four to five days after emergence, the pair remaining *in copula* for up to three hours. Following separation, females begin oviposition almost immediately, laying a few eggs on the first night and more on subsequent nights. Although this species is attracted to flowers and feeds avidly from beds of, for example, *Petunia*, *Nicotiana* or *Silene*, few come to light.

FLIGHT-TIME

Univoltine in its northern range, mainly from mid-June to the beginning of July, although individuals can be found as early as mid-May. Farther south, two generations are normal, one in May/June, the other in August/September. However, at higher altitudes in southern Europe there is only one generation during June and July (Ganev, 1984).

EARLY STAGES

OVUM: Almost spherical (1.1 × 1.0mm), light glossy blue-green. Deposited either singly or in small clusters at the tip of a shoot of the hostplant, hatching about ten days later. Interestingly, the eggs of this species are extremely hard; if dropped on to a hard surface they will rebound to a considerable height.

LARVA (Pl. 2, fig. 10): Full-fed, 70–80mm. Polymorphic.

The newly-hatched larva measures approximately 4mm and is off-white with a black head and horn. With feeding, this primary colour is soon replaced by dark olive black which then lightens with further feeding. At this stage it rests low down on the hostplant during the day, at dusk moving up the stem to feed during darkness in the company of others. After the first moult, the characteristic bright pattern appears, superimposed on a now light greenish yellow-brown background. With each successive moult this becomes more startling until, eventually, the fully-grown larva can feed quite openly, relying on its aposematic coloration for protection. Although the basic pattern remains the same, size and colour are just as variable in the larva as in the adult insect. The dorso-lateral line of eye-spots may be red, yellow, green, white or orange, set in a ground colour which varies from green through light yellowish brown to black.

At all stages, its feeding powers are prodigious. Vast quantities of leaves and also soft stems are consumed between spells of basking, which it does more frequently as it grows. If disturbed, a thick stream of dark green fluid is ejected from the mouth towards the attacker, accompanied by violent body twitching from side to side. Prior to pupation, the larva is very active and wanders a great distance at a very fast pace, finally pupating a considerable way from the hostplant.

Occurs in most areas throughout July and August; in June and August/September where two generations are present.

Hostplants. Principally herbaceous *Euphorbia* spp., especially *E. paralias* and *E. cyparissias*; also, but very occasionally, *Rumex*, *Polygonum*, *Vitis*, *Mercurialis* and *Epilobium*.

PUPA: 45–50mm. Light yellowish brown with fine, dark striations. Formed within a loosely spun, net-like cocoon, incorporating leaves, soil and other debris, on the surface of the soil. The time spent in this stage can vary considerably – from two weeks to two years. Overwinters as pupa.

PARASITOIDS

See Table 4, p. 54.

BREEDING

Pairing. Usually successful after keeping the moths for two or three nights in a moderate-sized, cylindrical, net cage (40 × 40cm), with at least two males to every female. Pairing rarely lasts more than a few hours and occurs just after sunset.

Oviposition. Easy to obtain in an empty net cage, though better results are likely if plant material is included so that gravid females can cling on and curl their abdomen underneath. Eggs are deposited in clusters, most over three or four nights. The moths will require hand-feeding during this period.

Larvae. Easy to rear if sleeved on growing hostplant and can be crowded under warm, dry, well-ventilated conditions. Sunlight is very beneficial in reducing the risk of disease and increasing growth rate. If reared in glass containers, each larva should be kept separate after the third instar and its hostplant changed every day. Never feed with wilted food or plants standing in water, as this is usually fatal. Furthermore, in years when rainfall is high, rearing in closed containers is almost impossible as the high water-content of the hostplant causes gastric problems. The larvae are very sensitive to photoperiod and temperature, and must be reared under long-day conditions of 17 hours light and 7 hours darkness and 18°C to prevent the pupae going into diapause (Harbich, 1976b).

Pupation. Easy to achieve in a cardboard box containing moss or dead leaves over dry sand. Pupae store well in clean, dry tins; when placed on damp peat for emergence the following year they should not be sprayed if kept at normal room temperatures.

DISTRIBUTION

The coast of North Africa from the Straits of Gibraltar to Tunis. Also across southern and central Europe to the Urals (Kumakov, 1977) and from Turkey (Daniel, 1939; de Freina, 1979) to Transcaucasia, northern Afghanistan (Ebert, 1969) and the western Tian Shan (Alphéraky, 1882; Derzhavets, 1980). Not found in Corsica, Sardinia or the Balearic Islands (Pittaway, 1983b). A migrant to the north.

Extra-limital range. The eastern Tian Shan (Bang-Haas, 1936). The U.S.A. and Canada as an introduced species for the control of pest species of *Euphorbia* (Hodges, 1971; Batra, 1983).

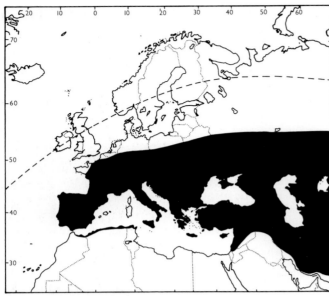

Map 38 – *Hyles euphorbiae*

HYLES EUPHORBIAE CONSPICUA
(Rothschild & Jordan, 1903)

Celerio euphorbiae conspicua Rothschild & Jordan, 1903, *Novit. zool.* **9** (Suppl.): 720.

Type locality: Beirut, Syria [Lebanon].

(*Taxonomic note.* There is a disjunct population of *H. euphorbiae* in the Asir Mountains of Saudi Arabia which may warrant subspecific status. Until more specimens are available, it is tentatively placed in *H. e. conspicua*. Larvae and adults from most of the Asir are almost indistinguishable from those from Lebanese montane populations, although most adults are slightly darker in colour. However, some adults from the area just south of Abha are, in appearance, quite clearly intermediate between *H. e. conspicua* and *H. tithymali himyarensis* (Meerman, 1988b) from Yemen. The larvae are also intermediate in colour (see Taxonomic note (ii), p. 141).)

ADULT DESCRIPTION AND VARIATION (Pl. 10, figs 3, 4)

Wingspan: 60–85mm. As illustrated, but those from drier and warmer areas are smaller and lighter in colour, resembling *H. tithymali deserticola*. Those from cool mountainous regions tend to be darker. Individuals may also be heavily flushed with pink.

ADULT BIOLOGY

An inhabitant of mountain chains running northwards from Israel through Lebanon into Turkey, where it frequents warm mountain-sides. Behaviour similar to that of subsp. *euphorbiae*.

FLIGHT-TIME

Multivoltine; April to October.

EARLY STAGES

OVUM: As subsp. *euphorbiae*.

LARVA (Pl. 2, fig. 9; Pl. 4, fig. 10): Full-fed, 75–90mm. Resembles that of subsp. *euphorbiae* from Greece and Spain, much of the black pigment being replaced by orange

or greenish yellow. Some may even more closely resemble those of *H. nicaea* (Pl. 3, fig. 4). At high altitudes, however, most larvae are blacker and more like central European forms of *euphorbiae*; such is the case with the Asir Mountain race from Saudi Arabia (Pl. 4, fig. 10) (see Taxonomic note above).

Occurs from April until November.

Hostplants. Herbaceous *Euphorbia* spp.

PUPA: Indistinguishable from subsp. *euphorbiae*. Overwinters as a pupa.

PARASITOIDS

See Table 4, p. 54.

BREEDING

As for subsp. *euphorbiae*.

DISTRIBUTION

Israel (Eisenstein, 1984), Lebanon (Zerny, 1933; Ellison & Wiltshire, 1939) Syria and southern Turkey (Daniel, 1932).

Extra-limital range. An isolated population exists in the south-western mountains of Saudi Arabia (Wiltshire, 1986), where, to a very limited extent, it has hybridized with *Hyles tithymali himyarensis*.

HYLES EUPHORBIAE ROBERTSI (Butler, 1880)
Stat. rev.

Deilephila robertsi Butler, 1880, *Proc. zool. Soc. Lond.* **1880**: 412.

Type locality: Kandahar, Afghanistan.

Deilephila peplidis Christoph, 1894, *Ent. Nachr. Berlin* **20**: 333. **Syn. rev.**

Hyles robertsi orientalis Ebert, 1969, *Reichenbachia* **12**: 46. **Syn. nov.**

(*Taxonomic note.* The eastern populations from the Pamirs, Afghanistan and ?Kashmir (subsp. *orientalis* Ebert, 1969) are often regarded as a separate subspecies of *H. robertsi*, which some authors consider to be a distinct species. Given the variability of *Hyles* spp., the arguments used to support this view are not tenable. Larvae from Shiraz, Iran, which produced typical adults of *H. e. robertsi*, were indistinguishable from individuals from southern Turkey which produced typical *H. e. euphorbiae*. Indeed, in north-western Iran, northern Iraq, south-eastern Turkey and across southern Turkmenistan, Uzbekistan and Tajikistan, the two subspecies intergrade, with many intermediate adult forms present.)

ADULT DESCRIPTION AND VARIATION (Pl. 10, figs 5, 6)

Wingspan: 65–85mm. Very similar to *H. e. euphorbiae* but usually distinguished by the silver-white, oblique median stripe of the forewing and by the fringes of the tergites being pure white. Females are often very much larger than the males.

ADULT BIOLOGY

A montane subspecies occurring on dry, rocky slopes with *Euphorbia*, e.g. open steppe. Behaviour similar to subsp. *euphorbiae*.

FLIGHT-TIME

End of March to June or later in a number of broods.

EARLY STAGES

OVUM: Very similar to that of subsp. *euphorbiae*.

LARVA (Pl. 4, fig. 1): Full-fed, 80–100mm. The ground colour varies from yellowish green to black; however, the general pattern remains the same for both forms and is very similar to that of some southern European forms of subsp. *euphorbiae*.

Hostplants. Herbaceous *Euphorbia* spp.

PUPA: Indistinguishable from subsp. *euphorbiae*. Overwinters as a pupa.

PARASITOIDS

None recorded.

BREEDING

As for subsp. *euphorbiae*.

DISTRIBUTION

Northern Iraq (Wiltshire, 1957), Iran (Daniel, 1971; Kalali, 1976), the Kopet Dağ Mountains of Turkmenistan (Derzhavets, 1984), eastward to central and eastern Afghanistan (Bell & Scott, 1937; Ebert, 1969; Daniel, 1971), ?Kashmir and the Pamirs (Ebert, 1969). As adults are easily confused with *H. e. euphorbiae*, *H. costata* and *H. nervosa* (see Taxonomic note (i), p. 135), the exact distribution of this subspecies is not fully known; records from Kashmir are unconfirmed.

Extra-limital range. Western Pakistan (Daniel, 1971)

HYLES TITHYMALI (Boisduval, 1832)   Map 39
E – Barbary Spurge Hawkmoth,
F – le Sphinx du tithymale

HYLES TITHYMALI TITHYMALI
(Boisduval, 1832)

Deilephila tithymali Boisduval, 1832, *Icones hist. Lépid.* 2: 30.
Type locality: Canary Islands.

> *Deilephila calverleyi* Grote & Robinson, 1865, *Proc. ent. Soc. Philad.* 5: 56.

ADULT DESCRIPTION AND VARIATION (Pl. 10, figs 7, 8)

Wingspan: 60–85mm. Many individuals resemble subsp. *mauretanica* (Pl. 10, fig. 9), from which it has undoubtedly descended. Quite distinct from *H. euphorbiae*: median stripe of forewing narrow and cream in female, broader and silver in male, many having superimposed silver venation. Not as variable as *H. euphorbiae*.

ADULT BIOLOGY

Restricted to the Canary Islands, where it is widespread, occurring from sea-level to 1000m in short-lived but well-defined colonies (Schurian & Grandisch, 1991). Commonest in the drier and warmer parts, such as steep-sided valleys and cultivated areas where its main hostplant is most abundant. Behaviour identical to that of *H. euphorbiae*; attracted to the flowers of *Bougainvillea* (van der Heyden, 1991).

FLIGHT-TIME

Continuous-brooded throughout the year, although scarce

from May to August (van der Heyden, 1990, 1991).

EARLY STAGES

OVUM: Very similar to that of *H. euphorbiae*, but with a noticeable bluish hue. Laid in clusters of up to 25 on the growing tips of smaller hostplants, hatching four to eight days later.

LARVA (not illustrated): Full-fed, 70–80mm. Almost identical to those of subsp. *mauretanica* and subsp. *deserticola* (Pl. 2, fig. 4)

Young larvae are black at first, turning to olive black after feeding, with the final colour pattern appearing in the third or fourth instar. Fully grown, it is quite unlike any colour form of *H. euphorbiae*, more closely resembling certain varieties of *H. livornica* (q.v., p. 154). In some individuals, the white eye-spots may be flushed with red.

Eggs and small larvae are generally found on small plants of *Euphorbia regis-jubae* (van der Heyden, 1991). Larger larvae move to and then feed quite openly on the taller, more mature bushes, alternating frenzied bouts of feeding with long stretches of basking (Pittaway, 1976). They may occur in such numbers as to defoliate large areas of hostplant. In midsummer some individuals become fully grown in 20 days.

Being continuous-brooded, various stages of development can be found simultaneously on the same plant, although larvae are scarce from July to October (van der Heyden, 1991).

Hostplants. *Euphorbia regis-jubae*; occasionally *E. paralias*.

PUPA: As *H. euphorbiae*, but with many pupae diapausing for two or more years, a characteristic of species adapted to desert conditions.

PARASITOIDS

None recorded.

BREEDING

As for *H. euphorbiae*.

DISTRIBUTION

The Canary Islands.
Extra-limital range. None.

HYLES TITHYMALI MAURETANICA
(Staudinger, 1871)

Deilephila mauretanica Staudinger, 1871, *in* Staudinger & Wocke, *Cat. Lepid. eur. Faunengeb.*: 36.
Type locality: Mauretania [Morocco and Algeria].

ADULT DESCRIPTION AND VARIATION (Pl. 10, fig. 9)

Wingspan: 60–85mm. Highly variable, often resembling the nominate subspecies but tending to become smaller and paler towards desert areas.

ADULT BIOLOGY

Frequents dry, rocky slopes and, in more open areas, dry, sandy river-beds with a good supply of *Euphorbia*.

FLIGHT-TIME

Trivoltine; April/May, June/July and August/September.

EARLY STAGES

OVUM: Identical to that of *H. euphorbiae*.

LARVA (not illustrated): Full-fed, 70–80mm. Almost identical to that of subsp. *tithymali* and subsp. *deserticola* (Pl. 2, fig. 4). Generally, the colours are less bright than either of these subspecies, the pure white eye-spots are slightly larger and are set in a slightly larger black surround. Examination of a large number of full-fed larvae, both in the wild and preserved in the Natural History Museum, London, clearly shows the relationship of this subspecies to subsp. *tithymali* and subsp. *deserticola*; it is quite distinct from *H. euphorbiae*.

In a year with good rainfall, larvae can be very abundant from April to October.

Hostplants. Herbaceous *Euphorbia* spp., especially *E. paralias*, *E. terracina* and *E. nicaeensis* (Rungs, 1981).

PUPA: Morphologically indistinguishable from that of *H. euphorbiae*. Overwinters as a pupa.

PARASITOIDS

None recorded.

BREEDING

As for *H. euphorbiae*.

DISTRIBUTION

Restricted to the mountains of North Africa from Morocco (Rungs, 1981; Speidel & Hassler, 1989) to Tunisia (Pittaway, 1983b); in desert areas, merges with subsp. *deserticola* to form a cline.

Extra-limital range. None.

HYLES TITHYMALI DESERTICOLA
(Staudinger, 1901)

Deilephila mauretanica deserticola Staudinger, 1901, *in* Staudinger & Rebel, *Cat. Lepid. palaearct. Faunengeb.*: 102. **Syn. rev.**
Type locality: 'Maur. desert. arenos. (Biskra etc.)' [North African desert areas].

> *Deilephila mauretanica* ab. *deserticola* Bartel, 1899, *in* Ruhl, *Groß-Schmetterlinge* 2: 79.
> *Celerio deserticola saharae* Günther, 1939, *Ent. Z., Frankf. a. M.* **53**: 261. **Syn. rev.**

(*Taxonomic notes*. (i) The taxon *deserticola* was first established as an aberration of *Deilephila mauretanica* by Bartel, therefore this name is unavailable. It was first raised to subspecfic level by Staudinger and is therefore attributable to him.
(ii) The population in the Cape Verde Islands, off West Africa, superficially resembles subsp. *deserticola* (Bauer & Traub, 1980), but more specimens are needed for examination before its taxonomic status can be determined.
(iii) Speidel & Hassler (1989) consider that subsp. *deserticola* and subsp. *mauretanica* should still be classified as subspecies of *H. euphorbiae*, ignoring the close relationship they obviously show to each other and to subsp. *tithymali*. Therefore the taxon *deserticola* is reinstated as a subspecies of *H. tithymali*.)

ADULT DESCRIPTION AND VARIATION (Pl. 10, fig. 10)

Wingspan: 45–75mm. Variable, especially in colour intensity and size; the smallest examples come from the driest areas and are usually referred to as f. *saharae* Günther,

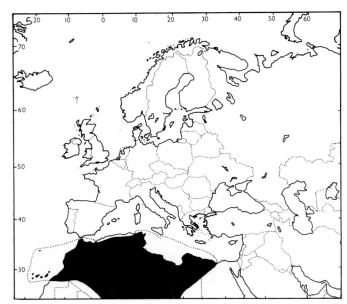

Map 39 – *Hyles tithymali*

1939, which are just small, pale replicas of larger individuals. In the southern Atlas Mountains, this subspecies merges with *H. t. mauretanica*.

ADULT BIOLOGY

Occurs in desert steppe, in dry, sandy river-beds and, especially in southern Algeria, the fringes of oases.

FLIGHT-TIME

Continuous-brooded; throughout the year but commonest in spring.

EARLY STAGES

OVUM: Cannot be distinguished from subsp. *tithymali*.

LARVA (Pl. 2, fig. 4): Cannot be distinguished from subsp. *tithymali*.

Occurs throughout the year but most commonly during April and May.

Hostplants. Herbaceous *Euphorbia* spp., especially *E. calyptrata*, *E. biglandulosa*, *E. terracina* (Rungs, 1981) and *E. guyoniana* (Speidel & Hassler, 1989).

PUPA: As *H. euphorbiae*. Most, but not all, overwinter.

PARASITOIDS

None recorded.

BREEDING

As for *H. euphorbiae*; however, many pupae will fail to overwinter.

DISTRIBUTION

North Africa south of the Atlas Mountains, from eastern Mauretania (Rungs, 1981), Morocco and southern Algeria (Speidel & Hassler, 1989) to eastern Egypt. During the last ice age this subspecies probably also occurred as far east as Yemen, where it has since given rise to a distinct and isolated subspecies – *H. t. himyarensis*.

Extra-limital range. Senegal. Possibly the Cape Verde Islands (Bauer & Traub, 1980).

HYLES TITHYMALI GECKI de Freina, 1991
Comb. nov.

Hyles euphorbiae gecki de Freina, 1991, *NachrBl. bayer. Ent.* **40**: 65–72.

Type locality: Madeira, 10 km east of Funchal.

(*Taxonomic note.* Studies of both larvae and adults demonstrate that *H. euphorbiae gecki* is a subspecies of *H. tithymali*. The three worn adults (1 ♂, 2 ♀♀) in the BM(NH) (collected by T. V. Wollaston) are referable to *H. tithymali*, a conclusion also reached in part by Baker (1891). Comparison of these specimens and a number of recently reared adults with a large number of *H. t. mauretanica* indicate that *H. t. gecki* is a valid subspecies, although some individuals of *H. t. mauretanica* do approximate to *H. t. gecki* in coloration and pattern. The larva of *H. t. gecki* closely resembles that of *H. t. mauretanica*, differing mainly in having a narrower, yellow dorso-lateral stripe and reduced yellow coloration on the prolegs in the final instar, and confirms the placement of this subspecies within *H. tithymali*. This masking of the yellow by black is also found in *H. t. himyarensis*.)

ADULT DESCRIPTION AND VARIATION (Pl. 10, fig. 11)

Wingspan: 70–85 mm. As illustrated. Similar in coloration and pattern to some individuals of *H. t. mauretanica* from Morocco. In some, the normally dark olive brown coloration of the forewing may be reddish brown. The median stripe may also be off-white rather than of the normal pale creamy yellow colour.

ADULT BIOLOGY

An inhabitant of steep cliffs with a good growth of *Euphorbia regis-jubae*.

FLIGHT-TIME

So far as is known, April to October in a number of broods. May be continuous-brooded.

EARLY STAGES

OVUM: Similar to that of subsp. *tithymali*, i.e. small, very hard and blue-green in colour. Laid in large clusters on the growing tips of the hostplant.

LARVA (not illustrated): Full fed: 75–100 mm.

According to de Freina (1991), the young larva is initially matt black, its colour changing with feeding to olive black. In the second instar very similar to that of *H. t. tithymali*: ground colour yellow; a dorso-lateral line of round, startling white eye-spots; dorsal and ventro-lateral lines greenish yellow; head dark orange; horn almost matt black. In the third and fourth instars the larva is very similar to that of *H. t. mauretanica*. Full-grown larvae are also similar to that subspecies but differ mainly in having a much narrower yellow dorso-lateral stripe and reduced yellow on the prolegs. As in *H. t. tithymali*, about 66 per cent of the larvae have a red flush to the normally white eye-spots.

In behaviour very similar to that of *H. t. tithymali* but, until the final instar, generally active only by day (de Freina, 1991).

Occurs from April to December.

Hostplant. Euphorbia regis-jubae.

PUPA: Indistinguishable from subsp. *tithymali*. Overwinters as a pupa.

PARASITOIDS

None recorded.

BREEDING

As for *H. euphorbiae*.

DISTRIBUTION

Confined to the island of Madeira (de Freina, 1991), where it was also found by M. Geck (pers. comm.), de Worms (1964) and Gardner & Classey (1960) who recorded it as '*Celerio euphorbiae*'. Cockerell (1923) noted that there were five species of Sphingidae on Madeira but did not name them. *Extra-limital range.* None.

HYLES TITHYMALI HIMYARENSIS
Meerman, 1988 Map 39d

Hyles tithymali himyarensis Meerman, 1988, *Ent. Ber., Amst.* **48**: 61–67.

Type locality: Dhufar, 10 km south-east of Yarim, Dhamar Province, northern Yemen.

(*Taxonomic notes.* (i) Harbich (1991) treats this taxon as a subspecies of *H. euphorbiae* on the basis of a number of adult and larval characters, but fails to take into account the findings of Wiltshire (1986) and Pittaway (1987).

(ii) During the last ice age it is probable that *H. euphorbiae* was resident in the western and central parts of the Arabian Peninsula and that *H. tithymali* had penetrated as far east as what is now Yemen; the present-day eastern limit of its main area of distribution is in western Egypt. Today, *H. t. himyarensis* forms a disjunct eastern population, with adults resembling dark forms of *H. t. deserticola* (Meerman, 1991). Some also resemble those of the disjunct, montane population of *H. euphorbiae* which occurs from Taif to Abha in south-western Saudi Arabia and which can tentatively be referred to *H. e. conspicua*, being almost indistinguishable from the high-altitude population of that subspecies in the Lebanon (q.v., p. 138). There is little overlap between these two populations of *H. tithymali* and *H. euphorbiae*, although what appear to be hybrids do occur on the Yemen border south of Abha. Fully-integrated hybrid populations between these two species can be found on Crete (Meerman & Smid, 1988) and Malta (see Pl. H, fig. 2) (pers. obs.) but *H. t. himyarensis* is certainly a valid subspecies, with little interbreeding between it and *H. e. conspicua*.

(iii) Larvae of *H. e. conspicua* from Baha in Saudi Arabia are very similar to those found in the mountains of Lebanon and are hardly distinguishable from central European examples (Pittaway, *in* Wiltshire, 1986). In larvae of *H. t. himyarensis* from Yemen, the white dorso-lateral eye-spots are smaller and a pronounced orange-yellow longitudinal line develops between them, a character of *H. tithymali*. The small, pale spots immediately below this line are pale yellow (as opposed to white), indicating the presence of a masked yellow band, also characteristic of *H. tithymali*. The extensive dark coloration of larvae of the Arabian populations of both *H. euphorbiae* and *H. tithymali* may be a protective adaptation to high ultra-violet radiation levels experienced at altitude. Alternatively, it may assist heat absorption. It is interesting to note that in hybrids between *H. tithymali* and *H. euphorbiae* obtained by Harbich (1988), the adults resemble *H. tithymali* whereas the larvae look like those of *H. euphorbiae*.)

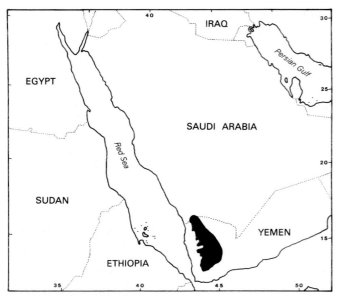

Map 39d – *Hyles tithymali himyarensis*

ADULT DESCRIPTION AND VARIATION (Pl. 10, fig. 12)

Wingspan: 60–80mm. As illustrated. Very similar in appearance to *H. t. deserticola* (Pl. 10, fig. 10), with which it could easily be confused. However, some individuals resemble those of the population of *H. e. conspicua* (Pl. 10, fig. 3) to the north.

ADULT BIOLOGY

Confined to the narrow, juniper-forest zone of the highlands of Yemen, at 2000–2500m. Males come readily to light and are avid visitors to flowers.

FLIGHT-TIME

April to September, in two or three broods.

EARLY STAGES

OVUM: As *H. t. tithymali*.

LARVA (Pl. D, fig. 2): Full-fed, 70–80mm. Similar in some respects to *H. dahlii* (Pl. 3, fig. 9) in that in the final instar it is sooty black and covered with fine white spots. Superimposed on this primary colour scheme is a dorso-lateral row of single, very white eye-spots set in an orange-yellow stripe. The small, pale spots immediately below this line are pale yellow (as opposed to the remainder, which are white), indicating the presence of a masked yellow band. The horn, head, prolegs, true legs and dorsal stripe are orange-yellow or red, and there is an orange and yellow chequered ventro-lateral line.

Occurs from April until October.

Hostplants. Herbaceous *Euphorbia* spp., especially *E. cyparissioides*, but also feeds on *E. peplus*.

PUPA: As *H. euphorbiae*, but with a much greater tendency to remain in this stage for two or more years.

PARASITOIDS

None recorded.

BREEDING

As for *H. euphorbiae*, but the pupal stage may last for several years, a characteristic of *H. tithymali*.

DISTRIBUTION

Restricted to the mountains of Yemen.
Extra-limital range. None.

HYLES DAHLII (Geyer, 1827) Map 40

E – Smoky Spurge Hawkmoth

Sphinx dahlii Geyer, [1827], *in* Hübner, *Samml. eur. Schmett.*, Sphingidae: pl. 36, figs 161–164.
Type locality: Cagliari, Sardenia [Sardinia].

> *Sphinx dahlii* Boisduval, 1828, *Eur. Lepid. Index meth.*: 33 [homonym].
> *Celerio euphorbiae balearica* Rebel, 1926, *Dt. ent. Z. Iris* **40**: 141. **Syn. nov.**

(*Taxonomic note*. Subsp. *balearica* (Rebel, 1926) is not tenable. Rebel described it from one specimen from Majorca, thinking it represented an isolated form; however, specimens from there cannot be distinguished from those from Corsica and Sardinia.)

ADULT DESCRIPTION AND VARIATION (Pl. 11, fig. 1)

Wingspan: 65–85mm. Superficially similar to a dark, heavily blotched *H. t. tithymali* (Pl. 10, fig. 7); not very variable.

ADULT BIOLOGY

A species of rocky mountainsides scattered with shrubs and pine trees.

FLIGHT-TIME

May/June and August/September, in two broods, with a partial third in warm years.

EARLY STAGES

OVUM: As *H. euphorbiae*.

LARVA (Pl. 3, fig. 9): Full-fed, 80–100mm.

On hatching, the 3–4mm-long larva is off-white. Fully grown it is dark grey with a profusion of small, whitish spots; head usually pink and horn orange. Instead of a solid dorso-lateral line, there are two oval, white eye-spots set in a velvet-black patch on each segment. Narrow orange dorsal line; laterally, a red and yellow chequered line runs between the white spiracles. Ventral surface yellowish.

Occurs from June to October.

Hostplants. Herbaceous *Euphorbia* spp., especially *E. paralias*, *E. characias* and *E. terracina*.

PUPA: Very similar in appearance to that of *H. euphorbiae*. Overwinters as a pupa.

PARASITOIDS

See Table 4, p. 54.

BREEDING

Exactly as for *H. euphorbiae*. Readily accepts *Epilobium* in captivity.

DISTRIBUTION

Corsica, Sardinia and the Balearic Islands. Considered to be restricted to these islands, although there are occasional reports of its occurrence in Sicily. In October 1975, several larvae of this distinctive species were also found on the Catalan coast, north-east Spain (Planas *et al.*, 1979).
Extra-limital range. None.

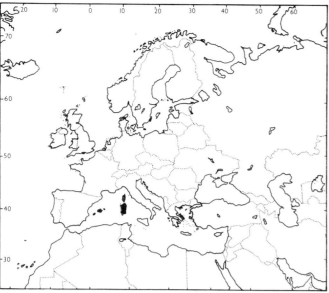

Map 40 – *Hyles dahlii*

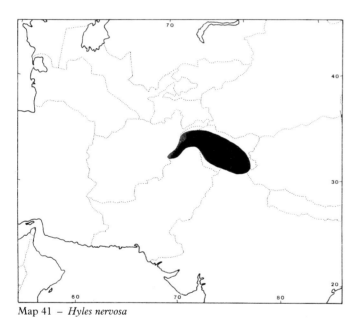

Map 41 – *Hyles nervosa*

HYLES NERVOSA (Rothschild & Jordan, 1903)

Map 41

E – Ladakh Hawkmoth

Celerio euphorbiae nervosa Rothschild & Jordan, 1903, *Novit. zool.* **9** (Suppl.): 721.
Type locality: Sabathu, [Kashmir,] N.W. India.

ADULT DESCRIPTION AND VARIATION (Pl. 11, fig. 2)

Wingspan: 68–87mm. In the male, the tegulae are fringed with white, the abdomen has only two black lateral spots and the fringes of the tergites are not white along the dorsal ridge. Dorsally, the forewing veins are outlined with white in the post-discal area.

ADULT BIOLOGY

A local species which can occur in very large numbers in a given area. In Kashmir, generally occurring at 2500–3000m (Bell & Scott, 1937) where stands of herbaceous *Euphorbia* are common.

FLIGHT-TIME

Apparently trivoltine; March to May, June/July and September.

EARLY STAGES

OVUM: Almost spherical, pale green and very like that of *H. euphorbiae*. Laid in small clusters of up to twenty on the young shoots of the hostplant.

LARVA (Pl. 4, fig. 6): Full-fed, 80mm.

In the first instar the 4mm-long larva is off-white with a black horn and head. The coloration of the final instar, which is primarily black with small white to pale yellow spots, begins to show in the second instar. A dorso-lateral line of single, white eye-spots, one per segment, runs from the white-speckled head to the jet black horn. The shield, anal claspers and prolegs are black and an unbroken dorsal line of the same colour stretches from head to horn. There is also a noticeable ventro-lateral line of medium-sized yellow spots. This colour scheme is constant, with little or no variation, unlike that of *H. euphorbiae*.

According to Bell & Scott (1937), the larvae are gregarious and, having stripped one plant of its leaves, move *en masse* to the next. They feed quite openly, although when nearly full grown they tend to rest along the stems of the hostplant close to the earth. If disturbed, they simultaneously regurgitate drops of green fluid. Such gregarious behaviour is probably a defence against tachinid parasitoids.

Occurs from April to October, most commonly from June to August.

Hostplants. Herbaceous *Euphorbia* spp.

PUPA: 45mm. Very similar to that of *H. euphorbiae*, but head, thorax and wings dull green. The overwintering stage.

PARASITOIDS

None recorded.

BREEDING

Nothing known.

DISTRIBUTION

Eastern Afghanistan (Ebert, 1969).

Extra-limital range. Northern-western India, northern Pakistan and the extreme west of Xizang Province/Tibet, China.

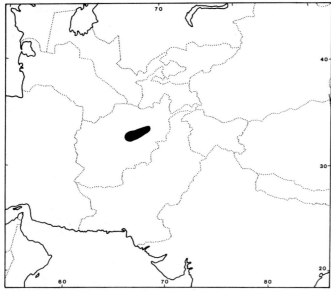

Map 42 – *Hyles salangensis*

HYLES SALANGENSIS (Ebert, 1969)   Map 42

E – Salang Hawkmoth
Celerio salangensis Ebert, 1969, *Reichenbachia* 12: 47–48.
Type locality: Salang Pass, eastern Afghanistan.

ADULT DESCRIPTION AND VARIATION (Pl. 11, fig. 3; Text fig. 47: HOLOTYPE)
Wingspan: 55–70mm. Although it resembles a dark hybrid between *Hyles hippophaes* and *H. euphorbiae*, the number of specimens caught over many years indicates that it is a valid species.

ADULT BIOLOGY
Little known. In the Salang Pass it flies at 2000–2700m.

FLIGHT-TIME
Early July.

EARLY STAGES
Unknown.

Text Figure 47 *Hyles salangensis* (Ebert) ♂ **HOLOTYPE**

PARASITOIDS
Unknown.

BREEDING
Nothing known.

DISTRIBUTION
Salang Pass, Afghanistan, and surrounding mountains.
Extra-limital range. None.

HYLES CENTRALASIAE (Staudinger, 1887)
Map 43

E – Foxtail-lily Hawkmoth

HYLES CENTRALASIAE CENTRALASIAE
(Staudinger, 1887)

Deilephila euphorbiae var. *centralasiae* Staudinger, 1887, *Stettin. ent. Ztg* 48: 64.
Type locality: Trans-Caspia [Transcaspia].
    Celerio centralasiae transcaspica O. Bang-Haas, 1936, *Ent. Z., Frankf. a. M.* 50: 256. **Syn. nov.**
(*Taxonomic note.* 'Subsp.' *transcaspica* is a local form of *H. c. centralasiae* and does not warrant subspecific status.)

ADULT DESCRIPTION AND VARIATION (Pl. 11, figs 4, 5)
Wingspan: 60–75mm. Resembles a very pale *H. e. euphorbiae*, but the pale median stripe extends to the costa and is broken only twice, basally and centrally. First segment of foretarsus shorter than in *H. euphorbiae*, with fewer, longer spines.

ADULT BIOLOGY
Restricted to areas where *Eremurus* (foxtail lily) is common on rock-strewn mountain-sides; in northern Iran, usually 1000–2500m; in Turkey, slightly lower.

FLIGHT-TIME
Bivoltine; between late April and early July, and July to early September, depending on altitude.

EARLY STAGES
OVUM: Almost spherical, pale, glossy green.

LARVA (Pl. 3, fig. 12): Full-fed 70–80mm. Dimorphic: principally pale grey or off-white.
    The newly-hatched 4mm-long larva is pale yellow with a blackish brown horn, legs, head and shield. The body bears a number of 'warts' of the same colour. In the second instar the primary body colour is olive green, with a pale yellow dorso-lateral line bearing rudimentary eye-spots. The third instar is similar, although the primary colour can vary from pale greyish green through greenish olive to almost black. The dorso-lateral line is more pronounced, the eye-spots pure white or white with yellow centres, and a whitish ventro-lateral line develops. In the fourth instar, the coloration ranges from pale grey to black; both forms are speckled white and bear a dorso-lateral row of round, black-ringed, startling white eye-spots. The dorso-lateral line is absent, but the ventro-lateral stripe becomes more evident, especially in the dark form. The head, legs, horn and shield are either dark brown or black.

Fully grown, either pale grey or off-white, sometimes with a pale rose dorsal suffusion. Along either side a dorso-lateral line of large, brilliant white, black-ringed eye-spots. Horn, head, legs and anal flap red or black. Spiracles white, ringed with red.

Feeds quite openly on the tall, columnar flower- and seed-spikes of its hostplant, alternating frenzied bouts of feeding with long spells of basking.

Most likely to be found in May, June, August and September. Grum-Grshimailo (1890) reported larvae to be common in southern Tajikistan during July, although he mistakenly identified them as *Hyles hippophaes bienerti*.

Hostplants. Principally the flower- and seedheads of *Eremurus*, especially *E. anigapterus* and, in Afghanistan, *E. stenophyllus* (Daniel, 1971); also *Asphodelus* seedheads.

PUPA: 40–50mm. Pale brown, in a loosely spun, net-like cocoon amongst ground debris. The overwintering stage.

PARASITOIDS

None recorded.

BREEDING

As for *H. euphorbiae*.

DISTRIBUTION

From central Turkey eastward across northern Iraq, northern Iran to southern Turkmenistan (Derzhavets, 1984), Uzbekistan (Bang-Haas, 1936), Tajikistan (Grum-Grshimailo, 1890; Bang-Haas, 1936) and Afghanistan (Ebert, 1969; Daniel, 1971).

Extra-limital range. Western Pakistan (Daniel, 1971), north-western China (most of the Tian Shan area) and Mongolia (Derzhavets, 1977). Although quite common, its distribution is not fully understood due to confusion with other related species.

HYLES CENTRALASIAE SIEHEI
(Püngeler, 1903)

Deilephila siehei Püngeler, 1903, *Berl. ent. Z.* **47**: 235–238, Pl. 3.
Type locality: Bulghar Dagh [Bolkar Dağları, Toros Mountains, Turkey].

ADULT DESCRIPTION AND VARIATION (Pl. 11, fig. 6)
Wingspan: 65–70mm. Central gap in the oblique median stripe of forewing (very apparent in subsp. *centralasiae*) much reduced, or absent. Most also have a distinct rosy hue to both wings and body.

ADULT BIOLOGY
Apparently confined to the upper elevations of the Toros Mountains in southern Turkey and in parts of northern Syria, where it is found in similar locations to subsp. *centralasiae*, i.e. rocky, herb-rich steppe.

FLIGHT-TIME
Univoltine, so far as known; late May and June.

EARLY STAGES
OVUM: Almost spherical; blue-green when freshly laid, becoming yellowish as the larva inside develops. Laid indiscriminately on the seed- and flower-heads of the host-

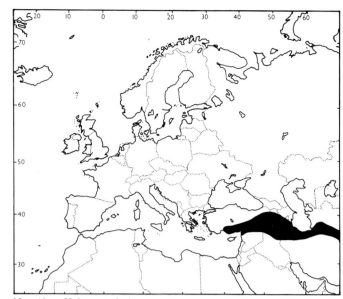

Map 43 – *Hyles centralasiae*

plant, with up to fifteen per plant. This stage lasts 5–6 days (Pelzer, 1982).

LARVA (Pl. D, fig. 3): Full-fed, 70–80mm. Polymorphic: brownish grey, off-white, greenish olive or black.

Early instars as *H. c. centralasiae*.

In the final instar, most larvae are pale brownish grey with a rose dorsal suffusion. The very large, single, dorso-lateral eye-spots are brilliant white and heavily ringed with black. The spiracles are also white, but ringed with red. The white speckling and the ventro-lateral stripe are absent (Pelzer, 1982). In black larvae, the eye-spots may be very reduced or absent.

The larvae feed quite openly on the seed- and flower-heads of their hostplant, alternating bouts of frenzied feeding with spells of basking.

Occurs mainly during June and July.

Hostplants. Principally the seed- and flower-heads of *Eremurus*, but also reported on *Asphodelus* (Pelzer, 1982); the leaves are never eaten.

PUPA: 40–50mm. As *H. euphorbiae*, but paler. Many spend several years in this stage.

PARASITOIDS

None recorded.

BREEDING

In captivity this subspecies will accept *Kniphofia* as a host-plant as well as all cultivated forms of *Eremurus*. Otherwise, treat as for *H. euphorbiae*.

DISTRIBUTION

The eastern Toros Mountains, southern Turkey (Daniel, 1932) and in northern Syria.

Extra-limital range. None.

HYLES GALLII (Rottemburg, 1775)   Map 44

E – Bedstraw Hawkmoth, **F** – le Sphinx de la garance,
G – Labkrautschwärmer, **NL** – Walstropijlstaart,
Sw – Brunsprötad skymningssvärmare,
R – Lodmarennikovyĭ Brazhnik, **H** – galajszender,
C – Lyšaj svízelový

HYLES GALLII GALLII (Rottemburg, 1775)

Sphinx gallii Rottemburg, 1775, *Naturforscher, Halle* 7: 107.

Type locality: Germany.

Sphinx galii [Denis & Schiffermüller], 1775, *Ankündung syst. Werkes Schmett. Wienergegend*: 42.

Celerio gallii chishimana Matsumura, 1929, *Insecta matsum.* 3: 166.

Celerio gallii sachaliensis Matsumura, 1929, *Insecta matsum.* 3: 166.

ADULT DESCRIPTION AND VARIATION (Pl. 11, fig. 7)

Wingspan: 65–85mm. The well-defined, pale oblique median stripe on the forewing distinguishes *H. gallii* from other similar members of the genus. There is very little variation: in f. *grisea* Tutt, the normally olive brown forewing areas are greyish and the red is absent from the hindwing discal band; in f. *pallida* Tutt, the forewing median stripe is white instead of yellowish.

ADULT BIOLOGY

Frequents rough commonland, roadside verges, meadow edges and, especially, clear-felled areas in mountain forests where *Epilobium* and *Galium* are present; occurs up to 2000m in the Alps.

A strong flyer, it becomes active at dusk and again at dawn, when it searches for flowers from which it feeds avidly. Pairing rarely takes place before early morning and lasts approximately two hours. After extra feeding, females commence egg-laying one or two days later, continuing to do so for up to ten days. Both sexes are attracted to light.

FLIGHT-TIME

May/June and again in August/September, the latter generation only partial. In more northerly latitudes and at high altitudes, only one generation, occurring in July, is normal.

EARLY STAGES

OVUM: Very small (1.1 × 1.0mm), almost spherical with a depression dorsally, glossy greenish blue. Deposited on the leaves of the hostplant or sometimes, as in the case of *Galium*, on the flowers, with up to four or five per plant.

LARVA (Pl. 3, fig. 11): Full-fed, 75–85mm. Polymorphic: one form mainly black.

Newly hatched, the 3–4mm-long larva is clear green with a number of yellow longitudinal lines. With growth, the green usually darkens to dark olive (or sometimes brown or black) and all lines except the dorso-lateral fade. This line gradually breaks up, to be replaced in the final instar by a row of startling, yellow or reddish eye-spots. The ventral surface in nearly all individuals is pinkish, and the head and horn red. In black larvae, both ventral and dorsal surfaces are of the same colour and the reduced eye-spots may be red. All colour forms bear small yellow spots laterally and ventro-laterally, which may be extensive or barely visible.

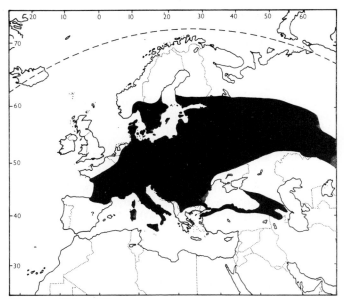

Map 44 – *Hyles gallii*

Initially, it feeds during the day and night, resting at intervals along the midrib under a leaf; when larger, larvae tend to retire low down on a plant by day, crawling up at dusk to feed. However, during wet weather fully-grown individuals may be seen in the daytime. As with *H. nicaea*, the number of larvae varies greatly from year to year, perhaps due to the weather or to the high rate of parasitism.

Occurs throughout most of its range during May, June and September; across its northern range and at high altitudes only from July to September.

Hostplants. Principally *Galium* and *Epilobium*, especially *E. angustifolium*. Also on *Betula*, *Vitis*, *Spiraea*, *Syringa*, *Geranium*, *Clarkia*, *Impatiens*, *Plantago*, *Asperula*, *Euphorbia* and, in central and southern Europe in particular, *Fuchsia*, *Rubia* and *Salix*.

PUPA: 40–45mm. Light brown with fine darker striations; encased in a loosely spun silk net amongst ground litter; very similar to that of *H. euphorbiae* (q.v., p. 137), although those produced by black larvae often have black wings and heavy black speckling on the abdomen. Those formed before mid-July usually hatch the same year, giving rise to a second generation. Overwinters as pupa.

PARASITOIDS

See Table 4, p. 54.

BREEDING

Pairing. As for *H. euphorbiae* (see p. 137), but occurring just after sunrise.

Oviposition. Conditions as for *H. euphorbiae*; however, between 300–400 ova are laid.

Larvae. As with all *Hyles* species, these should be started off on potted plants to avoid losses. Cut plants should be changed every day and the cages disinfected; if wilted or old food is given to last-instar larvae, disease can become a serious problem.

Crowding must be avoided; larvae tend to perish if disturbed by others when at rest on the midrib of a leaf. Losses can also be prevented if larvae which are moulting are not touched; they seem to be very delicate. Development can be considerably accelerated by warmth; at 30–35°C most will feed up in approximately twelve days with no loss in size.

Pupation. Best in an open box with a mixture of moss and dead leaves on clean, dry peat or sand. In Britain, pupae overwinter well between damp moss in a refrigerator. In central Europe, a sealed tin in an outdoor shed is adequate.

DISTRIBUTION

Resident throughout temperate Europe, apart from Britain (Newman, 1965), northern France and much of the Low Countries (Meerman, 1987), much of the Iberian Peninsula (Gómez Bustillo & Fernández-Rubio, 1976), Scandinavia, northern Russia and most of the Balkan Peninsula, where it occurs only as a migrant. It is, however, resident in the cooler north of Bulgaria (Ganev, 1984); also in northern Turkey and Transcaucasia. Absent from North Africa.

Extra-limital range. From central Siberia to China, the Kurile Islands, Sakhalin Island and northern Japan.

OTHER SUBSPECIES

Nepal as subsp. *nepalensis* (Daniel, 1960). China (Xizang Province/Tibet), Kashmir, Afghanistan (Ebert, 1969) and the Pamirs as subsp. *tibetanica* (Eichler, 1971). Canada and northern U.S.A. (Hodges, 1971) as subsp. *intermedia* (Kirby, 1837).

HYLES NICAEA (de Prunner, 1798)   Map 45
E – Greater Spurge Hawkmoth; Mediterranean Hawkmoth
F – Sphinx Nicéa, **G** – Nizzaschwärmer,
R – Nitstskiĭ Brazhnik, **H** – görög szender

HYLES NICAEA NICAEA (de Prunner, 1798)
Sphinx nicaea de Prunner, 1798, *Lepid. Pedemontana*: 86.
Type locality: Nice, Alpes-Maritimes [southern France].
  Sphinx cyparissiae Hübner, [1805], *Samml. eur. Schmett.*, Sphingidae: 115.

ADULT DESCRIPTION AND VARIATION (Pl. 11, fig. 8)

Wingspan: 80–100mm. As illustrated. Varies little except in the intensity of the coloration. A few may be pale yellowish white with very faint markings (f. *albina* Oberthür); others have a reddish tone (f. *rubida* Oberthür). As with *H. hippophaes*, some specimens found above 2000m in Kurdistan and Iran may be considerably larger than normal.

ADULT BIOLOGY

In the west of its range, mainly restricted to regions near the coast (above 500m in Spain) and, in the east, to high altitudes (2000–3000m, or sometimes higher, in Turkey (de Freina, 1979) and Iran). A local and scarce species, disappearing from one area for many years but appearing in another. Frequents very sunny, dry, stony slopes with scattered clumps of *Euphorbia*.

Unlike other members of this genus, *H. nicaea* has a very characteristic, fast, powerful flight reminiscent of that of *Daphnis nerii* (q.v., p. 115). This is well demonstrated by

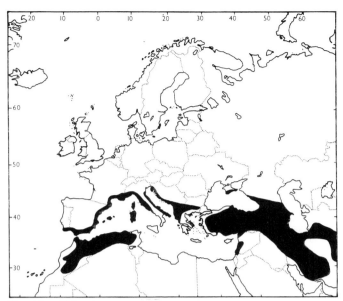

Map 45 – *Hyles nicaea*

individuals attracted to street lamps after dark.

FLIGHT-TIME

June; sometimes a partial second generation in August.

EARLY STAGES

OVUM: Very similar in size (1.6 × 1.4mm) and shape to that of *Smerinthus ocellatus* (q.v., p. 100), spherical and smooth; pale shiny green. The coloration changes to green and gold prior to hatching.

LARVA (Pl. 3, fig. 4): Full-fed, 100–120mm. Polymorphic: pale grey to black.

On hatching, the larva is 5mm long and a rich canary-yellow in colour, with a brownish black tail horn. With feeding, the yellow coloration is soon replaced by apple-green. This primary colour persists into the second instar, during which longitudinal rows of black dots appear; the horn and ventral surface also become black. In the third instar the final coloration of most individuals develops: pale grey with dorso-lateral and ventro-lateral rows of black-ringed yellow (fig. 4) or red eye-spots. The horn remains black and there is no red dorsal line. However, some larvae become totally black with red eye-spots and buff lateral patches; the amount of black pigmentation is very variable.

At first, the young larva rests along the midrib on the underside of a leaf, but growth is rapid and the large larva soon rests on the stem.

Occurs mainly during July and August, although a few may still be found feeding in September.

Hostplants. Principally herbaceous species of *Euphorbia*. Records of *Linaria* are erroneous.

PUPA: 45–50mm. Yellowish brown with fine black outlines to the legs and antennae. Enclosed in a loosely spun yellow cocoon amongst dead leaves and litter. The overwintering stage.

PARASITOIDS

Many larvae are parasitized by tachinid flies (pers. obs.) but the species are not recorded.

147

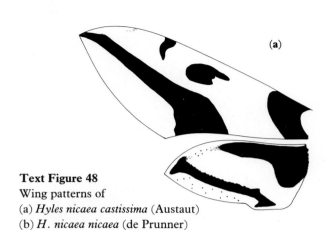

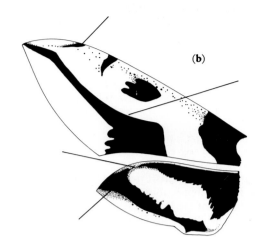

Text Figure 48
Wing patterns of
(a) *Hyles nicaea castissima* (Austaut)
(b) *H. nicaea nicaea* (de Prunner)

BREEDING

As for *H. euphorbiae*, but the larvae must be kept very dry and preferably reared in individual containers at above 30°C.

DISTRIBUTION

From southern Portugal (Gómez Bustillo & Fernández-Rubio, 1974) across southern Europe to Turkey (de Freina, 1979), including the Balearic Islands and south-western Bulgaria (Ganev, 1984). An English record (Newman, 1965) cannot be confirmed.
Extra-limital range. None.

HYLES NICAEA CASTISSIMA (Austaut, 1883)

Deilephila nicaea var. *castissima* Austaut, 1883, *Naturaliste* 5: 360.
Type locality: Sebdou, Algeria.

ADULT DESCRIPTION AND VARIATION (Pl. 11, fig. 9; Pl. H, fig. 3)

Wingspan: 76–110mm. Similar to subsp. *nicaea* but distal edge of the median band of forewing straight, not bowed as in subsp. *nicaea*; dark apical streak usually absent. The black post-discal band on the hindwing narrower and separated from the black basal area along the costa, unlike that in subsp. *nicaea* (cf. Text figs 48a,b).

ADULT BIOLOGY

Occurs on dry, stony slopes up to 2000m.

FLIGHT-TIME

Bivoltine; May/June, and again during late July and August.

EARLY STAGES

OVUM: As subsp. *nicaea*.

LARVA (Pl. D, fig. 4): As subsp. *nicaea*, pale grey to black.
  Occurs from June to September in two overlapping generations.
Hostplants. Herbaceous *Euphorbia* spp., especially *E. nicaeensis*.

PUPA: 50mm. Brown to pale buff, with a few fine markings. Otherwise as subsp. *nicaea*.

PARASITOIDS

None recorded.

BREEDING

As for subsp. *nicaea*.

DISTRIBUTION

The Atlas mountains in Morocco, northern Algeria and Tunisia (Rungs, 1981).
Extra-limital range. None.

HYLES NICAEA LIBANOTICA (Gehlen, 1932)

Celerio nicaea libanotica Gehlen, 1932, *in* Seitz, *Die Gross-Schmetterlinge der Erde* **2** (Suppl.): 153.
Type locality: Zahlé, Lebanon.

> *Celerio nicaea* var. *sheljuzkoi* Dublitzky, 1928, *Ent. Z., Frankf. a. M.* **42**: 38–40. **Syn. nov.**

(*Taxonomic note.* Apart from size, this subspecies has no clear characters separating it from subsp. *nicaea*, so it may well be only a form. *H. n. sheljuzkoi* (Pl. 11, fig. 11) is a local form of subsp. *libanotica* and certainly does not warrant subspecific status.)

ADULT DESCRIPTION AND VARIATION (Pl. 11, figs 10, 11)

Wingspan: 90–110mm. By far the largest subspecies of *H. nicaea*, but variable both in size and in the amount of dark speckling on the forewing; some specimens approach subsp. *crimaea* (q.v., p. 149).

ADULT BIOLOGY

Occurs on dry, stony slopes at high altitudes.

FLIGHT-TIME

May, July/August and sometimes September, in two or three generations, the third always partial.

EARLY STAGES

OVUM: As subsp. *nicaea*.

LARVA: As subsp. *castissima*.
Hostplants. Herbaceous *Euphorbia* spp.

PUPA: 50–55mm. Larger than but otherwise identical to that of subsp. *nicaea* (q.v., p. 147).

PARASITOIDS
None recorded.

BREEDING
As for subsp. *nicaea* (q.v., p. 148).

DISTRIBUTION
Lebanon (Zerny, 1933; Ellison & Wiltshire, 1939), southern Turkey (Hariri, 1971), northern Iraq (Wiltshire, 1957), Iran, Afghanistan (Daniel, 1971), Turkmenistan, Kazakhstan (Dublitzky, 1928) and the Pamirs (Derzhavets, 1984).
Extra-limital range. None.

HYLES NICAEA CRIMAEA (A. Bang-Haas, 1906)

Deilephila nicaea crimaea A. Bang-Haas, 1906, *Dt. ent. Z. Iris* **19**: 129.
Type locality: Sevastopol, Crimea [Ukraine].

Celerio nicaea var. *orientalis* Austaut, 1905, *Ent. Z., Frankf. a. M.* **18**: 143. **Syn. nov.**

(*Taxonomic note.* 'Subsp.' *orientalis* was described from a mixed collection of individuals from both the Crimea and northern Iran; however, the character used to justify this subspecies is found commonly throughout the range of *H. nicaea*. As the original description was based on a specimen from the Crimea and clearly indicated that it was a form of subsp. *crimaea*, the name *orientalis* is unavailable and therefore synonymized.)

ADULT DESCRIPTION AND VARIATION (Pl. 11, fig. 12)
Wingspan: 90–100mm. Immediately distinguished from other subspecies by the heavy, dark blotching on the wings; also less variable in colour and pattern.

ADULT BIOLOGY
Occurs on sunny, dry, stony slopes of hills.

FLIGHT-TIME
Univoltine; June/July.

EARLY STAGES
As subsp. *nicaea*.

PARASITOIDS
None recorded.

BREEDING
As for subsp. *nicaea* (q.v., p. 148).

DISTRIBUTION
The southern Crimean Peninsula (Ukraine) and Transcaucasia (Georgia).
Extra-limital range. None.

OTHER SUBSPECIES
Eastern Afghanistan (Ebert, 1969), north-west India (Kashmir) and China (Xizang Province/Tibet), as subsp. *lathyrus* (Walker, 1856).

HYLES ZYGOPHYLLI (Ochsenheimer, 1808)

Map 46

E – Bean-caper Hawkmoth, **G** – Doppelblattschwärmer, **R** – Tsarnolistnikovyĭ Brazhnik

HYLES ZYGOPHYLLI ZYGOPHYLLI

(Ochsenheimer, 1808)

Sphinx zygophylli Ochsenheimer, 1808, *Schmetterlinge Europa* **2**: 226.
Type locality: Southern Russia [Turkestan].

Celerio zygophylli jaxartis Froreich, 1938, *Ent. Rdsch.* **55**: 256. **Syn. nov.**

(*Taxonomic note.* In such a variable species, especially one which responds to local climatic conditions, the creation of a separate taxon, subsp. *jaxartis*, for pale individuals from the former Soviet Turkestan, is not justified.)

ADULT DESCRIPTION AND VARIATION (Pl. 12, fig. 1)
Wingspan: 60–75mm. Forewing yellowish brown with a narrow, yellowish white, median stripe running from base of inner margin to apex, from which one or two branches extend towards costa; wing margin yellowish. Forewing much narrower than in related species. Similar to some forms of *H. livornica* (q.v., p. 154). Variable in both colour intensity and size of markings; specimens from drier areas tend to be paler, with more yellow coloration.

ADULT BIOLOGY
Occurs on hot flats, sand-dunes and arid hillsides where *Zygophyllum fabago* (Syrian bean-caper) grows.

FLIGHT-TIME
End of April to mid-May, July and sometimes mid-September in two or three generations. In cooler mountainous areas, most are to be found during June/July, with a partial second generation in late September/October.

EARLY STAGES
OVUM: Size unrecorded but larger than most *Hyles* spp., spherical, bright green; very similar to that of *Laothoe populi*. Laid on the underside of a leaf, hatching from two to five days later.

LARVA (Pl. 2, fig. 12; Pl. 4, fig. 4): Full-fed, 70–80mm. Polymorphic.
Fully grown, primarily pale green or yellow, with a black reticulate pattern dorsally and laterally; ventro-lateral and lateral surface unmarked. A narrow yellow dorso-lateral band with small black-ringed yellow or white eye-spots of variable size. In some the head, dorsal line, horn and anal claspers are black; in others they are yellow or pale green. All forms have a yellow band beneath the orange or white spiracles.
On hatching, the larva takes up a position along the midrib on the lower surface of the leaf. During later stages, most feed fully exposed, clinging to a stem while avidly consuming leaves and flowers. From egg to pupa takes approximately 30 days.
Common between May and September, sometimes later in hot localities.
Hostplants. Principally *Zygophyllum fabago* in the region, but other species of *Zygophyllum* farther east. Possibly also the flower-heads of *Eremurus*.

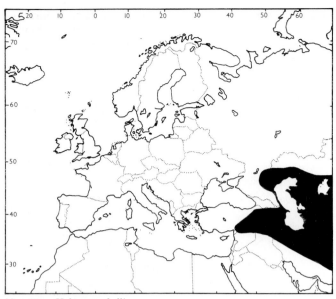

Map 46 – *Hyles zygophylli*

PUPA: 40–50mm. Light, sandy brown, with fine dark lines. In summer, this stage lasts no more than eighteen days. Overwinters as a pupa.

PARASITOIDS
None recorded.

BREEDING
Little information available; *Tribulus* will be accepted by most larvae as an alternative hostplant.

DISTRIBUTION
Eastern Turkey (de Freina, 1979), Armenia, northern Syria, eastern Transcaucasia with its main stronghold in the low-lying areas around the Caspian Sea (Eversmann, 1844); northern Iran, Turkmenistan, Kazakhstan, Uzbekistan, southern Tajikistan (Grum-Grshimailo, 1890) and northern Afghanistan (Ebert, 1969).
Extra-limital range. Northern Xinjiang Province, China.

OTHER SUBSPECIES
Mongolia as subsp. *xanthoxyli* (Derzhavets, 1977)

HYLES VESPERTILIO (Esper, 1779)   Map 47
E – Dusky Hawkmoth, **F** – le Cendré,
G – Fledermausschwärmer, **NL** – Vleermuispijlstaart,
H – denevérszender, **C** – Lyšaj metopýří
Sphinx vespertilio Esper, 1779, *Die Schmett.* **2**: 178.
Type locality: Verona, northern Italy.

ADULT DESCRIPTION AND VARIATION (Pl. 12, fig. 2)
Wingspan: 60–80mm. Forewing almost uniform slate-grey; hindwing reddish pink, with indistinct dark markings, which distinguish *H. vespertilio* from other species of *Hyles*. Some variation is exhibited: in f. *salmonea* Oberthür, no red on hindwing; in f. *flava* Black, hindwing yellow.

ADULT BIOLOGY
Unlike many members of this genus, *H. vespertilio* is not migratory and occurs in local, independent colonies up to 1600m in central Europe, but at higher altitudes in Asia Minor. Frequents warm, dry valleys, especially on well-drained, sunny, south-facing, gravel slopes where dry stream-channels are present. Shingle river-banks are a favourite haunt.

Whilst closely resembling *H. euphorbiae* in its feeding, copulating and activity behaviour (see p. 137), adults rest mostly on the ground, usually amongst rocks and pebbles (in a river-bed, for instance), with which their coloration blends.

FLIGHT-TIME
Over most of its range, May/June, with a partial second generation in August/September; at higher altitudes in central Europe only one emergence occurs, during June/July.

EARLY STAGES
OVUM: Oval, slightly dorso-ventrally flattened (1.1 × 1.0mm), pale, glossy green. Laid singly or in pairs on the stems, leaves and flower-shoots of *Epilobium*, often with up to ten to a plant. Very occasionally found on stones next to a potential hostplant.

LARVA (Pl. 3, fig. 3; Pl. D, fig. 5): Full-fed, 70–80mm.

Initially, the 3–4mm-long larva is pale green with a yellow, dorso-lateral line and small horn. It feeds by day from beneath the leaf on which it rests, rarely leaving the hostplant until the fourth instar. At this stage the primary colour is dark brown with a dorso-lateral line of pinkish grey eye-spots and a large number of small, dark brown spots covering the dorsal surface; ventral surface pale. It now feeds only at night, hiding by day on the ground amongst small stones at the base of the hostplant. Many curl up in an excellent imitation of a pebble. In the final instar the eye-spots and primary body colour are pale grey; very little change occurs in this grey coloration prior to pupation.

Most commonly found during June, July and September.

Hostplants. Principally *Epilobium* spp., especially *E. dodonaei* (also known as *Chamaenerion angustissimum*); also *Oenothera* and, occasionally, *Galium*.

PUPA (Text fig. 46): 35–40mm. Pale reddish brown with translucent green-brown wings. In shape, almost identical to that of *H. euphorbiae*. Contained in a fine-meshed, silk cocoon amongst debris on the ground. Overwinters as a pupa.

PARASITOIDS
See Table 4, p. 54.

BREEDING
As for *H. gallii* (q.v., p. 146), always using potted, under-watered hostplants. Can also be reared on *Fuchsia*.

DISTRIBUTION
Southern France (Herbulot, 1971; Rougeot & Viette, 1978) northwards to southern Germany (Forster & Wohlfahrt, 1960) and south through Austria to central Italy; thence south-eastwards through western Hungary, western Yugoslavia (Andus, 1986) to western and southern Bulgaria (Ganev, 1984) as one large isolated population. Thence

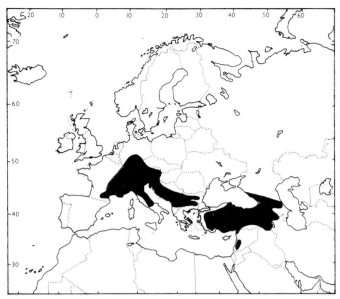

Map 47 – *Hyles vespertilio*

from western Turkey eastward to Transcaucasia (Milyanovskiĭ 1959) as another extensive population. A further small disjunct population also occurs in the mountains of Lebanon (pers. obs.).

Extra-limital range. None. Endemic to the western Palaearctic region.

HYLES HIPPOPHAES (Esper, [1793])   Map 48

E – Seathorn Hawkmoth, **F** – le Sphinx de l'argousier, **G** – Sanddornschwärmer, **NL** – Duindoornpijlstaart, **R** – Obletsikhovyĭ Brazhnik; Yuzhnyĭ Brazhnik, **H** – déli szender

HYLES HIPPOPHAES HIPPOPHAES
(Esper, [1793])

Sphinx hippophaes Esper, [1793], *Die Schmett.* (Suppl.) (Abschnitt 2): 6, pl. 38, figs 1–3.
Type locality: Wallachei, Milkowfluss bei Foxan [Wallachia region, southern Romania].

> *Celerio hippophaes kiortsii* Koutsaftikis, 1974, *Annls Mus. Goulandris* 2: 89–91. **Syn. nov.**

(*Taxonomic note.* The ranges of subsp. *hippophaes* and subsp. *bienerti* overlap in western and southern Turkey to produce intermediate hybrids, such as f. *malatiatus* Gehlen, 1934, and f. *kiortsii* Koutsaftikis (Pl. 12, fig. 3). The Aegean population is most certainly subsp. *hippophaes*, but with a trace of subsp. *bienerti* in many examples.)

ADULT DESCRIPTION AND VARIATION (Pl. 12, figs 3, 4)
Wingspan: 60–80mm. Very unlikely to be confused with other *Hyles* species. Most individuals from the Aegean population are slightly darker than those from central Europe; however, there is sufficient overlap in colour variation not to split the two groups into separate taxa. The original description and illustrations by Esper ([1793]) clearly indicate that the Romanian population is indis-

tinguishable from that in the Aegean. Where multiple climatic conditions are present, such as along mountain chains, adults of this species are very variable in wingspan, markings and colour intensity.

ADULT BIOLOGY
Within its range, populations can be somewhat isolated. However, as this species is prone to wander, individuals may turn up at great distances from known breeding grounds, leading to confusing records. Frequents river valleys in mountainous areas (up to 500m in Spain and Switzerland), mountainous steppe and sand-dunes. River islands overgrown with *Hippophae rhamnoides* are a favoured haunt in central Europe. In some western European localities, as a result of river flood-control measures, *H. hippophaes* is becoming increasingly rare as its hostplant cannot compete with riverine shrubs and trees which take over stabilized riverbanks.

As with nearly all members of this genus, pairing is a short affair lasting not more than three hours. Afterwards most females spend a few nights feeding, mainly after 23.00 hours, before egg-laying commences. Whilst strongly drawn to flowers, light holds little attraction.

FLIGHT-TIME
Late April to early July, with a peak in mid-June. A partial second brood in August often occurs. It is not unusual for only three weeks to elapse between the two broods.

EARLY STAGES
OVUM: Almost spherical (1.1 × 1.0mm), pale greenish grey. Deposited on both the upper and under surface of leaves, usually near the edge, on the lower branches of the hostplant. Thicket-edge or isolated shrubs are preferred, most eggs being laid in late June.

LARVA (Pl. 3, fig. 10): Full-fed, 75–80mm. Dimorphic: unstriped or striped.

On hatching, the eggshell is ignored, the 3–4mm-long larva proceeding to find a resting place below a leaf, a site to which it returns after each spell of feeding. At this stage it is dark green, thickly speckled with white and dark grey. The final instar has two colour forms. One is dark green (in some cases suffused with pink), thickly speckled with white and grey; superimposed on this are an off-white dorso-lateral line, often with orange eye-spots, and a broader, white, ventro-lateral stripe running just above the legs. Horn long, thin, orange below, black above, with two elongated orange spots at its base; head green, with two brown lines. The other colour form is silvery grey, with a black, broken dorso-lateral line from which emanate black, equally broken oblique lateral stripes with white, red, or yellow patches often present in between. Head brown and grey; horn as above. In a very rare colour form, all green coloration is replaced by pinkish brown.

Larvae frequently sun themselves openly on the upper branches, amongst those they have already stripped of leaves. There is a very heavy mortality due to parasitoids. Those that survive eventually become light brown before descending to find a pupation site, often after hours of perambulation on the ground.

Most common during late June and July; in some areas also during early September.

Hostplants. Principally *Hippophae rhamnoides*; also *Elae-*

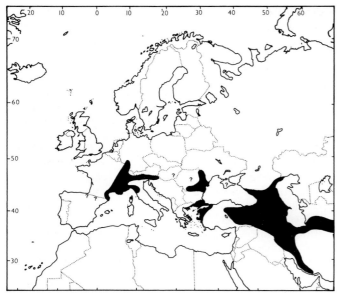

Map 48 – *Hyles hippophaes*

agnus angustifolia, an introduced oleaster from central and eastern Turkey now established over much of southern Europe. This is the main hostplant of the Aegean population (Pittaway, 1982a).

PUPA (Pl. E, fig. 15): 40–50mm. Pale yellowish brown, or light grey-brown, with dark brown striations. More elongated than others of the genus. Enclosed in a flimsy cocoon amongst roots or under stones. The overwintering stage.

PARASITOIDS

See Table 4, p. 54.

BREEDING

Pairing. Difficult but possible with force-feeding, two to three nights after emergence. Most pair around 01.00 hours and remain *in copula* for between thirty minutes and two hours.

Oviposition. Best achieved by placing cut shoots of the hostplant so that the tips touch the top of the cage. A female in captivity rarely lays more than 150 ova.

Larvae. Wilted food is unsuitable; as *Hippophae rhamnoides* wilts very quickly when cut, sprigs should be changed daily or a potted plant should be used. However the larvae will thrive on ornamental *Elaeagnus* and will also accept some *Epilobium* spp. They can be started off in glass jars but, from the third instar, they should be placed in well-ventilated, sunny cages.

Pupation. Best achieved by placing the larvae in a cardboard box containing a layer of dry leaves over dry sand. Pupae are most successful if stored cool and dry throughout the winter.

DISTRIBUTION

Separated into two main areas which seem to be the remnants of a much larger, post-glacial range. From northern Spain (Pittaway, 1983b) across southern France (Frionnet, 1910), Germany (Bavaria) (Heinemann, 1859; Forster & Wohlfahrt, 1960) and northern Italy to Yugoslavia. Then, as a separate population, Romania (Esper, [1793]), Mold-

ova (Derzhavets, 1984), northern Greece (Koutsaftikis, 1970; 1973; 1974) and western Turkey (Pittaway, 1982a). This species will very probably also be found in Bulgaria and Hungary (countries with large areas of potential hostplant); it has been recorded as a vagrant in England (Gilchrist, 1979) and the Crimea (Derzhavets, 1984).

Extra-limital range. None.

HYLES HIPPOPHAES BIENERTI
(Staudinger, 1874)

Deilephila bienerti Staudinger, 1874, *Stettin. ent. Ztg* **35**: 91. Type locality: Shahrud [Imamrud], north-east Iran.

Deilephila insidiosa Erschoff, 1874, *in* Fedtshenko, *Reise in Türkestan*: 25.

Celerio hippophaes ornatus Gehlen, 1930, *Ent. Z., Frankf. a. M.* **44**: 174–176. **Syn. nov.**

Celerio hippophaes transcaucasica Gehlen, 1932, *in* Seitz, *Die Gross-Schmetterlinge der Erde* 2 (Suppl.): 153. **Syn. nov.**

Celerio hippophaes anatolica Rebel, 1933, *Z. öst. EntVer.* **18**: 23–24. **Syn. nov.**

Celerio hippophaes bucharana Sheljuzhko, 1933, *Mitt. münch. ent. Ges.* **23**: 43. **Syn. nov.**

Celerio hippophaes shugnana Sheljuzhko, 1933, *Mitt. münch. ent. Ges.* **23**: 43. **Syn. nov.**

Celerio hyppophaes malatiatus Gehlen, 1934, *Ent. Z., Frankf. a. M.* **48**: 61. **Syn. nov.**

(*Taxonomic note*. This subspecies can be very variable in both coloration and size where numerous climatic conditions occur in close proximity to each other, such as in mountainous areas. Many of these forms were described as distinct subspecies but this is not warranted. Subspecies *ornatus*, *transcaucasica*, *anatolica*, *bucharana*, *shugnana* and *malatiatus* are therefore synonymized with subsp. *bienerti*.)

ADULT DESCRIPTION AND VARIATION (Pl. 12, figs 5, 6)

Wingspan: 65–80mm. Considerably paler and browner than related subspecies. A pale, oblique median line is noticeable on the underside of the forewing; hindwing patches more orange than red. Some large specimens found above 2000m in north-west Iran and Kashmir tend to resemble subsp. *caucasica* in coloration (q.v.). (See Taxonomic note above.)

ADULT BIOLOGY

Often common in mountainous, arid steppe, especially along rivers overgrown with *Hippophae* or *Elaeagnus*. Although found at any altitude from 400–3000m, most populations occur from 1000–2000m where *H. rhamnoides* often forms discrete thickets away from rivers. Attracted to the flowers of *Cistanche* at dusk (Shchetkin, 1956).

FLIGHT-TIME

April to September, in two or three overlapping generations.

EARLY STAGES

OVUM: As subsp. *hippophaes* (q.v., p. 151), with up to 500 being laid by each female.

LARVA (not illustrated): Full-fed, 75–85mm. Dimorphic: unstriped or striped.

In early stages, very similar to subsp. *hippophaes* (q.v.). Fully grown, usually also very similar to those of subsp. *hippophaes*; however, some are dark green with a dorsal lilac tint on the anterior segments and a broken, white ventro-lateral streak.

This stage lasts as little as 28 days, during which the larva basks quite openly on the topmost branches of its hostplant.

Often abundant from April to August.

Hostplants. Principally *Hippophae rhamnoides* and *Elaeagnus* spp., especially *E. angustifolia* and *E. hortensis* in Tajikistan (Shchetkin, 1956); possibly also *Daphne* spp. (Thymelaeaceae).

PUPA: Similar to subsp. *hippophaes*; in the summer months it remains in this stage for no more than 20 days. Formed in a chamber in the soil, often up to 10cm deep (Chu & Wang, 1980b). Overwinters as a pupa.

PARASITOIDS

None recorded.

BREEDING

As for subsp. *hippophaes* (q.v., p. 152).

DISTRIBUTION

The central Anatolian plateau forms the western limit of this subspecies. Central (Rebel, 1933), south-eastern and eastern Turkey (Daniel, 1932, 1939), northern and central Iran (Bienert, 1870; Barou, [1967]; Kalali, 1976), Turkmenistan and Uzbekistan (Derzhavets, 1984), Tajikistan (Shchetkin, 1956), Afghanistan (Ebert, 1969; Daniel, 1971) to the western Tian Shan.

Extra-limital range. Provinces of Xinjiang, Ningxia and Nei Mongol (Inner Mongolia), China (Chu & Wang, 1980b), and Mongolia (Derzhavets, 1977).

HYLES HIPPOPHAES CAUCASICA
(Denso, 1913)

Celerio hippophaes caucasica Denso, 1913, *Dt. ent. Z. Iris* 27: 35–37.

Type locality: Aresh, Transcaucasia [Areshperani, Georgia (41°50'N, 46°02'E)].

Celerio hippophaes caucasica Clark, 1922, *Proc. New Engl. zool. Club* 8: 19 [homonym]. **Syn. nov.**

(*Taxonomic note.* May be a hybrid population between subsp. *hippophaes* and subsp. *bienerti* and thus not merit subspecific status.)

ADULT DESCRIPTION AND VARIATION (Pl. 12, fig. 7)

Wingspan: 68–75mm. Intermediate in colour between subsp. *hippophaes* and subsp. *bienerti* (qq.v., Pl. 12, figs 4, 5).

ADULT BIOLOGY

Found on mountainsides at altitudes where *Hippophae rhamnoides* forms a distinct zone of vegetation.

FLIGHT-TIME

Late May to mid-July, depending on altitude.

EARLY STAGES

As subsp. *bienerti*.

PARASITOIDS

None recorded.

BREEDING

As for subsp. *hippophaes* (q.v., p. 152).

DISTRIBUTION

Mainly the Caucasus in Georgia: on the north side of the Bolshoi Kavkaz (Greater Caucasus) Mountains around Derbent and the south side as far as Tbilisi; and on the north face of the Malyi Kavkaz (Lesser Caucasus) Mountains around Areshperani. Also in small numbers over eastern Anatolia and southern Russia.

Extra-limital range. None.

OTHER SUBSPECIES

Kashmir and north-west India as subsp. *baltistana* (O. Bang-Haas, 1939).

HYLES CHAMYLA (Denso, 1913)   Map 49
E – Dogbane Hawkmoth, **R** – Kendyrnyĭ Brazhnik

Celerio hippophaes chamyla Denso, 1913, *Dt. ent. Z. Iris* 27: 37–39.

Type locality: Chamyl [Hami/Kumul], western Gobi [eastern Tian Shan, Xinjiang Province, China].

Celerio chamyla apocyni Shchetkin, 1956, *Izv. Akad. Nauk tadzhik. SSR*, (Otdel. Estest Vennkh Nauk) (16): 143–156. **Syn. nov.**

(*Taxonomic note.* Subspecies *apocyni* is not tenable. In many *Hyles* species which inhabit desert and semi-desert biomes, specimens from the more arid and hotter areas tend to be paler and smaller than those from less hostile environments; such is the case with *H. chamyla*. Shchetkin (1956) himself points out that many of the specimens he obtained from the more fertile southern areas of Tajikistan were virtually indistinguishable from the type series from arid Hami/Kumul, China, although most were larger and darker in colour, as one would expect.)

ADULT DESCRIPTION AND VARIATION (Pl 12, figs 8, 9)

Wingspan: 52–75mm. Varies considerably, with some resembling a pale, creamy *H. hippophaes bienerti* (Pl. 12, figs 5, 6), others *H. centralasiae siehei* (Pl. 11, fig. 6). A few even look like *H. centralasiae centralasiae* (Pl. 11, figs 4, 5) in having a large discoidal spot in the pale median stripe of the forewing, which itself can be very faint or pronounced. The pink area of the hindwing can be intense or faint, or even ochreous yellow.

ADULT BIOLOGY

A species of *Elaeagnus/Apocynum* thickets along riverbanks and on river flood-plains. However, with the cultivation of *Apocynum* (dogbane) as a fibre-plant and the introduction of extensive irrigation projects, *H. chamyla* has spread along irrigation canals and has become a local pest of cultivated dogbane in Uzbekistan and Tajikistan.

An avid visitor to *Cistanche* flowers at dusk. By day, most adults rest among grass tussocks.

FLIGHT-TIME

Trivoltine; late April to mid-May, mid-June to mid-July,

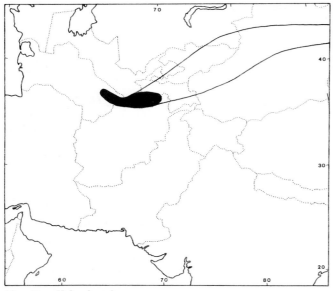

Map 49 – *Hyles chamyla*

DISTRIBUTION

Recorded only from southern Turkmenistan, southern Uzbekistan and southern Tajikistan (Shchetkin, 1956; Derzhavets, 1984). It probably also occurs in river valleys along the entire southern slopes of the Tian Shan into China, and in southern Turkmenistan as far west as the Kopet Daǧ (Ashkhabad), as its hostplant is common in these areas (Shchetkin, 1956). Damage to *Apocynum* plantations in Kyrgyzstan by larvae of '*H. euphorbiae*' was probably caused by this species, but this needs confirmation. Due to confusion with *H. hippophaes* and *H. euphorbiae*, the exact distribution of *H. chamyla* is unclear.

Extra-limital range. Recorded only from Hami/Kumul (the type locality) in northern Xinjiang Province, China.

and late July to late August in Tajikistan (Shchetkin, 1956).

EARLY STAGES

OVUM: Unknown, but presumably as *H. euphorbiae*.

LARVA (Pl. 4, fig. 7) Full-fed, 70–80mm. Dimorphic: dark green or bluish grey.

Similar to that of *Hyles hippophaes* (Pl. 3, fig. 10), from which the species undoubtedly evolved, but with significant differences. The coloration and pattern tend to remain the same for all instars. In the final instar the primary body colour is grass-green, although some have a bluish grey suffusion. The legs, prolegs, anal claspers, head and shield are of the same colour. The body is covered with small yellowish white dots, but these are usually larger and fewer in number than in *H. hippophaes*, and may even be absent. The horn is yellow with a black tip. Unlike *H. hippophaes*, no dorso-lateral stripe is present and there is no elongated yellow spot at the base of the horn. A ventro-lateral yellowish white band runs from thoracic segment 1 to abdominal segment 8. In some individuals the dorsal surface has a slight cinnamon hue, whilst others bear large but regular black patches.

The larvae are voracious feeders and grow very quickly, particularly in the fourth and fifth instars.

Occurs from May to June, in July, and from mid-August to mid-September.

Hostplants. Apocynum scabrum and *A. lancifolium*.

PUPA: Very similar to that of *H. euphorbiae* (q.v., p. 137), but slightly smaller. Formed in a chamber in the soil, as with *H. hippophaes bienerti*. Summer pupae remain at this stage for only nine to fourteen days. Overwinters as a pupa.

PARASITOIDS

In Tajikistan, larvae succumb to tachinid flies (Shchetkin, 1956), but the species are not recorded.

BREEDING

Nothing known.

HYLES LIVORNICA (Esper, 1779)   Map 50

E – Striped Hawkmoth, F – le Livournien,
G – Linienschwärmer, NL – Gestreepte Pijlstaart,
Sp – Esfinge de la vid; Oruga de esteva,
R – Lineĭchatyĭ Brazhnik, C – Lyšaj vinný,
Sw – Vitribbad skymningssvärmare, H – sávos szender

HYLES LIVORNICA LIVORNICA (Esper, 1779)

Sphinx livornica Esper, 1779, *Die Schmett.* 2: 88.
Type locality: Germany.

> *Phinx koechlini* Fuessly, 1781, *Arch. Insektengeschichte* 1: 1.
> *Celerio lineata saharae* Stauder, 1921, *Dt. ent. Z. Iris* 35: 179.

(*Taxonomic note. H. lineata* (Fabricius, 1775), under which *H. livornica* used to be listed as a subspecies, is a different species restricted to the New World (Harbich, 1980a, 1982).)

ADULT DESCRIPTION AND VARIATION (Pl. 12, figs 10, 11)

Wingspan: 60–85mm. Similar to many other species of the genus, but with distinctive, white forewing venation. Although extremely variable in size, with some individuals dwarfing others, it exhibits very little other variation except in the intensity of coloration and degree of pattern. In the southern Sahara, small pale individuals occur which are referable to f. *saharae* Stauder (Pl. 12, fig. 11).

ADULT BIOLOGY

A noted migrant, generally found in open ground with few trees and shrubs, such as rough grazing land, parched hillsides and sand-dunes, or in vineyards. In semi-desert areas, huge numbers can build up during winter and spring, especially after heavy rains.

An extremely active species, normally flying towards evening, when considerable numbers are often attracted to sweet-smelling flowers and to light. Pairing always takes place towards dawn over a period of two or three hours. Thereafter, females can cover considerable distances whilst egg-laying. In southern Europe and North Africa, many are also active during daylight hours, especially when on migration.

FLIGHT-TIME

Migrant and multivoltine in its southern range from late February to October, with a peak in March or April. Farther north, migrants and their offspring have been recorded from late May to October.

EARLY STAGES

OVUM: Slightly oval (1.1 × 1.0mm), glossy and pale green. Laid on the upper and underside of leaves of the hostplant, with four or five to a small plant.

LARVA (Pl. 3, fig. 8; Pl. D, fig. 6): Full-fed, 65–80mm. Polymorphic.

Newly hatched, 4mm long; greenish white, with black head and horn. With feeding, body darkens to olive green; in second instar the final pattern appears, thereafter becoming more brilliant at each moult. Fully-grown larvae are usually of the colour form depicted (fig. 8). Some, however, have very extensive black markings; green stripes instead of yellow; and a plum-coloured ventral surface. A very different colour form exists commonly in North Africa, i.e. pale apple-green with yellow eye-spots and speckling (fig. 6).

At all stages, larvae feed quite openly on their low-growing hostplants, alternating bouts of frenzied feeding, when large quantities of food are consumed, with spells of basking. When disturbed, young individuals will drop from the plant; older larvae will twitch violently from side to side while regurgitating food.

Found in North Africa from February until October – sometimes in hundreds of thousands in semi-desert areas; farther north from June until September.

Hostplants. Principally *Rumex*, *Polygonum* and, in North Africa and the Middle East, the flower- and seed-heads of *Asphodelus* (Pittaway, 1979b; Rungs, 1981); also *Vitis*, *Fuchsia*, *Galium*, *Antirrhinum*, *Plantago*, *Zygophyllum fabago*, and, in Asia Minor, the flower-heads of *Eremurus*. It has also been known to feed on various other plants, including *Pelargonium*, *Boerhavia*, *Asparagus*, *Acacia* and *Cicer*. In recent years this species has become a minor pest of *Arbutus unedo* (the strawberry tree) in the Mediterranean area.

PUPA: 30–45mm. More elongate than most *Hyles* species and variable in the amount of brown coloration present, some being yellowish or even translucent. Like others of the genus, encased in a flimsy, silk cocoon amongst ground litter or in a grass tussock. Overwinters as pupa.

PARASITOIDS

See Table 4, p. 54.

BREEDING

Pairing. As for *H. euphorbiae* (q.v., p. 137), but with most pairings taking place during the early morning (Heinig, 1981b).

Oviposition. As for *H. euphorbiae*.

Larvae. Must be kept dry and well ventilated. Succulent food may cause heavy losses, so under-watered potted plants should ideally be used or, alternatively, sprigs from water-starved plants. Cool conditions and a lack of sunlight produce abnormally small and often infertile adults.

Pupation. Easy to achieve in a mixture of dry leaves and moss on dry sand, kept dry and well ventilated.

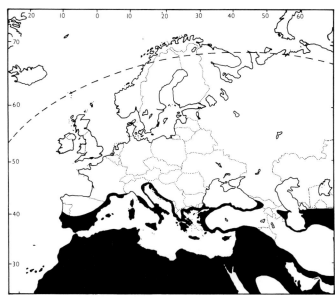

Map 50 – *Hyles livornica*

DISTRIBUTION

Occurs throughout the region, but resident only in the south; a migrant elsewhere, often in very large numbers, except the extreme north.

Extra-limital range. Africa to southern India, China and Japan.

OTHER SUBSPECIES

Madagascar as subsp. *malgassica* (Denso, 1944). Australia as subsp. *livornicoides* (Lucas, 1891). Xizang Province/Tibet, China, as subsp. *tatsienluica* Oberthür, 1916.

DEILEPHILA [Laspeyres], 1809

Deilephila [Laspeyres], 1809, *Jena. allg. Lit.-Ztg* **4**(240): 100.

Type species: *Sphinx elpenor* Linnaeus, 1758.

‡*Eumorpha* Hübner, [1806], *Tentamen determinationis digestionis . . .* [1].

‡*Elpenor* Oken, 1815, *Okens Lehrbuch Naturgesch.* **3**(1): 760.

‡*Choerocampa* Duponchel, 1835, *in* Godart & Duponchel, *Hist. nat. Lépid. Papillons Fr.* (Suppl.) **2**: 159.

Metopsilus Duncan, 1836, *in* Jardine, *Naturalist's Library* (Edn 1) **14** (Brit. Moths): 154.

Elpenor Agassiz, 1846, *Nomencl. zool.* (Nom. syst. Generum Anim.) Lepid.: 24.

Cinogon Butler, 1881, *Trans. ent. Soc. Lond.* **1881**: 1.

A Palaearctic genus containing four species.

IMAGO: Labial palpus rough, with long lateral hairs; scaling on inner surface as in *Hyles* (q.v., p. 136). Eyelashes much more distinct than in *Hyles*. Female antenna clubbed, male almost filiform; terminal segment short. Spines on abdomen numerous, less strongly chitinized than in *Hyles*. First row of spines of the first protarsal segment doubled at base. Pulvillus always present.

Genitalia. Uncus slender, much narrower than gnathos; the latter flat, or slightly convex beneath, not keeled or boat-shaped, rounded-truncate or rounded at end. Valva broadly sole-shaped, with a dozen or more friction scales. Sacculus ending in a more or less spatulate process, which is concave on upperside and slightly curved upwards. Aedeagus without apical process, but with a strong, subapical, oblique, dentate ridge.

OVUM: Ovoid, pale glossy green.

LARVA: Not typically sphingiform: anal horn reduced or absent; thoracic segments and head retractile into the eyespot-bearing first and second abdominal segments. Oblique lateral bars present.

PUPA (Text fig. 49): Proboscis fused with body, weakly carinate. Similar to those of *Hippotion* and *Theretra* (cf. Text figs 49–51) but with a dorso-lateral transverse row of fine, caudad-pointing hooks on each motile abdominal segment. Cremaster broadly triangular and down-curving, without terminal spines.

HOSTPLANT FAMILIES: Herbaceous plants mainly of the Onagraceae and Rubiaceae.

Text Figure 49 Pupa: *Deilephila e. elpenor* (Linnaeus)

DEILEPHILA ELPENOR (Linnaeus, 1758)

Map 51

E – Elephant Hawkmoth, **F** – le Sphinx de la vigne, **G** – Mittlerer Weinschwärmer, **Sp** – Esfinge de la vid, **NL** – Avondrood; (larva) Olifantsrups, **Sw** – Allmän snabelsvärmare, **R** – Sredniĭ vinnyĭ Brazhnik, **H** – szőlőszender, **C** – Lyšaj vrbkový

DEILEPHILA ELPENOR ELPENOR

(Linnaeus, 1758)

Sphinx elpenor Linnaeus, 1758, *Syst. Nat.* (Edn 10) **1**: 491.
Type locality: Unspecified [Europe].

Sphinx porcus Retzius, 1783, *Genera et Species Insect.*: 34.

‡*Elpenor vitis* Oken, 1815, *Okens Lehrbuch Naturgesch.* **3**(1): 760.

Chaerocampa lewisii Butler, 1875, *Proc. zool. Soc. Lond.* **1875**: 247. **Syn. rev.**

(*Taxonomic note*. According to I. J. Kitching (pers. comm.), the characters which separate *D. e. lewisii* from *D. e. elpenor* are not confined to that subspecies. Individuals of what are typical subsp. *lewisii* can be found within the range of subsp. *elpenor*; the reverse is also true. Subsp. *lewisii* is therefore synonymized with subsp. *elpenor*.)

ADULT DESCRIPTION AND VARIATION (Pl. 13, fig. 1; Pl. H, fig. 4)

Wingspan: 60–75mm. A very distinct species and difficult to confuse with any other in the region except *D. rivularis* (Boisduval, 1875) (Pl. 13, fig. 2). Variation is confined to the extent of or, as in the case of f. *unicolor* Tutt, absence of pink coloration on the forewing.

ADULT BIOLOGY

Frequents the flood-plains of streams and rivers, but also occurs in damp forest clearings, town wastelands or railway cuttings; up to 1500m in the Alps.

Hiding away amongst foliage by day, adults do not take flight, unlike many other hawkmoths, until well after dark, when flowers such as *Lonicera*, *Silene*, *Buddleja* and *Valeriana* are avidly visited. Towards midnight, pairing takes place, with pairs rarely remaining *in copula* for more than two hours. Following separation, females immediately begin to lay eggs, continuing to do so over a number of nights until approximately 100 have been deposited. It is strongly attracted to light, with large numbers often coming to mercury-vapour lamps between 22.00 and 24.00 hours.

FLIGHT-TIME

Late May to early July, with a partial second generation in southern Europe from mid- to late August. In exceptional years, some adults may also be found in August in southern England (Tutt, 1904).

EARLY STAGES

OVUM: Almost spherical (*c.* 1.5 × 1.2mm), pale glossy green. Laid singly or in pairs underneath the leaves of its hostplant, hatching about ten days later.

LARVA (Pl. 3, fig. 5): Full-fed, 70–80mm. Dimorphic: brown or green.

Newly-hatched larva (4–5mm long) pale green, cylindrical, with a small, narrow horn. In the second instar the head is disproportionately small. With a further moult, the first and second abdominal segments enlarge and have large and extremely realistic eye-spots. It is during this instar that most change to the final dark form (fig. 5), although some remain green. In between feeding, both by day and at night, the young larva rests stretched out beneath a leaf, where it is extremely well camouflaged. Later, larger individuals feed fully exposed at the top of a plant by day (Theunert, 1990) or night, preferring flowers and seed-heads to leaves. When not feeding, it often retires to hide at the base of the plant where its dark coloration is of greater advantage. Some larvae, especially those on *Galium*, may feed openly only at night, often in the company of those of *Macroglossum stellatarum* and *Deilephila porcellus*.

A striking feature of this species is its defensive behaviour. When alarmed, the head and the three thoracic segments are withdrawn into the first and second abdominal segments, which expand greatly, enlarging the startling eye-spots. Even quite large birds have been known to flee at this sight. The coloration of these 'eyes' remains bright until pupation. The larva can also swim if it drops from emergent aquatic hostplants into the water below.

Occurs between early June and late September, in overlapping generations.

Hostplants. Principally *Epilobium* and, to a lesser extent, *Galium*; also *Fuchsia*, *Impatiens*, *Lythrum*, *Calla*, *Menyanthes* and *Lonicera* in northern Europe. In its southern range, also on *Vitis* (it is sometimes a pest of grapevines), *Parthenocissus*, *Circaea*, *Oenothera*, *Arisaema*, *Polygonum*, *Daucus*, *Lysimachia* and *Rumex*.

PUPA (Pl. E, fig. 16): 35–47mm. Similar to those of *Hippotion celerio* and *Theretra alecto* (Pl. E, figs 17, 18), but easily distinguished by having a strongly reduced proboscis; brown coloration streaked with even darker brown; and a single row of spines on each mobile, abdominal segment. The function of these has yet to be ascertained, but Brock (1990) has demonstrated that if cocoons of this species are flooded or exposed to high humidity, up to 90 per cent of pupae rapidly work their way out of the cocoon by using these spines. This behaviour may be an adaptation to avoid drowning during flooding of the habitat. Even so, pupae are very active, often working their way out of their cocoon prior to emergence. Pupation takes place very close to the hostplant in a strongish mesh cocoon amongst debris on the ground. Overwinters as a pupa.

PARASITOIDS

In Britain, the larvae of *D. e. elpenor* are heavily parasitized by *Amblyjoppa proteus* (Christ) (Ichneumonidae), a fact which was recorded by Albin (1720), Dutfield [1749], Wilkes [1749] and Harris (1766); the adult ichneumon is easily recognized from their excellent colour plates.

See also Table 4, p. 54.

BREEDING

Pairing. Easy to achieve in a medium-sized flight-cage or netting-covered, wooden box. Better results can be obtained if this is placed outside in a sheltered location. Adults need not be fed prior to copulation.

Oviposition. Relatively easy to achieve in a flight-cage with potted hostplant, which should touch the top of the cage. Many more eggs can be obtained if the adults are hand-fed as many times as possible.

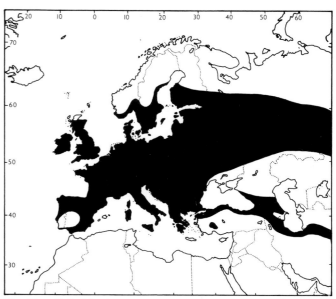

Map 51 – *Deilephila elpenor*

Larvae. Easily reared on well-ventilated, potted hostplant and can be crowded without loss. However, some losses can be expected on cut food in water or when rearing in glass jars, though, in the latter case, these should be minimal if crowding is avoided. Heat and sunlight are not required.

Pupation. Straightforward in an open box with damp moss on damp peat or soil. Pupae are easily stored in sealed tins, from which they should be removed in May and placed on damp moss or peat. As they benefit from humid conditions, spraying at this stage is advisable.

DISTRIBUTION

A non-migrant; resident throughout temperate Europe (Newman, 1965), including Sicily (Mariani, 1939). Turkey (de Freina, 1979); common in the Alborz Mountains of northern Iran (Sutton, 1963; Barou, [1967]; Ebert, 1976); possibly also north-eastern Afghanistan (Ebert, 1969), and recorded once only from the western Tian Shan (Alphéraky, 1882). Absent from northern Scandinavia, northern Russia, most of the Middle East, and North Africa (Rungs, 1981).

Extra-limital range. From western Siberia (Eversmann, 1844) to Amurland (Derzhavets, 1984) and Japan. Also northern and central China as far south as the Tian Shan, Xizang Province/Tibet, Sichuan Province and Nanjing (Chu & Wang, 1980b).

OTHER SUBSPECIES

Nepal, Sikkim, Bhutan, Assam (India), northern Burma and Yunnan Province, China, as subsp. *macromera* (Butler, 1875). Sichuan Province, China, as subsp. *szechuana* (Chu & Wang, 1980a), although the status of this subspecies is uncertain. See Taxonomic note (iii) under *D. rivularis*.

DEILEPHILA RIVULARIS (Boisduval, [1875])
Map 52

E – Chitral Elephant Hawkmoth

Chaerocampa rivularis Boisduval, [1875], *in* Boisduval & Guenée, *Hist. nat. Insectes* (Spec. gén. Lépid. Hétérocères) **1**: 280.

Type locality: Simlah [Simla], India.

>*Chaerocampa fraterna* Butler, 1875, *Proc. zool. Soc. Lond.* **1875**: 247.

(*Taxonomic notes.* (i) Kernbach (1958) regards this species as a subspecies of *D. elpenor*, an opinion he shares with many others. However, Ebert (1974) provides convincing evidence that *D. rivularis* is a valid species.

(ii) The **LECTOTYPE**, here designated, is a female in the Carnegie Museum, Pittsburg, Pennsylvania, U.S.A., bearing two two labels One states '*rivularis* B., Darjiling', the other '*epilobis* Bv., Simlah'. In his explicit description of *D. rivularis*, Boisduval stated quite clearly that this specimen was from Simlah [Simla], but added that a second specimen was reared by a Captain Shervill from an '*elpenor*-like' larva collected at Darjeeling. The latter was probably the source of the incorrect labelling. The correct type locality is therefore 'Simlah [Simla]'. *D. rivularis* does not occur farther east than Uttar Pradesh, India. The individual reared by Captain Shervill was probably *D. elpenor macromera* (Butler).

(iii) A possible relationship between *D. elpenor szechuana* (Chu & Wang, 1980a) from Sichuan Province, China, and *D. rivularis*, as suggested by their illustration (Chu & Wang, 1980b), needs to be fully investigated, but it appears to be doubtful.)

ADULT DESCRIPTION AND VARIATION (Pl. 13, fig. 2)

Wingspan: 64–82mm. Very like *D. elpenor* (Pl. 13, fig. 1), but with the rosy red parts of the body and wings heavily suffused with cinnamon, the red coloration being far less bright than in *D. elpenor*, especially on the wings. Marginal area of hindwing broad. In the male genitalia, the processes of the sacculi are bent claw-like and strongly sclerotized. Aedeagus anteriorly with a noticeably strong, subapical, oblique, dentate ridge, much more so than in *D. elpenor*. Even the single cornutus is more pronounced.

ADULT BIOLOGY

Nothing known, except that it tends to occur at 2000–4000m.

FLIGHT-TIME

Bivoltine; February/March, and late June and July, so far as known.

EARLY STAGES

As *D. elpenor* (Bell & Scott, 1937).

Hostplants. Arisaema and *Impatiens* in India (Bell & Scott, 1937).

PARASITOIDS

Unknown.

BREEDING

Nothing known.

DISTRIBUTION

Known only from eastern Afghanistan (Safed Koh Mountains, Kotkai) at 2350m (Ebert, 1974).

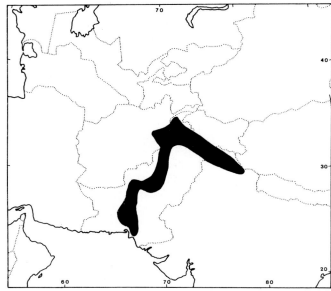

Map 52 – *Deilephila rivularis*

Extra-limital range. Pakistan, as far south as Karachi (Bell & Scott, 1937; Ebert, 1974), and northern India as far east as Dehra Dun, Uttar Pradesh, India. Records from Sikkim are almost certainly erroneous, even though there is a specimen in the Natural History Museum, London, attributed to this locality (see Taxonomic Note (ii) above).

DEILEPHILA PORCELLUS (Linnaeus, 1758)
Map 53

E – Small Elephant Hawkmoth,

F – le Petit Sphinx de la vigne; le Petit Pourceau,

G – Kleiner Weinschwärmer,

NL – Klein Avondrood; (larva) Kleine Olifantsrup,

Sw – Liten snabelsvärmare, **R** – Rozovyĭ malyĭ Brazhnik,

H – piros szender, **C** – Lyšaj kyprejový

DEILEPHILA PORCELLUS PORCELLUS
(Linnaeus, 1758)

Sphinx porcellus Linnaeus, 1758, *Syst. Nat.* (Edn 10) **1**: 492.

Type locality: Unspecified [Europe].

>*Pergesa porcellus porca* O. Bang-Haas, 1927, *Horae Macrolepidopt. Reg. palaearct.* **1**: 80.

ADULT DESCRIPTION AND VARIATION (Pl. 13, figs 3, 4)

Wingspan: 40–55mm. A distinctive species over most of its range but, in Asia Minor where it intergrades with *D. p. suellus*, easily confused with *D. p. suellus* f. *rosea* Zerny, though the wing pattern differs slightly (see p. 160). To add to the confusion, *D. p. porcellus* itself is highly variable in coloration: f. *indistincta* Tutt, in which the red is replaced by pinkish grey, is very similar to a normal *D. p. suellus*, especially when the yellow coloration is also subdued. Such examples tend to be more frequent in arid areas. Controlled temperature-breeding experiments produce specimens similar to *D. p. suellus* (fig. 4). (see Taxonomic Note, p. 160.)

ADULT BIOLOGY

Frequents coastal areas, heaths and meadowland, especially uncut field verges and roadsides where *Galium* is present, usually in local colonies where it may be abundant. Up to 1600m in the Alps, but in Turkey and Iran it is restricted to northward-facing mountain slopes at high altitudes (Ebert, 1976).

Although its general coloration is very noticeable against an artificial background, amongst low vegetation, where it rests by day, it breaks up the outline of the moth very well. At dusk, *D. porcellus* sallies forth in search of flowers such as *Valeriana, Iris, Echium, Silene* and, especially, *Rhododendron*, ceasing to feed 2–3 hours later. After midnight, the species is once again active, but this time in search of a mate. Pairing takes only a short time, after which females alternate feeding with egg-laying as soon as dusk falls, depositing their small green ova amongst *Galium* shoots whilst hovering. Readily attracted to light, large numbers often arriving in a short space of time.

FLIGHT-TIME

Late May to early July, depending on the weather, and, in its southern range, again in August.

EARLY STAGES

OVUM: Small (1.20 × 1.05mm), oval, with a dorsal depression, clear green. Changes to yellowish green before hatching. Laid singly near the growing tips of the hostplant, often two or three to a clump of shoots.

LARVA (Pl. 3, fig. 6): Full-fed, 60–70mm. Dimorphic: brown and green.

On hatching, the larva is approximately 3mm long, pale greyish green, with a tinge of yellow ventrally and two small tubercles in place of a tail horn. In the second and third instars the primary green colour darkens, a pale dorsolateral line appears and the caudal tubercles are pink. In the next moult the eye-spots and a dark bloom develop on the body, the bloom becoming dominant in the final instar when most assume a grey-brown coloration (fig. 6); only a few remain green. If the larva is alarmed at this stage, it retracts its head and thoracic segments causing the eye-spots on abdominal segment 1 to become prominent. If this fails to deter the 'attacker', *D. porcellus* then feigns death, becoming completely limp and flaccid. Fully grown, nearly all feed by night, resting during daylight hours low down on the hostplant. Although a mixed diet of flowers and leaves is initially preferred, when feeding on *Galium*, many make do with leaves only with no ill-effects. Prior to pupation the skin darkens, an event more noticeable in the rare green form; the larva may then travel a considerable distance in search of a suitable pupation site.

Most common between July and September.

Hostplants. Principally *Galium* spp.; also *Epilobium, Impatiens, Asperula, Lythrum, Vitis* and *Parthenocissus*.

PUPA: 25–31mm. Light brown with streaks and dashes of darker brown; prominent eyes, a cariniform proboscis and a pointed, triangular, downward-curving cremaster. As in *D. elpenor*, there is a semi-circle of small hooks on each movable abdominal segment. Pupation takes place in a loosely spun cocoon amongst litter on the ground or sometimes under a stone. The overwintering stage.

PARASITOIDS

See Table 4, p. 54.

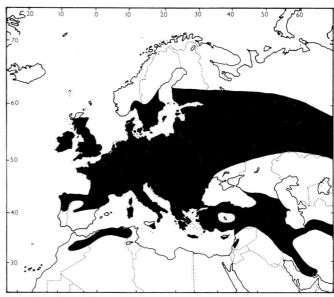

Map 53 – *Deilephila porcellus*

BREEDING:

As for *D. elpenor* (q.v., p. 157), but with the following differences:

(i) it does not tolerate mass rearing;
(ii) adults must be fed before pairing will take place;
(iii) once started on *Epilobium*, it is very difficult to change to another hostplant.

DISTRIBUTION

With the exception of most of the Iberian Peninsula and northern Scotland and northern Scandinavia, widespread throughout the region, from Ireland (Lavery, 1991) and the rest of western Europe (Newman, 1965) to western Siberia (Eversmann, 1844) and Kazakhstan. Also northern Turkey (de Freina, 1979), Transcaucasia and northern Iran (Ebert, 1976). In Turkey and Iran it intergrades with *D. p. suellus* (q.v. p. 160).

Extra-limital range. Central Siberia (Russia) as far east as Lake Baikal (Derzhavets, 1984).

DEILEPHILA PORCELLUS COLOSSUS
(A. Bang-Haas, 1906)

Metopsilus porcellus colossus A. Bang-Haas, 1906, *Dt. ent. Z. Iris* **19**: 129.

Type locality: Teniet-el-Haad, northern Algeria.

ADULT DESCRIPTION AND VARIATION (Pl. 13, fig. 5)

Wingspan: 55–62mm. Like a large, brightly-coloured example of subsp. *porcellus* (Pl. 13, fig. 3).

ADULT BIOLOGY

Occurs in similar habitats to those of subsp. *porcellus*, but at 1500–2000m. Strongly attracted to light.

FLIGHT-TIME

Univoltine; late April/May/June, depending on the location; the Moroccan populations mainly in mid-June.

EARLY STAGES

OVUM: As subsp. *porcellus*.

LARVA: As subsp. *porcellus*.

Hostplants. Principally *Galium* and *Epilobium* spp.

PUPA: As subsp. *porcellus*.

PARASITOIDS

None recorded.

BREEDING

As for *D. elpenor* and subsp. *porcellus* (qqv., pp. 157, 159).

DISTRIBUTION

Restricted to the Middle Atlas and Rif Mountains of Morocco (Ifrane, Jaba Forest, Mischliffen, Dayet-Aaoua, Ketama, etc. (Rungs, 1981)), and the Atlas Mountains of Algeria and western Tunisia.

Extra-limital range. None.

DEILEPHILA PORCELLUS SUELLUS
Staudinger, 1878

Deilephila porcellus var. *suellus* Staudinger, 1878, *Horae Soc. ent. ross.* **14**: 298.

Type locality: Amasia [Amasya, northern Turkey].

Pergesa suellus sus O. Bang-Haas, 1927, *Horae Macrolepidopt. Reg. palaearct.* **1**: 80. **Syn. nov.**

Pergesa suellus rosea Zerny, 1933, *Dt. ent. Z., Iris* **47**: 60. **Syn. nov.**

Pergesa suellus kuruschi O. Bang-Haas, 1938, *Ent. Z., Frankf. a. M.* **52**: 180.

Deilephila suellus kashgoulii Ebert, 1976, *J. ent. Soc. Iran* **3**: 91. **Syn. nov.**

Pergesa porcellus sinkiangensis Chu & Wang, 1980, *Acta zootaxon. sin.* **5**: 421. **Syn. nov.**

Eumorpha suellus gissarodarvasica Shchetkin, 1981, *Izv. Akad. Nauk tadzhik. SSR*, (Otdel. Biol.) (4): 90–92. **Syn. nov.**

(*Taxonomic note*. The status of *D. p. suellus* presents some problems. Many regard it as a valid species; however, it is almost certainly a subspecies of *D. porcellus* (see also Kernbach, 1958; Ebert, 1976; de Freina, 1979), which diverged from the parent stock in an isolated refugium during the last ice age, and it is treated as such in this work. The two subspecies have since come together again and are now intergrading in Turkey, northern Iran and the Tian Shan. Further evidence for the subspecific status of subsp. *suellus* is supplied by temperature-breeding experiments with subsp. *porcellus*: heat applied to developing pupae of subsp. *porcellus* can result in adults which are externally almost identical to those of subsp. *suellus* (cf. Pl. 13, figs 4, 6). There are no differences between the male genitalia and early stages of subsp. *porcellus* and subsp. *suellus* (de Freina, 1979). For these reasons, all attempts to describe 'subspecies' and forms of *suellus* (e.g. fig. 7) should be abandoned.)

ADULT DESCRIPTION AND VARIATION (Pl. 13, figs 6, 7)

Wingspan: 45–51mm. Very similar to *D. p. porcellus* (q.v., p. 158) but normally all pink coloration replaced by sandy buff. However, a confusing aberration with pink coloration

(f. *rosea* Zerny) occurs quite commonly in its south-western range where it intergrades with *D. p. porcellus*. Even this 'form' is variable, with some individuals having very little pink.

ADULT BIOLOGY

Frequents open, dry montane forest, or scrub with a good ground cover of herbs. Usually found at about 2500m, rarely below 2000m.

Apart from the fact that it is attracted to the flowers of *Lonicera caprifolium* and to light, little is known about the behaviour of this subspecies.

FLIGHT-TIME

Univoltine; mid-May to mid-June, depending on altitude.

EARLY STAGES

OVUM: Unknown.

LARVA (not illustrated): Full-fed, 55mm.

A study of the very detailed description given by Degtyareva & Shchetkin (1982) of a larva of *Eumorpha suellus gissarodarvasica* Shchetkin (=*D. porcellus suellus*) taken in the Darai-Nazarak Gorge, Gissar Mountains, Tajikistan, reveals no discernible difference between the larva of *D. p. suellus* and that of *D. p. porcellus*.

Occurs during June and July.

Hostplants. *Galium* spp.

PUPA: 38mm. Indistinguishable from that of *D. p. porcellus* (Degtyareva & Shchetkin, 1982). The overwintering stage.

PARASITOIDS

None recorded.

BREEDING

Nothing known.

DISTRIBUTION

From Sinop in northern Turkey (de Freina, 1979) eastward across Transcaucasia, the Caucasus (Bang-Haas, 1938; Derzhavets, 1984), north-east Iraq (Wiltshire, 1957), northern and eastern Iran (Ebert, 1976), southern Turkmenistan and Uzbekistan to Tajikistan (Alphéraky, 1882; Bang-Haas, 1927; Degtyareva & Shchetkin, 1982), and southwards to the Lebanese mountains (Zerny, 1933; Ellison & Wiltshire, 1939). In Turkey and northern Iran it intergrades with *D. p. porcellus*.

Extra-limital range. The Tian Shan and other parts of Xinjiang Province, China (Chu & Wang, 1980a; 1980b).

HIPPOTION Hübner [1819]

Hippotion Hübner [1819], *Verz. bekannter Schmett.*: 134.
Type species: *Sphinx celerio* Linnaeus, 1758.

    Panacra Walker, 1856, *List Specimens lepid. Insects Colln. Br. Mus.* **8**: 77 (key), 154.

A genus consisting of about 22 species restricted to the tropics and subtropics of the Old World.

IMAGO: Forewing very elongate, pointed, as is the abdomen. Female antenna clubbed, male antenna rod-like; terminal segment short. Labial palpus smoothly scaled on outer side, but lacking erect hairs present in *Deilephila* (q.v., p. 156). Scales at apex of first segment regularly arranged on inner side to form an even border; no apical tuft on second segment. Hindtibia with two pairs of spurs, the inner spurs longer than the outer.

Genitalia. Differ little from those of *Deilephila* and *Theretra* (qq.v., pp. 156, 163).

OVUM: Ovoid, small, glossy green.

LARVA: Anteriorly tapering, with retractile head. Abdominal segments 1 and 2 swollen, bearing eye-spots. Horn weakly curved backwards.

PUPA (Text fig. 50): Proboscis fused with body, laterally compressed and cariniform, similar to that of *Theretra*. Cremaster long, pointed, without apical spines on the sharp, glossy tip.

HOSTPLANT FAMILIES: Mainly herbaceous plants and climbers of the Vitaceae, Araceae, Nyctaginaceae, Rubiaceae, Polygonaceae and Onagraceae.

Text Figure 50 Pupa: *Hippotion celerio* (Linnaeus)

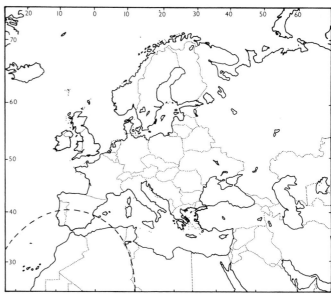

Map 54 – *Hippotion osiris*

HIPPOTION OSIRIS (Dalman, 1823)   Map 54

E – Large Silver-striped Hawkmoth,
F – le Deiléphile Osyris

Deilephila osiris Dalman, 1823, *Analecta Entomologica*: 48.
Type locality: Africa.

ADULT DESCRIPTION AND VARIATION (Pl. 13, fig. 8)

Wingspan: 89–90mm. Very similar to *Hippotion celerio* (Pl. 13, fig. 9), but distinguished by its larger size, heavier markings and the absence of any black coloration along the hindwing veins. Shows little variation; all European specimens have been of the typical form.

ADULT BIOLOGY

Frequents gardens and public parks with flower-beds or walls and shrubs overgrown with *Vitis* and *Parthenocissus*; also, small, private vineyards which are not sprayed with insecticide. Attracted to light.

FLIGHT-TIME

In the region, individuals have been captured at light mainly between September and November.

EARLY STAGES

OVUM: Very like that of *H. celerio*, but larger.

LARVA (Pl. 2, fig. 2): Full-fed, 80–90mm. Dimorphic: brown or green.

    No authenticated records exist from the region; however, if present, most would probably be of the brown form rather than the green; the former predominates in Africa.

    In the brown form (fig. 2), dorsal surface to spiracles medium brown with fine, darker brown cross-lines and a fine, black dorsal stripe. Intersegmental areas pale yellow, with two yellow spots on each segment. Fine subdorsal and dorso-lateral yellow lines also present, which merge into a broad, yellow, lateral line in which the white spiracles are located. On abdominal segment 1, a large, black-ringed, purplish brown eye-spot present, containing white-pupilled, purple spots; a smaller eye-spot is sometimes also present on abdominal segment 2, with black patches replacing it on subsequent segments. Ventral surface of body pale brown. Head pale brown. Horn very small (1mm), pale yellow, with a dark tip. The other form has similar black markings but the brown coloration is replaced by green.

Hostplants. In Africa, principally *Vitis* and *Parthenocissus* spp.; also *Rumex, Polygonum, Impatiens, Cissus, Ipomoea, Spathodea, Fuchsia* and various Rubiaceae.

PUPA: 55–61mm. In shape, very similar to that of *H. celerio*. Pale greyish brown with black specks. Proboscis projecting forward, laterally compressed and cariniform. Encased in a light brown, loosely spun cocoon on the ground. Does not overwinter in the region.

PARASITOIDS
None recorded.

BREEDING
Nothing known.

DISTRIBUTION
A migrant to North Africa (Rungs, 1981) and southern Spain (Gómez Bustillo & Fernández-Rubio, 1976; Vives Moreno, 1981). Significant numbers have been recorded from southern Spain and Gibraltar but the frequency with which this species visits Europe is not known due to its confusion with *H. celerio*.

Extra-limital range. Sub-Saharan Africa and Madagascar.

HIPPOTION CELERIO (Linnaeus, 1758)   Map 55
E – Silver-striped Hawkmoth, **F** – le Sphinx Phaenix, **G** – Großer Weinschwärmer, **Sp** – Esfinge de la parra, **NL** – Wijnstokpijlstaart, **R** – Bolshoĭ vinnyĭ Brazhnik, **H** – csíkos szender, **C** – Lyšaj révový

Sphinx celerio Linnaeus, 1758, *Syst. Nat.* (Edn 10) **1**: 491. Type locality: Unspecified [Europe].

> *Sphinx tisiphone* Linnaeus, 1758, *Syst. Nat.* (Edn 10) **1**: 492.
> *Phalaena inquilinus* Harris, 1776, *An exposition of English Insects*: 93.
> ‡*Elpenor phoenix* Oken, 1815, *Okens Lehrbuch Naturgesch.* **3**: 760.
> *Hippotion ocys* Hübner, [1819], *Verz. bekannter Schmett.*: 135.
> *Deilephila albo-lineata* Montrouzier, 1864, *Annls Soc. linn. Lyon* (2) **11**: 250.

ADULT DESCRIPTION AND VARIATION (Pl. 13, fig. 9; Pl. H, fig. 5)
Wingspan: 60–80mm. Very similar in appearance to *H. osiris*, but distinguished by its smaller size and black venation on the hindwing. There is greater variation, however, in *H. celerio*. In f. *pallida* Tutt, a pale terracotta ground coloration is present; in f. *rosea* Closs, the wings are suffused with red and, in f. *brunnea* Tutt, with deep brown. In f. *augustei* Trimoul, the black markings are so extensive as to cover the entire wings; in f. *luecki* Closs, the silver markings are absent and in f. *sieberti* Closs, the oblique stripe on the forewing is yellowish instead of silver.

ADULT BIOLOGY
No particular habitat preference is shown by the adult which may occur wherever flower-beds are plentiful; however, for breeding colonies to become established, the presence of cultivated or wild grapevines is essential.

Its cryptic coloration, whatever the variation, makes this species difficult to see as it rests during daylight hours on stones, walls, tree-trunks, or amongst foliage. At dusk it takes flight in search of tubular nectar flowers. Although active for only short periods, its powerful and rapid flight enables it to cover great distances; it is frequently attracted to light. Pairing commences a few hours after dusk and lasts only a short time, the pair remaining *in copula* 1–3 hours.

FLIGHT-TIME
Migrant and multivoltine in its resident range, throughout the year, with up to five well-defined generations. In the Mediterranean area, two or three migrant-induced generations are normal between June and October; individuals from these migrate farther north, to be found during August, September and October.

EARLY STAGES
OVUM: Variable in size and shape, ranging from near spherical to distinctly oval; clear, glossy, bluish green, assuming a greenish yellow hue prior to emergence. Laid singly on both the upper and lower surfaces of leaves near a growing tip, with rarely more than one to a shoot. The lowest shoots, usually of a vine growing along a wall or fence or along the ground, are preferred.

Incubation lasts for 5–10 days.

LARVA (Pl. 3, fig. 1): Full-fed, 80–90mm. Dimorphic: brown or green.

On hatching, the 4mm-long larva is pale yellow with a disproportionately long black horn. It immediately consumes its eggshell, then moves off to find a resting place on the lower surface of a leaf. A period of several hours elapses before any plant material is consumed, after which the body becomes glossy green. In the second instar, eye-spots appear on the first and second abdominal segments; the long dark horn becomes bifurcated at the tip and waves up and down as the larva moves. In the third instar the eye-spots assume their final coloration and a yellow, dorso-lateral line appears, running from thoracic segment 3 to the base of the horn. In the final instar most assume a mid- to dark brown coloration, while a few remain green (fig. 1). However, unlike most other species there is very little change in colour prior to pupation, even in the green form.

As with most larvae exhibiting anterior eye-spots, the head is retracted when the larva is alarmed, expanding the large eye-spots on the first abdominal segment. When feeding, it rarely consumes the whole of a leaf; shoots with quarter- or half-eaten leaves often indicate the presence of a larva. Whereas young larvae may be found beneath a leaf, fully-grown specimens usually rest away from the feeding area, farther down the stem.

In southern Europe, most occur from July to September; in more northerly localities, during late summer. In the coastal region of Saudi Arabia it occurs throughout the year (Pittaway, 1979b).

Hostplants. Principally *Vitis* and *Parthenocissus* spp.; also *Galium, Fuchsia, Epilobium, Beta, Impatiens, Convolvulus, Scrophularia, Verbascum, Syringa, Rumex, Begonia, Arum, Mirabilis, Cissus* and *Caladium*.

PUPA (Pl. E, fig. 17): 45–51mm. Pale greyish brown with

extensive dark brown specks. Proboscis projecting, laterally compressed and cariniform. Cremaster long, glossy, terminating in a very sharp point. Formed in a loosely spun brown cocoon, either on the ground amongst litter, or just below the surface of the soil. Does not overwinter in the region.

PARASITOIDS

See Table 4, p. 54.

BREEDING

Pairing. In general as for *Acherontia atropos* (q.v., p. 83), very difficult to achieve, though not impossible in a large flight-cage with plenty of nectar flowers; if flowers are not available, force-feeding is essential. Pairing usually takes place on the third or fourth night after emergence.

Oviposition. Not difficult to achieve under the above conditions, with the inclusion of vine shoots. A dull, orange light suspended in the centre of the cage amongst the twigs of its hostplant not only encourages egg-laying, but also reduces damage to the adults, as they are drawn into the cage centre, away from the netting. Again, force-feeding may be necessary. Egg-laying commences on the second or third night after mating and can go on for up to two weeks. The eggs hatch in 5–6 days at 21°C (Friedrich, 1986).

Larvae. Very easy to rear if sleeved outside, or placed in glass containers where they should be kept clean and dry and never overcrowded. Adults of *H. celerio* reared in Europe are often sterile; this can be remedied to some extent by providing a direct heat source and warm surroundings for the larvae, though sunlight is not essential. The houseplant, *Cissus antarctica* (kangaroo vine), is a very good substitute hostplant during the winter months.

Pupation. Easy to achieve by placing the fully-grown larva in a cardboard box with dead leaves over a thin layer of sand. Emergence of the adults is approximately three weeks later.

NOTE. This species is continuous-brooded under short-day conditions and can produce several generations in a year.

DISTRIBUTION

A notable migrant in most years from tropical Africa and India to the western Palaearctic region. In warm years, colonies may even be established in North Africa and Europe, so the delineation between resident and migrant ranges cannot be clearly defined. It is, however, resident in the Canary Islands and probably also along the Atlantic coast of Morocco. It is certainly resident in many areas of the Levant and the Arabian Peninsula (Pittaway, 1979b), and Egypt (Badr *et al.*, 1985).

Extra-limital range. Tropical Africa, Asia and Australia.

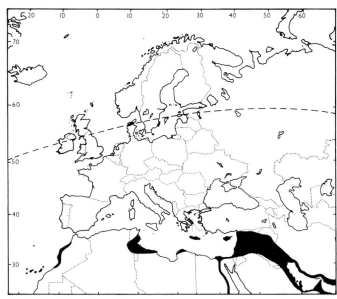

Map 55 – *Hippotion celerio*

THERETRA Hübner, [1819]

Theretra Hübner, [1819], *Verz. bekannter Schmett.*: 135.
Type species: *Sphinx equestris* Fabricius, 1793 (=*S. nessus* Drury, 1773).

> *Oreus* Hübner, [1819], *Verz. bekannter Schmett.*: 136.
> *Gnathostypsis* Wallengren, 1858, *Öfvers. K. Vetensk-Akad. Förh. Stockh.* **15**: 137.
> *Hathia* Moore, [1882], *Lepid. Ceylon* **2**: 19.
> *Florina* Tutt, 1903, *Entomologist's Rec. J. Var.* **15**: 76.
> *Lilina* Tutt, 1903, *Entomologist's Rec. J. Var.* **15**: 101.

A genus consisting of about 30 species occurring in the Oriental and Australasian regions, with some penetrating into the southern Palaearctic and Afrotropical regions.

IMAGO: Labial palpus smoothly scaled; first segment with an apical cavity on outer side bounded by this scaling, with the joint visible in it. On the inside, the apical scaling on segment 1 is regular as in *Hippotion* (q.v., p. 161), but unlike other genera, segment 2 bears at its apex a tuft of inwardly directed scales. Antenna in both sexes rod-like, long, with terminal segment short. Hindtibia with two pairs of spurs, the inner spurs longer than the outer. Very closely allied to *Hippotion*. As in that genus, wings and abdomen elongate.

Genitalia. In male, uncus rather narrow, underside weakly concave apically. Gnathos obtusely pointed, apex curved upwards, strongly convex beneath. Valva with numerous friction scales; sacculus slender, S-shaped, pointed. Aedeagus with a short multidentate process on right side, and a long oblique row of teeth on a raised ridge on the left. In the female, the ridge before the ostium bursae rather thin, but well chitinized and smooth, forming a half-moon shape.

OVUM: Broadly ovoid, smooth, shiny green or whitish.

LARVA: Like that of *Hippotion*. Head small and rounded, with body tapering forward from swollen, ocellate abdominal segments 1 and 2. Rest of body cylindrical. Horn erect, short or of medium length. Always with a pale dorso-lateral line in which there may be numerous eye-spots.

PUPA (Text fig. 51): Very similar to that of *Hippotion*. Proboscis fused with body, but projecting forward, laterally compressed and carinate. A coxal piece always present. Ante-spiracular ridges almost always present. Cremaster long, pointed, without apical spines.

HOSTPLANT FAMILIES: Usually herbaceous plants and vines of the Vitaceae, Araceae, Onagraceae, Lythraceae, Rubiaceae and Dilleniaceae.

Text Figure 51 Pupa: *Theretra alecto* (Linnaeus)

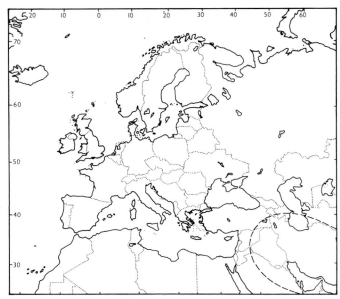

Map 56 – *Theretra boisduvalii*

THERETRA BOISDUVALII (Bugnion, 1839)
Map 56

E – Boisduval's Hawkmoth

Sphinx boisduvalii Bugnion, 1839, *Annls Soc. ent. Fr.* **1839**: 115.
Type locality: Île de Candie [Crete].

> *Chaerocampa punctivenata* Butler, 1875, *Proc. zool. Soc. Lond.* **1875**: 248.
> *Cechenena sumatrensis* Joicey & Kaye, 1917, *Ann. Mag. nat. Hist.* (8) **20**: 307.

ADULT DESCRIPTION AND VARIATION (Pl. 13, fig. 10)
Wingspan: 85–110mm. Immediately distinguished from *T. alecto* (Pl. 13, fig. 11) by its dark brown forewing and blackish hindwing; not very variable.

ADULT BIOLOGY
Frequents edges of woodland and cultivated areas where species of Vitaceae grow.

Very little is known about the behaviour of this species except that it is strongly attracted to light and flowers.

FLIGHT-TIME
Most likely during late summer as a migrant to the region.

EARLY STAGES
OVUM: Large (2mm), oval, pale glossy green.

LARVA (not illustrated): Very like that of *T. alecto* (Pl. 3, fig. 7).

Hostplants. Vitis and *Parthenocissus* spp.

PUPA: 53–70mm. Very similar to that of *T. alecto* (q.v., p. 165). Does not overwinter in the region.

PARASITOIDS
None recorded.

BREEDING
As for *T. alecto* (q.v.).

DISTRIBUTION
A rare migrant to south-eastern Iran, and thence possibly westwards as a very rare vagrant to Turkey and Greece. No individuals have been recorded from Europe in recent years. As *T. boisduvalii* and *T. alecto* have both shared the name *cretica*, it may be that many records of this species from Europe are due to misidentification.

Extra-limital range. From south-eastern Iran eastwards to Sri Lanka, the Himalayan foothills (to 400m), and across South East Asia to Borneo.

THERETRA ALECTO (Linnaeus, 1758)
Map 57

E – Levant Hawkmoth, **F** – le Deiléphile Candiote,
G – Riesenweinschwärmer, **R** – Alekto Brazhnik

Sphinx alecto Linnaeus, 1758, *Syst. Nat.* (Edn 10) **1**: 492.
Type locality: India.

> *Sphinx cretica* Boisduval, 1827, *Annls Soc. linn. Paris* **6**: 118.
> *Theretra freyeri* Kirby, 1892, *Synonymic Cat. Lepid. Heterocera* **1**: 650.
> *Theretra alecto transcaspica* O. Bang-Haas, 1927, *Horae Macrolepidopt. Reg. palaearct.* **1**: 80. **Syn. nov.**
> *Theretra alecto intermissa* Gehlen, 1941, *Ent. Z., Frankf. a. Main* **55**: 185–186. **Syn. nov.**

(*Taxonomic note*. The coloration of adults of this species can be influenced greatly by environmental conditions experienced during the larval stage. This can give rise to environmentally produced forms, such as 'subsp.' *transcaspica* and 'subsp.' *intermissa*; neither warrants subspecific status.)

ADULT DESCRIPTION AND VARIATION (Pl. 13, fig. 11; Pl. H, fig. 6)
Wingspan: 80–100mm. As illustrated, with very little variation, apart from the intensity of coloration. However, a

number of forms exist: f. *transcaspica* O. Bang-Haas bears an orange-red, oblique, submarginal line on the forewing; f. *cretica* Boisduval is paler than normal, with a buff tint to the forewing and orange-red hindwing. These forms are geographical, but not exclusively: the former occurs in Turkmenistan; the latter seems to be found mainly in the drier and hotter regions of south-eastern Europe and the Middle East.

ADULT BIOLOGY

Occurs in areas where Vitaceae are grown; in Europe, on ornamental vines rather than in vineyards due to the widespread commercial use of pesticides; in Greece, up to 1200m. The commonest sphingid in Lebanon, apart from *Macroglossum stellatarum* (Ellison & Wiltshire, 1939).

Little is known about the behaviour of this species except that it is attracted to flowers and light.

FLIGHT-TIME

April/May, June/July and August/September, usually in three overlapping generations, with sometimes a partial emergence in October and November.

EARLY STAGES

OVUM: Large (2mm), oval, smooth, pale glossy green; not unlike a large egg of *Laothoe populi* (q.v., p. 102). Up to five in a cluster may be laid on both the upper and lower surfaces of young leaves, each female depositing 150–250 eggs.

LARVA (Pl. 3, fig. 7): Full-fed, 90–110mm. Dimorphic: brown or green.

On hatching, the 6mm-long larva is bright yellow with a very long, straight, black horn, which remains black until the final instar. With feeding, the body acquires a vine-green coloration upon which a paler, dorso-lateral line is superimposed. In the second instar, several eye-spots of diminishing size may appear along this; that on abdominal segment 1 is very bold – black above and white below – and, unlike some of the others, always present. Although it begins feeding from beneath a vine leaf whose colour it matches, in the fourth or fifth instar most larvae have assumed the colour and pattern of an old, gnarled vine stem (fig. 7), on which most now rest. Those that remain green resemble a third-instar larva of *Deilephila elpenor* (q.v., p. 157), speckled yellow, with a yellow dorso-lateral line in which, from abdominal segments 1–8, yellow-ringed eye-spots are usually present. There are also several pale yellow, oblique, lateral streaks and a blue, slightly curved, stumpy horn.

It grows very rapidly and achieves its final length and a diameter of 10–12mm in 15–25 days. With its preference for young vine shoots and its abundance in certain localities, it can cause considerable damage. Having finished feeding, it descends very quickly from the hostplant in search of a suitable place to pupate.

Most common between May and late September.

Hostplants. Principally *Vitis* and *Parthenocissus* spp.; also *Rubia* and *Gossypium*.

PUPA (Pl. E, fig. 18): 50–60mm. Light brown and noticeably dorso-ventrally flattened, with a granulose, keel-shaped proboscis projecting beyond the head. Cremaster short, broadly triangular, downward curving. Pupation may take place in a variety of sites: most larvae construct a

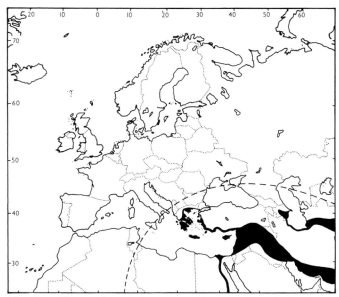

Map 57 – *Theretra alecto*

loosely spun cocoon amongst dead leaves on the ground; others may pupate under stones or beneath the bark of a nearby tree, without forming a cocoon; during the summer months many pupate above ground, forming a cradle of vine leaves pulled together, usually in the notch of a stem. This stage lasts from fifteen days to five months, depending on the climate. Overwinters as a pupa.

PARASITOIDS

See Table 4, p. 54.

BREEDING

Pairing. As for *Deilephila elpenor* (q.v., p. 157).

Oviposition. As for *Deilephila elpenor*.

Larvae. As for *Deilephila elpenor* but must be reared in bright conditions with temperature above 25°C or many resulting adults may be sterile.

Pupation. As for *Deilephila elpenor* but do not spray with water; cannot tolerate dampness nor temperatures below 0°C over long periods during the winter. If possible, pupal emergence should be stimulated by maintaining well-ventilated heat (over 25°C).

DISTRIBUTION

A species which is partially migrant, individuals having been found as far west as Sicily (Spuler, 1908, although Mariani, 1939, doubts this) and north to Romania (Fleck, 1901); resident in south-western Bulgaria (Ganev, 1984). It is regularly taken in Corfu, where it may be a resident. Mainly from Greece (Koutsaftikis, 1970; 1973; 1974) across Turkey (Daniel, 1932; de Freina, 1979) to most of Iran (Bienert, 1870; Kalali, 1976), Turkmenistan and Uzbekistan (Grum-Grshimailo, 1890; Derzhavets, 1984) and Afghanistan (Ebert, 1969; Daniel, 1971), and south to Iraq (Wiltshire, 1957), Lebanon (Zerny, 1933; Ellison & Wiltshire, 1939), Israel (Eisenstein, 1984) and the more fertile areas of Egypt (Badr *et al.*, 1985).

Extra-limital range. India to Taiwan and the Indonesian Islands.

Appendix 1 – REARING HAWKMOTHS

Hawkmoths are bred to obtain perfect specimens, for scientific study, or simply for pleasure. For successful breeding, it must be remembered that hawkmoths will thrive only if kept under conditions as close as possible to those they would experience in the wild (Sokoloff, 1984; Friedrich, 1986). These comprise plenty of good quality food, fresh air, warmth and daylight. They should also be afforded protection from predators and parasitoids.

Obtaining livestock

One method of obtaining livestock is to search for eggs, caterpillars or chrysalids, but this is productive for only a limited number of species and requires a good knowledge of the ecology of the species sought.

1. Eggs (*Ova*)

The eggs of most species are difficult to find, and the amount of effort needed to discover them is rarely worthwhile. Exceptions include those of *Smerinthus ocellatus* and *Laothoe populi*, which can easily be seen at the right time of year (late May to early July in southern England) by gently twisting over the top shoots of small sallow bushes or the lower shoots of poplar trees and examining the underside of the leaves. The same method applies to the conspicuous eggs of *Rethera komarovi* which are laid on the flowering tips of *Rubia*, or those of *Hemaris tityus* or *H. fuciformis* on the underside of leaves of *Knautia* and *Lonicera* respectively. Where colonies are known to exist, upward of 20 eggs can be found in a day.

2. Caterpillars (*Larvae*)

Larvae are much easier to find than eggs. For species that feed on bushes and small trees, such as *Sphinx ligustri*, *Laothoe populi* and *Smerinthus* species, young (early-instar) larvae may also be found under leaves by twisting over the shoots on suitable plants. Small larvae of all *Hemaris* species chew very characteristic holes on either side of a leaf midrib. Larger larvae leave denuded twigs and, sometimes, very noticeable deposits of frass (droppings) on the ground. Species that feed on herbaceous plants may require a slightly different approach. Many, such as *Rethera komarovi*, *Deilephila elpenor* and *Hyles vespertilio*, hide away at the base of the hostplant by day and are best sought out by looking for denuded shoots or searching with a torch at night. The former method is best for finding the caterpillars of *Acherontia atropos* in potato fields. By contrast, the bright warning colours of aposematic caterpillars, such as those of *Hyles euphorbiae* and *H. nicaea*, make them much easier to see and several may be found by a thorough search of a small area. However, it should be remembered that although larger caterpillars are more conspicuous, as a result of predation there will be fewer of them, and many may produce parasitoids rather than the desired moth.

3. Chrysalids (*Pupae*)

In suburban areas, pupa-digging can be a most worthwhile winter activity, particularly for chrysalids of *Mimas tiliae*, *Smerinthus ocellatus* and *Laothoe populi*. These species generally pupate within 50cm of the base of the host-tree (*Tilia*, *Salix*, *Populus* or *Ulmus*) and can be dug out with a trowel. The best trees are those growing in soft earth covered with a layer of loose turf. Pupae of *M. tiliae* are found in or just under the turf, those of the other two species will be a little deeper. Unfortunately, this method is not very productive in rural areas nor for those species which feed on herbaceous plants.

4. Adults (*Imagines*)

The most satisfactory way of obtaining good breeding stock in quantity is to capture a gravid female moth. A mercury-vapour light-trap operated in the right area at the right time of year will capture most nocturnal species. Some individuals may be caught while feeding at flowers or after having been attracted to street- or other bright lights. Very few are discovered by accident while at rest. The sexes can readily be distinguished: female moths are generally more heavy-bodied, bear smooth filiform antennae and have an abdomen terminating in an anal pore; males are of a lighter build, have antennae whose lower surface is covered with bristle-like platelets and have two terminal abdominal claspers.

Although less satisfying and more expensive than finding one's own livestock, purchasing it is often the only way of obtaining rare or foreign species. Several entomological dealers and individual breeders sell hawkmoths at all stages, and many species can be found by scanning the lists of suppliers and entomological societies. A list of sources of western Palaearctic material is given below (p. 177).

Rearing livestock

Hygiene is of vital importance when rearing livestock in captivity. In all cases and for all stages, cages, sleeves and containers should always be sterile. They should be washed, disinfected and dried thoroughly before use.

1. *Pairing*

Most western Palaearctic species, being nocturnal, are relatively easy to pair if certain conditions are observed. Having sexed the moths (see above), select up to three pairs of the same species; do not mix species unless hybrids are required. Nearly all will pair in a suspended cylindrical net cage of 60 × 80cm (Text fig. 52). Net cages such as those used for rearing larvae are also suitable (see p. 170); for species such as *Sphinx ligustri* and *Acherontia atropos*, a 60 × 60cm rough wooden box covered with netting will be quite satisfactory. If the cage is placed outside, make sure that cats and birds cannot reach it and avoid brightly lit or extremely dark locations. An optimum temperature of 10–20°C should be maintained. Most species of Sphinginae remain paired (*in copula*) for about 20 hours but the Macroglossinae generally separate after 2–3 hours.

Diurnal species are much more difficult to pair, but not impossible. A large flight cage (minimum 2 × 2 × 2m) is required, stocked with plenty of hostplants and nectar-bearing flowers, and with access to sunshine. The plants can be grown either in the ground or in pots. A moth-house (as described on p. 172) is ideal for the *Hemaris* species and also for many of the nocturnal Macroglossinae.

The following advice and information may be useful.

(a) Never separate or disturb pairs *in copula*. If they are separated prematurely the resulting eggs may be infertile.

(b) Never use males for pairing more than twice – their 'spermbank' becomes depleted. Old and exhausted males often pair for much longer than fresh males, i.e. up to three days.

(c) Always make sure that a weak directional wind is blowing through the cage. Females 'call' the males using sex pheromones (see below under Assembling).

(d) Species which feed do not generally pair until the second or third night after emergence, or even later.

(e) There is no need to include sprigs of hostplant in the cage. A few flowers for the feeding species will suffice and even these are usually not required.

(f) In confined cages it is best to force-feed those adults that feed as they rarely feed themselves. Commence this on the second night after emergence (see p. 169).

2. Assembling

If a male is required for an unpartnered, virgin female, for the cabinet, or to obtain new blood for stock, 'assembling' is advocated. Female hawkmoths attract a mate by means of wind-borne sex pheromones. A virgin female will, if placed in an air-current out-of-doors at the right time of night and year, draw many males to her. This also applies to the day-flying *Hemaris* species. Some breeders tether the female to a piece of wood or gauze using a cotton thread looped around her waist, and then rise before sunrise to bring their hopefully paired captive to safety and shelter – sparrows are very partial to hawkmoths! Another method (not to be recommended as it is extremely drastic) is sometimes employed for the Smerinthini and entails clipping off most of the female's wings and placing her in an open window or dense shrub to call. The ideal method is to use special 'assembling cages' with one-way cones which allow males to enter but prevent females from escaping. However, hanging cylinder cages (see Text fig. 52, p. 171) with the zipper left open can be equally successful.

3. Oviposition

In general, most hawkmoths will lay eggs given the same conditions as for pairing, i.e., in cylindrical emergence cages, larval rearing cages or outside flight-cages. In the last, moths can be left both to pair and lay eggs in the same cage. With net cages, it is best to transfer mated females to another cage where they can oviposit undisturbed. The non-feeding Smerinthini will deposit all their eggs over a period of four or five nights in almost any net cage or net-lined container. They do not require the presence of the hostplant. However, the nocturnal feeding Sphingini and Macroglossinae require rather different conditions for egg-laying. Use a net cage at least 60 × 80cm, preferably larger for the tropical species (2 × 1m). Include a few sprigs of hostplant, with some touching the top of the cage, or better still, a potted hostplant. Do not overcrowd the females as the stress caused will reduce the numbers of eggs laid.

These moths will need to be force-fed nightly. Prepare a 10 per cent solution of white sugar laced with a dash of honey. This can be placed in a small bottle-top with a piece of absorbent paper underneath for the moth to rest on. Alternatively, a cotton-wool ball placed in a saucer of solution can be used. Grasp the moth on each side of the thorax with the wings held vertically over the back. Gently unroll the proboscis with a long pin and place the end of it in the sugar solution while allowing the moth to 'taste' the sugar with its sensitive forelegs. The moth will indicate it is feeding by probing the solution. At this stage it can be released. It will signal it has finished either by coiling up its proboscis or by walking away. Wash the moth after feeding by dipping it in fresh water, especially the forelegs.

Species of tropical origins, such as *Agrius convolvuli*, *Acherontia atropos*, *A. styx*, *Daphnis nerii* and *Hippotion celerio*, generally require temperatures above 20°C to oviposit successfully. These species usually lay their eggs over 1–2 weeks, so patience is required. Some, such as *D. nerii*, rarely lay more than a few dozen eggs. Once the eggs are laid they can be handled as recommended below.

4. Keeping eggs

For most species, keeping eggs is fairly straightforward. Small clear acetate (plastic) boxes make the best storage receptacles, although glass specimen-tubes can also be used. However, plastic containers are always better than glass; the latter are colder and, as the larvae cannot obtain a grip on their smooth surfaces, a piece of absorbent paper must always be inserted with the eggs for the larvae to climb. *Never* include plant material as this often leads to the formation of fatal mould. This can also arise if too many eggs are placed in one container as eggs give off water-vapour. An excessive number of eggs in a container will also result in overcrowding among the newly-hatched larvae, causing them to damage each other.

If possible, *always* remove the eggs from whatever surface they were laid on (except a potted hostplant) and place them in a container. Within a few days of being laid they can quite safely be eased off netting, a leaf or paper with a fingernail. However, one western Palaearctic species has to be treated differently. In its first instar, the larva of *Marumba quercus* eats nothing except, sometimes, the egg-shell. The eggs of this species need to be stuck (one per leaf) with wood glue on the underside of the leaves of a potted hostplant, or on a sheet of paper, 3cm apart. If using the latter substrate, it is essential that newly-emerged larvae are transferred to individual leaves within twelve hours of hatching to allow them to moult. If overcrowded, they will lash about quite violently and may severely injure other larvae.

Fertile eggs will change colour before hatching and the young caterpillars can be seen through the shell. They will hatch in from four to thirty days, depending on the species and the temperature. Where possible, avoid handling eggs. Use a paint brush to move them around. An infertile egg usually collapses and dries out after a week. Such eggs have a very characteristic appearance which should not be confused with the shallow depression that develops naturally on the eggs of some species.

5. *Rearing larvae*

After hatching, the larvae of most species pass through five instars before pupating; *Marumba quercus*, however, has six instars and species of *Laothoe* often have only four. This is essentially the feeding and growing stage in the insect's life and nearly all of its existence is dedicated to finding food while itself avoiding becoming the food of a predator.

Although under natural conditions many species are confined to only a few hostplants, in captivity they will often accept substitutes, e.g. *Daphnis nerii* will accept *Ligustrum* and *Hyles euphorbiae* will accept *Epilobium* and *Vitis*. Alternatively, larvae can be fed on synthetic diets (Gardiner, 1978; 1986), but these are beyond the scope of most amateurs and are principally used to rear pathogen-free livestock for research.

Newly-hatched larvae may be started off in small plastic containers (i.e. ice-cream or sandwich boxes) or glass jars (Text fig. 54) but are best transferred to cages or sleeved (see below) midway through the second instar. Small numbers can be reared in closed plastic containers but no container of any kind should be exposed to direct heat or sunlight.

(i) Net Cages

These may be either wood- or metal-framed containers with textile netting or wire-mesh walls and top, a solid plywood bottom and a door. They provide good ventilation, allow cut food to be changed with the minimum of fuss and can double as emerging/pairing cages. They also restrain larvae which leave the plant by day. A good size is 60 × 45 × 35cm, but there are larger and smaller sizes available. Several firms manufacture them, although they are not difficult to make. Both cages and sleeves should be disinfected after each use.

In net cages, use either potted plants or cut twigs in a bottle of water. The former will require daylight to keep them healthy, and should be under-watered, under-fertilized and never used more than twice a year – replacement leaves often contain high concentrations of natural toxins to deter or kill would-be browsers (see p. 47). If using cut twigs, do not expose them to direct sunlight and never put so many twigs in a bottle that the water supply becomes exhausted while unattended. Wilted twigs rarely recover. Some cut plants wilt very rapidly, even in water, e.g. *Betula* and *Hippophae*. However, their freshness (and that of many plants) can be extended by using distilled instead of tap water. To avoid larvae crawling down into the bottle and drowning themselves, plug the top with cotton wool or tissue paper. This also stops frass contaminating the water. To recharge the bottle, one can either unplug it and pour in more water or place a large straw, topped with a small plastic funnel, among the twigs and pour water through that, which saves time and irritation.

Some species, such as *Sphinx pinastri*, *Rethera komarovi* and many of the *Hyles* genus, will not do well if their hostplant is standing in water and, if fed in this way, may perish through

Text Figure 52 (top left)
Hanging cylinder cage
for pairing or emerging

Text Figure 53 (top right)
A brown polyester sleeve
on a poplar sapling

Text Figure 54 (left)
Individual glass and
plastic rearing containers

'waterlogging'. This applies to a lesser extent to most of the other Macroglossini.

Laothoe populi and *Deilephila elpenor*, among other species, need to drink if kept too dry, so spray the hostplant regularly.

To transfer larvae to a new batch of twigs, lay the old twigs on top of the new ones and the larvae will transfer themselves. Once again, *M. quercus* is an exception. Under artificial conditions, its larvae do not recognize leaves which have dried out, continuing to feed on badly desiccated material, so transfer them as necessary. Extra twigs can be stored in a plastic bag in the refrigerator for up to five days to avoid having to collect food frequently.

(ii) Sleeving

The other method of keeping larvae – sleeving – is far superior and should be used whenever possible. A sleeve is simply a large tube of netting placed over a growing branch, twig or potted plant and tied at both ends (Text fig. 53). It allows larvae to be reared under the most natural conditions possible. Many materials may be used. Cotton-muslin is the cheapest, but it is dense and retains too much water after rain, which can severely chill the atmosphere inside. Nylon degrades when exposed to sunlight, but it is reasonably cheap and does not retain water. Best of all is polyester (Terylene). It is light, tough, durable, non-absorbent and allows a great deal of sunlight through it. If possible, use black or brown material in colder areas as this absorbs heat thereby raising the internal temperature by several degrees. Also, the finer the mesh the warmer and less windy it will be inside. Ladies' tights (pantyhose) or stockings, suitably cut up, make excellent small sleeves.

When using sleeves there are various points to remember. Leave enough slack in the sleeve to accommodate growing plants and use fine cotton or hessian string to tie it. Plastic string slips and may become undone, and elastic bands perish and break too readily. Always place the sleeve in a sunny, sheltered location, but not under a tree. Always change a sleeve before the food reserves are depleted. When changing sleeves on branches, transfer larvae into the new sleeve on a leaf or small piece of twig. If using potted herbaceous plants, it may be necessary to support the sleeve internally with canes pushed into the soil. Do not allow too much frass to accumulate in the sleeve.

Commercial or dedicated amateur breeders may use walk-in net aviaries and net-lined greenhouses planted with hostplants to rear large numbers of larvae. Provided they are well protected against spiders and parasitoids, these 'moth-houses' are excellent, especially for tropical or sub-tropical species.

(iii) General

Whether in a cage or sleeve, most larvae require warm, dry conditions or even direct sunshine. This is particularly important for such southern species as *Marumba quercus*, *Rethera komarovi* and *Hyles nicaea* which come from dry environments. These will become sluggish, cease growing and feeding, and often waste away and die if kept cool and humid. A dry atmosphere will also allow them to lose any excess water they take up from the hostplant. Many such species will thrive if sleeved on a potted hostplant in a sheltered and sunny location out-of-doors, even during a normal British summer. However, it is better still to rear them in a greenhouse or place the potted plant in a sunny window.

All larvae should be protected from predators and parasitoids (see p. 176). Most of these are large and can be kept out of sleeves and cages by using 1–2mm mesh. Make sure all spiders, predatory bugs, earwigs and lacewing-fly larvae are removed from twigs, potted plants and cages before they are used. Even in a greenhouse, larvae should be caged or sleeved as most glasshouses contain a large population of spiders. Also, beware of tachinid flies; they can lay their eggs on a sleeve and the resulting parasitic larvae will crawl in and seek out the caterpillars.

The approach of pupation is heralded by a change in larval behaviour: the larva stops feeding and remains quiescent for up to a day. If touched, most will twitch violently from side to side, with head almost touching the tail horn. Green larvae usually acquire a brownish hue which darkens dorsally to form a broad purple band. All, whatever their colour, shrink a little in size and most void their gut contents as a brown, sticky fluid. At this stage, the larva begins to wander, first off the hostplant and then, in most species, in search of a suitable pupation site some distance from the hostplant.

Most larvae are relatively easy to rear provided certain principles are followed.

(a) Always maintain good hygiene. If using a closed container, line the bottom with newspaper or absorbent paper to absorb condensation every time the hostplant is changed.

(b) Never handle caterpillars. Always transfer small larvae with a sable-hair paint-brush and larger individuals on a twig or leaf. Moulting larvae must never be removed from their base. If it is necessary to move one, lay the larva and its attached base (be it a leaf or twig) on the cage-floor and leave it to change and eat its cast-off skin undisturbed.

(c) Never crowd larvae. Many react violently when disturbed and can injure other individuals; moreover, the stress caused often results in smaller than normal adults. Overcrowded larvae may also rapidly deplete their food source and resort to cannibalism, particularly of the tail horns. Crowding also promotes disease.

(d) If rearing in closed containers or cages, always ensure a constant supply of good quality food, replacing it every one to three days with just enough new material; too many leaves will create unhealthy, humid conditions. Prepare twigs and leaves on a clean sheet of newspaper to avoid contamination. Never collect twigs from, or sleeve larvae under, other trees or bushes. Such twigs are often contaminated with harmful viruses, bacteria and/or honeydew from above. Similarly, never gather twigs from plants bordering agricultural fields as they may have become contaminated with pesticides. If in doubt, wash and dry the leaves before use.

(e) Never use wilted food.

(f) Never keep more larvae than you have hostplant for. Be ruthless – dispose of any surplus in the first instar.

(g) Destroy any sick larvae immediately (see p. 176).

6. *Pupation*

When full-grown, almost without exception, hawkmoth caterpillars pupate in soil or just below the surface litter and provision has to be made for this.

For pupation, the larva must be transferred to a suitable container in which it can burrow. Use an open-topped (well-ventilated) box or plastic bag half filled with damp loam (a 2 : 2 : 1 mix of soil, peat or leaf-mould, and washed sand) and covered with a layer of moss, absorbent paper or dry peat. Bags of potting compost from a garden centre are ideal. Place six to eight caterpillars to a 3-litre bag and they will quickly burrow from sight.

Some species can be induced to pupate in individual soil-less tubs or foam-polystyrene cups. Put the larva into the sealed container (lined daily with fresh absorbent paper or it will become wet and soggy) and then, when the larva has shrunk and become quiescent, remove all the paper and make sure the larva is lying on a flat, even surface. It will pupate in this position. However, any irregularities in the container's surface will imprint themselves on the pupa and may cripple it. This technique works well for some species, such as *Acherontia atropos*, but humidity variations are a problem. *Marumba quercus* will not tolerate dampness, whereas newly-formed pupae of *Laothoe populi*, *Smerinthus ocellatus* and *Mimas tiliae* dehydrate severely in a dry atmosphere. Appendages often fail to extend to their full length, leaving shiny, bald regions.

Three to four weeks after larvae have pupated they should be dug up and checked to see that all are alive before they are stored. (The exception to this is *Proserpinus proserpina* which suffers from disturbance and should be left in the pupation medium until spring.) Many species have very motile pupae which will wriggle if touched. Others will respond only if a fingernail is pressed gently into the soft intersegmental cuticle of the abdomen. Before storing, or if an infection is found, sterilize all pupae in a mild solution of disinfectant for five minutes and then dry them thoroughly. Apart from this, try to avoid handling chrysalids but if it is necessary to do so use cotton gloves or soft paper. Some species, such as *S. ocellatus*, will die if handled excessively. The most suitable storage containers for pupae are sealed plastic boxes or tins, lined with absorbent paper, which should be kept in a cool, shady place. The contents should be checked at least once a week and any dead individuals removed, the remainder then being transferred to a sterile container. Any severely malformed pupae should also be discarded but mildly deformed pupae usually produce healthy adults. Never spray pupae stored in closed containers with water.

Most western Palaearctic species diapause as pupae and emerge in the spring. The major exception is *Macroglossum stellatarum*, which overwinters as a hibernating adult and requires specialized treatment (see p. 134). However, those species which produce more than one generation a year also produce non-diapausing pupae which will hatch after 7–30 days. Some species also diapause for only short periods of less than two months, e.g. *A. atropos*. Non-diapausing pupae should be checked more frequently and, if soft and dark and still alive, be placed in an emerging cage.

Avoid sudden changes in temperature. Diapausing pupae should be allowed to cool naturally in the autumn in a cool shed, greenhouse or sheltered spot in the garden. Make sure, however, that the container is proof against mice, birds, hedgehogs and children! Most species can be safely left in such locations all winter and will benefit from exposure to frosts. However, the more sensitive species of tropical origin, such as *Agrius convolvuli*, *Acherontia atropos* and *Hyles livornica*, cannot diapause for very long and will die if exposed to severe cold spells (i.e. lower than −4°C); therefore they should be protected. These species may also die under prolonged damp conditions. *Daphnis nerii* and *Hippotion celerio* do not overwinter but may enter diapause during the hot summer months. Other species, especially those from arid areas, may diapause for up to three years or more years.

Important points to remember are as follows.

(a) Never overcrowd pupating larvae. Many will try to pupate in the same corner and will end up crushing each other. Larger individuals should be given their own small containers, e.g. a glass jar.

(b) Never use a closed plastic container for pupation. Many pupae will perish of fungal diseases in these.

(c) Provide at least 20cm depth of loam for pupation.

(d) Ensure that the loam is damp, but not wet. Conversely, do not allow it to dry out.

(e) Place the pupation container in a cool, shaded place and do not move containers that are not rigid, as this may cause the earthen pupal chambers to disintegrate and result in malformed pupae.

(f) Secure the top of the pupation containers with netting to prevent the escape of any restless larvae.

(g) Resist the temptation to look before three weeks have elapsed. The larvae need time to construct a chamber, shrink, change into a pupa and then harden off, before being handled.

(h) When the pupae are fit to be handled, their sex may be determined by examination of the ventral side of the abdominal segments (see Text fig. 55, p. 175).

7. *Emergence*

With the arrival of warmer weather in the spring, most pupae will develop and the moths emerge in April or May under natural climatic conditions. This may, however, be inconvenient for the breeder and some species can be 'forced' to emerge earlier than usual by exposing them to warmth. This is a successful technique for the *Hemaris* species but not for *Smerinthus*. Most pupae *must* undergo a set period of cooling before they will emerge and will not do so if warmed too early or if kept too warm during the winter. The more tropical species, however, will develop and emerge at any time if warmed up. Remember, however, that forced adults are nearly always sterile.

Alternatively, it may be desirable to delay emergence. In this case, place the pupal container in a refrigerator (at 4–6°C) during December/January and bring them into the warmth (15–29°C) about three weeks before the moths are required. However, do not expose these pupae to instant heat. Acclimatize them for a week in a cool location.

About eight days before emergence, the chrysalids will darken in colour and, after four or five more days, become soft. At this stage, or even earlier, they should be placed in an emergence cage. The net cages in which larvae are reared, or suspended cylindrical net cages (see Text fig.

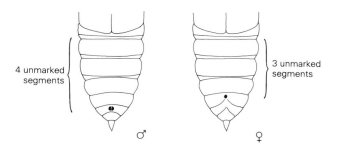

Text Figure 55
Ventral view of male and female pupae. Between the wings and the marks indicating future genital openings are, in the male four, and in the female three, unmarked segments (after Friedrich, 1986)

4 unmarked segments

3 unmarked segments

♂ ♀

52, p. 171), which can also double as pairing cages, are best for this purpose. Emergence cages must have the following features.

(i) Absorbent paper, peat/leaf-mould, or moss placed in the base to absorb any excess moisture or voided meconium.

(ii) Rough walls of netting or unplaned wood or, alternatively, a supply of dead twigs.

(iii) Good ventilation and a shaded position.

(iv) Easy access via a removable top, side or zipper.

Under humid, cool conditions pupae may be laid on the moss or absorbent paper at the bottom of the cage and sprayed once a day with water. Under dry conditions the pupae will need to be buried in the moss or laid between layers of paper, which should then be wetted every day. It is better still to leave pupae in their storage container until they soften. Remember that species which burrow in soil or live in damp areas are more prone to desiccation than surface-pupating species from dry zones.

To emerge, the imago splits the pupal skin around the antennae and across the thorax and wriggles out. Just as it clears the pupal case it may void a quantity of meconium and then crawl up a suitable surface to expand its wings. During this process, which generally takes 30–90 minutes, the moth should not be disturbed. The wings then take a further 3–4 hours to dry and harden. Most species emerge in the late morning or early afternoon so that their wings have enough time to dry before the evening flight. Most also prefer to emerge when the temperature is between 10–15°C.

The following suggestions may be useful.

(a) Never put too many pupae in one cage as the wings of the emerging adults can become damaged due to overcrowding.

(b) Never use the same cage simultaneously for emerging and pairing moths.

(c) Having emerged, most adult moths will retain a great deal of meconium in their body which they will squirt at any 'attacker'. To avoid any mess they can be held firmly, with the thorax gripped between forefinger and thumb, and 'milked' over absorbent paper.

(d) Do not kill any specimens for the collection until they are at least 12 hours old. This allows the wings to harden.

(e) Moths which have failed to escape properly from the pupal skin and become crippled can still be used for breeding.

Mailing livestock

The established method of mailing eggs is to place them in a length of rigid or thick plastic tubing, sealed at each end with cotton-wool. This can be taped to a letter inside an envelope and posted. Larger quantities of eggs can be sent in small plastic or metal containers packed in padded bags or boxes.

Caterpillars are more difficult to mail and it is recommended that only first- or second-instar larvae be sent by post. Always use a metal container (e.g. a tobacco tin) lined with absorbent paper which will accommodate enough hostplant to last the trip. Tape the box shut and place it in a well-padded bag or box. Always employ an express postal service for long distances. If sending larvae from hot climates, remember to ensure that the box or bag is not left in the sun or subjected to high temperatures even for short periods. If possible, place the package in a refrigerator until it can be posted.

Pupae are almost as easy to send as eggs. Use the same method as recommended for larvae, but fill the box with cotton-wool or soft paper and wrap each pupa individually. Make sure that the pupae cannot rattle or slide around – 50 per cent of a consignment of *Acherontia styx* were once lost by the author in this manner. As with larvae, diapausing pupae of cold-tolerant species can also be stored in a refrigerator until posting. Conversely, for cold-sensitive species, such as *Daphnis nerii* and *Hippotion celerio*, insulate the package against low winter temperatures that may be experienced in transit.

Never send live adults through the post. Unlike butterflies, adult hawkmoths cannot be adequately restrained without killing them.

Parasitoids, predators and diseases

The larval stage is the most vulnerable to parasitism and disease. Many wild-collected larvae will be parasitized and there is little that one can do about this. The parasitoid should, however, be preserved for identification, with full data of its host-species, date of emergence and place found. Parasitoids and predators can be excluded from bred material by using sufficiently fine netting or by rearing the livestock indoors.

Disease is the main cause of losses to breeders, particularly of caterpillars. An outbreak of viral or bacterial disease can be rapid and catastrophic, leading to the loss of an entire stock if not treated immediately. However, many species are partially immune to the common diseases and tend to succumb only if subjected to stress, liable to be caused by one or more of the following factors:

   (i) overcrowding;

   (ii) excessive handling;

  (iii) poor ventilation and/or excessive humidity;

  (iv) lack of warmth;

   (v) too much rain;

  (vi) poor quality food;

 (vii) contaminated food;

(viii) poor hygiene, i.e. lack of cleaning;

  (ix) inbreeding and genetically weak stock;

   (x) contact with diseased stock.

The symptoms of disease vary. Eggs, kept under very humid conditions, develop a layer of mould on them which good ventilation and a rubbing normally cures. In larvae, of the two types of disease which occur, viral infections are relatively rare and not very contagious, unlike most bacterial infections. A larva infected with a polyhedrosis virus tends to stop feeding, shrink and harden; when infected by bacteria, a larva not only generally stops feeding but also loses its turgidity and ability to grip the hostplant tightly. A pale brown, watery diarrhoea appears around the anal flap and drops of highly contagious liquid may be voided. Thereafter the larva may either crawl around the sleeve or cage voiding dark fluid or simply turn into a bag of dark liquid and drop from the plant. Bacterial disease can run through stock like wildfire. Similar infections occasionally occur at the pupal stage and are indicated by an unpleasant smell and/or darkening of the cuticle. They tend to strike at moments of stress.

Fungal infection is nearly as big a problem, especially with pupae kept under poorly-ventilated, humid conditions. The first sign is usually a dark patch around one of the spiracles. The pupa then becomes lighter in colour and rigid. When broken open it will be found to be filled with a mass of white or blue-grey fungus.

The treatment of any infection must be immediate and ruthless if the loss of the entire stock is to be avoided. Certain steps can be taken to prevent infections taking hold or spreading.

(a) Always sterilize containers, forceps, paint-brushes, etc. at the end of the season or after handling each species.

(b) Wash your hands before and after handling livestock and always prepare cut hostplants or transfer larvae/pupae on clean newspaper.

(c) Destroy any larvae or pupae at the first sign of an infection by burning them or dropping them into a strong solution of disinfectant. Never throw them outside as sporolated pathogens can survive long periods in the soil.

(d) Immediately change all the hostplant and/or sterilize the container. Prune potted plants severely to remove contaminated leaves.

(e) Isolate the remaining stock and repeat the above if required.

(f) Always soak bought pupae in a mild solution of disinfectant for five minutes on arrival to kill any disease spores on their exterior. Dry them afterwards.

(g) Never store pupae in a big mass. Always put them in single layers (up to a maximum of four) between absorbent paper.

(h) With rare or prized species it is a good idea to divide stock into a number of containers or groups to reduce the risk of losing them all.

Suppliers of sphingid livestock

The addresses in the following list of sources of livestock are correct at the time of publication:

Amateur Entomologists' Society, 22 Salisbury Road, Feltham, Middlesex TW13 5DP (quarterly list of wants and exchanges)

The Entomological Livestock Group, c/o John Green, 11 Rock Gardens, Aldershot, Hampshire GU11 3AD (bi-monthly wants and exchange lists)

Herr Martin Geck, Söllereckstraße 9, D-8900 Augsburg, Germany
(annual list of livestock)

'Insektenbörse', *Entomologische Zeitschrift*, Alfred Kernen Verlag, Husmannshofstraße 10, Postfach 10 20 43, D-4300 Essen, Germany (bi-monthly list)

The Living World, Seven Sisters Country Park, Exceat, Seaford, East Sussex BN25 4AD (annual catalogues)

Nigel South, Butterfly Connections, Caribana, Silver Street, Misterton, Somerset TA18 8NH (several lists and catalogues per year)

Worldwide Butterflies Ltd., Compton House, Over Compton, Sherborne, Dorset DT9 4QN (annual catalogues)

Appendix 2 – HOSTPLANTS

Plant names are taken from Polunin (1969), with nomenclature updated in accordance with Mabberley (1987).

A. *Genera of sphingid hostplants*

This appendix lists the principal sphingid hostplants by genus, with English name, family and order for each:

GENUS	ENGLISH NAME	FAMILY	ORDER
Acacia	mimosa	Leguminosae	Fabales
Acer	maple, sycamore	Aceraceae	Sapindales
Actinidia	kiwifruit/Chinese gooseberry	Actinidiaceae	Theales
Adenium	desert rose	Apocynaceae	Gentianales
Aesculus	horse chestnut	Hippocastanaceae	Sapindales
Alnus	alder	Betulaceae	Fagales
Ampelopsis	ampelopsis	Vitaceae	Rhamnales
Amsonia	blue star	Apocynaceae	Gentianales
Antirrhinum	snapdragon	Scrophulariaceae	Scrophulariales
Apocynum	dogbane	Apocynaceae	Gentianales
Arisaema	arum	Araceae	Arales
Arum	arum	Araceae	Arales
Arbutus	strawberry tree	Ericaceae	Ericales
Asclepias	milkweed, swallow-wort	Asclepiadaceae	Gentianales
Asperula	woodruff, squinancywort	Rubiaceae	Rubiales
Asparagus	asparagus	Liliaceae	Liliales
Asphodelus	asphodel	Liliaceae	Liliales
Begonia	begonia	Begoniaceae	Violales
Beta	beet	Chenopodiaceae	Caryophyllales
Betula	birch	Betulaceae	Fagales
Boerhavia	—	Nyctaginaceae	Caryophyllales
Buddleja	buddleia	Loganiaceae	Gentianales
Caladium	elephant's ear	Araceae	Arales
Calla	water arum	Araceae	Arales
Carissa	Natal plum	Apocynaceae	Gentianales
Castanea	sweet chestnut	Fagaceae	Fagales
Cedrus	cedar	Pinaceae	Pinopsida
Centranthus	red valerian	Valerianaceae	Dipsacales
Cephalaria	—	Dipsacaceae	Dipsacales
Chamaenerion (see *Epilobium*)			
Chrysanthemum	chrysanthemum	Compositae	Asterales
Cicer	chick-pea	Leguminosae	Fabales
Cinchona	quinine tree	Rubiaceae	Rubiales
Circaea	enchanter's nightshade	Onagraceae	Myrtales
Cissus	kangaroo vine	Vitaceae	Rhamnales
Clarkia	clarkia/godetia	Onagraceae	Myrtales
Clerodendrum	pagoda flower	Verbenaceae	Lamiales
Convolvulus	convolvulus	Convolvulaceae	Solanales
Cornus	dogwood	Cornaceae	Cornales
Corylus	hazel	Betulaceae	Fagales
Cucurbita	pumpkin, gourd, marrow	Cucurbitaceae	Violales
Daphne	daphne, mezereon	Thymelaceae	Myrtales
Daucus	carrot	Umbelliferae	Apiales
Deutzia	deutzia	Hydrangeaceae	Rosales
Dipsacus	teasel	Dipsacaceae	Dipsacales
Duranta	—	Verbenaceae	Lamiales

GENUS	ENGLISH NAME	FAMILY	ORDER
Elaeagnus	oleaster, Russian olive	Elaeagnaceae	Proteales
Epilobium	willow-herb, fireweed	Onagraceae	Myrtales
Eremurus	foxtail lily	Liliaceae	Liliales
Euonymus	spindle-tree	Celastraceae	Celastrales
Euphorbia	spurge	Euphorbiaceae	Euphorbiales
Forsythia	forsythia	Oleaceae	Scrophulariales
Fraxinus	ash	Oleaceae	Scrophulariales
Fuchsia	fuchsia	Onagraceae	Myrtales
Galium	bedstraw	Rubiaceae	Rubiales
Gardenia	gardenia	Rubiaceae	Rubiales
Geranium	geranium	Geraniaceae	Geraniales
Gossypium	cotton	Malvaceae	Malvales
Helianthus	sunflower	Compositae	Asterales
Hippophae	sea buckthorn	Elaeagnaceae	Proteales
Ilex	holly	Aquifoliaceae	Celastrales
Impatiens	balsam	Balsaminaceae	Geraniales
Ipomoea	morning glory, sweet potato	Convolvulaceae	Solanales
Jasminum	jasmine	Oleaceae	Scrophulariales
Jaubertia	—	Rubiaceae	Rubiales
Juglans	walnut	Juglandaceae	Juglandales
Knautia	scabious	Dipsacaceae	Dipsacales
Kniphofia	red-hot poker	Liliaceae	Liliales
Kohautia	—	Rubiaceae	Rubiales
Lantana	lantana	Verbenaceae	Lamiales
Larix	larch	Pinaceae	Pinopsida
Laurus	bay laurel, bay tree	Lauraceae	Laurales
Ligustrum	privet	Oleaceae	Scrophulariales
Lonicera	honeysuckle	Caprifoliaceae	Dipsacales
Lychnis	campion	Caryophyllaceae	Caryophyllales
Lysimachia	yellow loosestrife	Primulaceae	Primulales
Lythrum	purple loosestrife	Lythraceae	Myrtales
Malus	apple	Rosaceae	Rosales
Mangifera	mango	Anacardiaceae	Sapindales
Menyanthes	bogbean	Menyanthaceae	Solanales
Mercurialis	mercury	Euphorbiaceae	Euphorbiales
Mirabilis	four o'clock plant, marvel of Peru	Nyctaginaceae	Caryophyllales
Nerium	oleander	Apocynaceae	Gentianales
Oenothera	evening primrose	Onagraceae	Myrtales
Olea	olive	Oleaceae	Scrophulariales
Parthenocissus	virginia creeper	Vitaceae	Rhamnales
Pelargonium	geranium	Geraniaceae	Geraniales
Phaseolus	bean	Leguminosae	Fabales
Physocarpus	ninebark	Rosaceae	Rosales
Picea	spruce	Pinaceae	Pinopsida
Pinus	pine	Pinaceae	Pinopsida
Pistacia	pistachio, mastic tree	Anacardiaceae	Sapindales
Plantago	plantain	Plantaginaceae	Plantaginales
Polygonum	redshank, knotgrass, persicaria	Polygonaceae	Polygonales
Populus	poplar, aspen	Salicaceae	Salicales
Prunus	cherry, plum, peach, almond, blackthorn, cherry laurel	Rosaceae	Rosales
Pseudotsuga	Douglas fir	Pinaceae	Pinopsida
Pyrus	pear	Rosaceae	Rosales
Quercus	oak	Fagaceae	Fagales
Rhazya	—	Apocynaceae	Gentianales
Rhus	—	Anacardiaceae	Sapindales
Ribes	currant	Grossulariaceae	Rosales

GENUS	ENGLISH NAME	FAMILY	ORDER
Rubia	madder	Rubiaceae	Rubiales
Rubus	bramble, blackberry	Rosaceae	Rosales
Rumex	dock, sorrel	Polygonaceae	Polygonales
Salix	sallow, willow	Salicaceae	Salicales
Sambucus	elder	Caprifoliaceae	Dipsacales
Saurauia	—	Actinidiaceae	Theales
Scabiosa	scabious	Dipsacaceae	Dipsacales
Scrophularia	figwort	Scrophulariaceae	Scrophulariales
Sesamum	sesame	Pedaliaceae	Scrophulariales
Solanum	nightshade, potato, aubergine	Solanaceae	Solanales
Sorbus	whitebeam, service-tree, mountain ash/rowan	Rosaceae	Rosales
Spathodea	flame tree	Bignoniaceae	Scrophulariales
Spiraea	spiraea	Rosaceae	Rosales
Stellaria	chickweed, stitchwort	Caryophyllaceae	Caryophyllales
Succisa	scabious	Dipsacaceae	Dipsacales
Symphoricarpos	snowberry	Caprifoliaceae	Dipsacales
Syringa	lilac	Oleaceae	Scrophulariales
Tabernaemontana	mock gardenia	Apocynaceae	Gentianales
Tecoma	tecoma	Bignoniaceae	Scrophulariales
Tilia	lime/linden	Tiliaceae	Malvales
Thevetia	yellow oleander	Apocynaceae	Gentianales
Trachelospermum	—	Apocynaceae	Gentianales
Tribulus	Maltese cross/caltrops	Zygophyllaceae	Sapindales
Ulmus	elm	Ulmaceae	Urticales
Uncaria	—	Rubiaceae	Rubiales
Verbascum	mullein	Scrophulariaceae	Scrophulariales
Viburnum	viburnum, laurustinus, guelder-rose, wayfaring tree	Caprifoliaceae	Dipsacales
Vinca	periwinkle	Apocynaceae	Gentianales
Vitex	chaste tree	Verbenaceae	Lamiales
Vitis	grapevine	Vitaceae	Rhamnales
Wendlandia	—	Rubiaceae	Rubiales
Zygophyllum	bean-caper	Zygophyllaceae	Sapindales

B. *Hostplants with associated sphingid species*

This appendix lists the principal hostplant families and genera with the species of Western Palaearctic hawkmoth larvae which feed on them:

FAMILY (ORDER): GENERA	SPHINGID SPECIES
Aceraceae (Sapindales)	
Acer	*Mimas tiliae*
Actinidiaceae (Theales)	
Actinidia	*Acosmeryx naga*
Saurauia	*Acosmeryx naga*
Anacardiaceae (Sapindales)	
Mangifera	? *Daphnis nerii*
Pistacia	? *Akbesia davidi*
Rhus	? *Akbesia davidi*
Apocynaceae (Gentianales)	
Adenium	*Daphnis nerii*
Amsonia	*Daphnis nerii*
Apocynum	*Hyles chamyla*
Carissa	*Daphnis nerii*
Nerium	*Acherontia atropos, Daphnis nerii*
Rhazya	*Daphnis nerii*
Tabernaemontana	*Daphnis nerii*
Thevetia	*Daphnis nerii*
Trachelospermum	*Daphnis nerii*
Vinca	*Daphnis nerii*
Aquifoliaceae (Celastrales)	
Ilex	*Sphinx ligustri*
Araceae (Arales)	
Arisaema	*Deilephila elpenor, D. rivularis*
Arum	*Hippotion celerio*
Caladium	*Hippotion celerio*
Calla	*Deilephila elpenor*
Asclepiadaceae (Gentianales)	
Asclepias	*Daphnis nerii*
Balsaminaceae (Geraniales)	
Impatiens	*Hyles gallii, Deilephila elpenor, D. rivularis, D. porcellus, Hippotion osiris, H. celerio*
Begoniaceae (Violales)	
Begonia	*Hippotion celerio*
Betulaceae (Fagales)	
Alnus	*Mimas tiliae, Smerinthus ocellatus, Laothoe populi*
Betula	*Mimas tiliae, Smerinthus ocellatus, Laothoe populi, Hyles gallii*
Corylus	*Mimas tiliae*
Bignoniaceae (Scrophulariales)	
Spathodea	*Acherontia styx, Hippotion osiris*
Tecoma	*Acherontia styx*
Caprifoliaceae (Dipsacales)	
Lonicera	*Sphinx ligustri, Hemaris ducalis, H. fuciformis, Hemaris tityus, Deilephila elpenor*
Sambucus	*Sphinx ligustri*
Symphoricarpos	*Sphinx ligustri, Hemaris fuciformis, H. tityus*
Viburnum	*Sphinx ligustri, Smerinthus ocellatus, Laothoe populi*
Caryophyllaceae (Caryophyllales)	
Lychnis	*Hemaris tityus*
Stellaria	*Macroglossum stellatarum*
Celastraceae (Celastrales)	
Euonymus	*Sphinx ligustri*
Chenopodiaceae (Caryophyllales)	
Beta	*Acherontia atropos, Hippotion celerio*
Compositae (Asteraceae) (Asterales)	
Chrysanthemum	*Agrius convolvuli*
Helianthus	*Agrius convolvuli*

FAMILY (ORDER): GENERA	SPHINGID SPECIES
Convolvulaceae (Solanales)	
Convolvulaceae *spp.*	*Agrius cingulatus, A. convolvuli*
Convolvulus	*A. convolvuli, Hippotion celerio*
Ipomoea	*Agrius cingulatus, A. convolvuli, Daphnis nerii, Hippotion osiris*
Cornaceae (Cornales)	
Cornus	*Sphinx ligustri*
Cucurbitaceae (Violales)	
Cucurbita	*Acherontia styx*
Dipsacaceae (Dipsacales)	
Cephalaria	? *Hemaris dentata, H. fuciformis, H. croatica*
Dipsacus	*Hemaris tityus*
Knautia	*Hemaris fuciformis, H. tityus*
Scabiosa	*Hemaris croatica, H. tityus*
Succisa	*Hemaris tityus*
Elaeagnaceae (Proteales)	
Elaeagnus	*Hyles hippophaes*
Hippophae	*Hyles hippophaes*
Ericaceae (Ericales)	
Arbutus	*Hyles livornica*
Euphorbiaceae (Euphorbiales)	
Euphorbia	? *Rethera komarovi, Hyles euphorbiae, H. tithymali, H. dahlii, H. nervosa, H. gallii, H. nicaea*
Mercurialis	*Hyles euphorbiae*
Fagaceae (Fagales)	
Castanea	*Mimas tiliae*
Quercus	*Marumba quercus, Mimas tiliae, Laothoe populi*
Geraniaceae (Geraniales)	
Geranium	*Hyles gallii*
Pelargonium	*Hyles livornica*
Grossulariaceae (Rosales)	
Ribes	*Sphinx ligustri*
Hippocastanaceae (Sapindales)	
Aesculus	*Mimas tiliae*
Hydrangeaceae (Rosales)	
Deutzia	*Hemaris fuciformis*
Juglandaceae (Juglandales)	
Juglans	*Mimas tiliae*
Lauraceae (Laurales)	
Laurus	*Laothoe populi*
Leguminosae (Fabales)	
Leguminosae *spp.*	*Acherontia styx*
Acacia	*H. livornica*
Cicer	*H. livornica*
Phaseolus	*Agrius convolvuli*
Liliaceae (Liliales)	
Asparagus	*Hyles livornica*
Asphodelus	*Hyles centralasiae, H. livornica*
Eremurus	*Hyles centralasiae,* ? *H. zygophylli, H. livornica*
Kniphofia	*Hyles centralasiae* (in captivity)
Loganiaceae (Gentianales)	
Buddleja	*Acherontia atropos*
Lythraceae (Myrtales)	
Lythrum	*Proserpinus proserpina, Deilephila elpenor, D. porcellus*
Malvaceae (Malvales)	
Gossypium	*Theretra alecto*
Menyanthaceae (Solanales)	
Menyanthes	*Deilephila elpenor*
Nyctaginaceae (Caryophyllales)	
Boerhavia	*Hyles livornica*
Mirabilis	*Hippotion celerio*

FAMILY (ORDER): GENERA	SPHINGID SPECIES
Oleaceae (Scrophulariales)	
Oleaceae *spp.*	*Acherontia atropos, A. styx*
Forsythia	*Sphinx ligustri*
Fraxinus	*Acherontia atropos, Sphinx ligustri, ? Dolbinopsis grisea, ? Kentrochrysalis elegans, Mimas tiliae, Laothoe populi*
Jasminum	*Daphnis nerii*
Ligustrum	*Acherontia atropos, Sphinx ligustri, ? Kentrochrysalis elegans, Smerinthus ocellatus*
Olea	*Acherontia atropos* (Stavrakis, 1976), *? Kentrochrysalis elegans*
Syringa	*Sphinx ligustri, ? Kentrochrysalis elegans, Hyles gallii, Hippotion celerio*
Onagraceae (Myrtales)	
Circaea	*Deilephila elpenor*
Clarkia	*Hyles gallii*
Epilobium	*Proserpinus proserpina, Macroglossum stellatarum, Hyles euphorbiae, H. gallii, H. vespertilio, Deilephila elpenor, D. porcellus, Hippotion celerio*
Fuchsia	*Hyles gallii, H. livornica, Deilephila elpenor, Hippotion osiris, H. celerio*
Oenothera	*Proserpinus proserpina, Hyles vespertilio, Deilephila elpenor*
Pedaliaceae (Scrophulariales)	
Sesamum	*Acherontia styx*
Pinaceae (Pinopsida)	
Cedrus	*Sphinx pinastri*
Larix	*Sphinx pinastri*
Picea	*Sphinx pinastri*
Pinus	*Sphinx pinastri*
Pseudotsuga	*Sphinx pinastri*
Plantaginaceae (Plantaginales)	
Plantago	*Hyles gallii, H. livornica*
Polygonaceae (Polygonales)	
Polygonum	*Hyles euphorbiae, H. livornica, Deilephila elpenor, Hippotion osiris*
Rumex	*Agrius convolvuli, Hyles euphorbiae, H. livornica, Deilephila elpenor, Hippotion osiris, H. celerio*
Primulaceae (Primulales)	
Lysimachia	*Deilephila elpenor*
Rosaceae (Rosales)	
Malus	*Mimas tiliae, Smerinthus ocellatus, Laothoe populi*
Physocarpus	*Sphinx ligustri*
Prunus	*Mimas tiliae, Smerinthus ocellatus*
Pyrus	*Sphinx ligustri, Mimas tiliae*
Rubus	*Sphinx ligustri*
Sorbus	*Mimas tiliae*
Spiraea	*Sphinx ligustri, Hyles gallii*
Rubiaceae (Rubiales)	
Rubiaceae *spp.*	*Daphnis hypothous, Sphingonaepiopsis gorgoniades, S. nana, Macroglossum stellatarum, Hippotion osiris*
Asperula	*Hemaris croatica, Hyles gallii, Deilephila porcellus*
Cinchona	*Daphnis hypothous* (in India)
Galium	*Hemaris fuciformis, H. tityus, Rethera komarovi, ? R. brandti, Sphingonaepiopsis gorgoniades, S. nana, Macroglossum stellatarum, Hyles gallii, H. vespertilio, H. livornica, Deilephila elpenor, D. porcellus, Hippotion celerio*
Gardenia	*Daphnis nerii*
Jaubertia	*Sphingonaepiopsis nana, Macroglossum stellatarum*
Kohautia	*Sphingonaepiopsis nana*
Rubia	*Rethera komarovi, Sphingonaepiopsis nana, Macroglossum stellatarum, Hyles gallii, Theretra alecto*
Uncaria	*Daphnis hypothous* (in India)
Wendlandia	*Daphnis hypothous* (in India)
Salicaceae (Salicales)	
Populus	*Smerinthus kindermanni, S. caecus, S. ocellatus, Laothoe populi, L. austauti, L. philerema, L. amurensis*
Salix	*Smerinthus kindermanni, S. caecus, S. ocellatus, Laothoe populi, L. austauti, L. amurensis, Hyles gallii*

FAMILY (ORDER): GENERA	SPHINGID SPECIES
Scrophulariaceae (Scrophulariales)	
Antirrhinum	*Hyles livornica*
Scrophularia	*Hippotion celerio*
Verbascum	*Hippotion celerio*
Solanaceae (Solanales)	
Solanaceae *spp.*	*Acherontia atropos, A. styx*
Solanum	*Acherontia atropos*
Thymelaeaceae (Myrtales)	
Daphne	*? Hyles hippophaes bienerti*
Tiliaceae (Malvales)	
Tilia	*Mimas tiliae, Smerinthus ocellatus*
Ulmaceae (Urticales)	
Ulmus	*Mimas tiliae, Laothoe populi*
Umbelliferae (Apiaceae) (Apiales)	
Daucus	*Deilephila elpenor*
Valerianaceae (Dipsacales)	
Centranthus	*Macroglossum stellatarum*
Verbenaceae (Lamiales)	
Verbenaceae *spp.*	*Acherontia atropos, A. styx*
Clerodendrum	*Acherontia styx*
Duranta	*Acherontia styx*
Lantana	*Acherontia styx*
Vitex	*Acherontia styx*
Vitaceae (Rhamnales)	
Ampelopsis	*Clarina kotschyi, Acosmeryx naga*
Cissus	*Hippotion osiris, H. celerio*
Parthenocissus	*Clarina kotschyi, Deilephila elpenor, D. porcellus, Hippotion osiris, H. celerio, Theretra boisduvalii, T. alecto*
Vitis	*Daphnis nerii, Clarina kotschyi, Acosmeryx naga, Hyles euphorbiae, H. gallii, H. livornica, Deilephila elpenor, D. porcellus, Hippotion osiris, H. celerio, Theretra boisduvalii, T. alecto*
Zygophyllaceae (Sapindales)	
Tribulus	*Hyles zygophylli* (in captivity)
Zygophyllum	*Hyles zygophylli, H. livornica*

Appendix 3 – GAZETTEER

Regions, places and geographical features mentioned in the text, many of which are not shown on the maps (Text figs. 1, 2, pp. 12, 13).

Alborz Mountains – mountain range in northern Iran

Alps – mountain range in central Europe

Almeria – province of southern Spain

Altai – mountain range bridging China, Russia, Kazakhstan and Mongolia

Amurland – portion of eastern Siberia and northern China bordering the Amur River

Anatolia – most of central Turkey

Asia Minor – area ranging from Turkey to the Caspian Sea

Asir Mountains – mountain range in S.W. Saudi Arabia

Ascension Island – island in the South Atlantic Ocean

Atlas Mountains – mountain range in N.W. Africa

Azores – group of islands in the North Atlantic Ocean

Baikal, Lake – freshwater lake in southern Siberia

Balearic Islands – include Majorca, Minorca and Ibiza; east of Spain in W. Mediterranean

Baluchistan – land of the Baluch people, split between Iran and Pakistan

Balkans – Yugoslavia, Albania, Greece, Bulgaria and Romania; also a mountain range in Bulgaria

Bavaria – southern state of Germany

Bessarabia – south-western area of Ukraine bordering Romania; includes part of Moldova (Moldavia)

Bukhara – area and town to the west of Samarkand, Uzbekistan

Canary Islands – include Gran Canaria, Tenerife and Lanzarote; off west coast of Morocco

Cape Verde Islands – islands off the west coast of Africa

Caucasus – mountain range between the Caspian and Black Seas

Crimea – peninsula in the northern Black Sea, Ukraine

Cuenca – province in eastern Spain

Dagestan – eastern portion of the Caucasus

Fergana – large valley and town along the western edge of the Tian Shan (east of Bukhara)

Galapagos Islands – islands off north-west coast of South America

Granada – province in southern Spain

Heilongjiang Province – province in N.E. China

Himalaya – mountain range between India and China

Hindu Kush – mountainous area in E. Afghanistan and N. Pakistan

Iberian Peninsula – Spain and Portugal

Ji Lin Province – province in N.E. China

Komiland – area of north-east European Russia inhabited by the Komi people

Kopet Dağ Mountains – mountain range between Turkmenistan and Iran

Kurdistan – land of the Kurds, split between Iran, Iraq and Turkey

Kurile Islands – chain of islands between Hokkaido (Japan) and the Kamchatka Peninsula, eastern Russia

Kuwait – Arab state at the northern end of the Persian Gulf between Iraq and Saudi Arabia

Levant – Syria, Lebanon, Israel, Jordan

Liaoning Province – province in N.E. China

Low Countries – the Netherlands, Belgium and Luxembourg

Macedonia – former Yugoslav republic, bordered by Serbia, Bulgaria, Greece and Albania. Also northern area of Greece

Mauretania – old name for area of N.W. Africa roughly comprising Tunisia, Algeria and Morocco, as used in this work; also a modern independent state in southern N.W. Africa, outside the scope of this book

Mesopotamia – plain of the Tigris and Euphrates rivers in Iraq

Middle East – Syria, Lebanon, Israel, Jordan, Iraq, Iran and the Arabian Peninsula

Nei Mongol Province – Inner Mongolia Autonomous Region of China

New World – North and South America; western hemisphere

Ningxia Province – Ningxia Hui Autonomous Region of China

Old World – Europe, Africa and Asia; eastern hemisphere

Pamirs – mountain range south of the Tian Shan range; western Himalaya

Patagonia – southern Argentina (New World)

Pontic Mountains – mountain range in northern Turkey (adj. **Pontine**)

Pyrenees – mountain range separating France and Spain

Qinghai Province – province in central northern China containing Lake Qinghai

Rif Mountains – mountain range in northern Morocco

Sakhalin Island – island off the eastern coast of Russia, north of Japan

Salang Pass – mountain pass north of Kabul, Afghanistan

Scandinavia – Denmark, Norway, Sweden and Finland

Senegal – westernmost country in Africa

Siberia – region of Russia east of the Ural Mountains

Sichuan Province – province in central southern China

Sierra de la Yedra – mountains in Granada province, southern Spain

Teruel – province in eastern Spain

Tian Shan – mountain range bridging Kyrgyzstan, Kazakhstan and western China

Toros (Taurus) – mountain range across southern Turkey

Transbaikal – area to the east of Lake Baikal, Russia

Transcaspia – area between the Caspian and Aral Seas

Transcaucasia – area south of the Caucasus Mountains and between the Black and Caspian Seas

Turkestan – land of the Turkic peoples, stretching from the Caspian Sea into China (Xinjiang Province)

Urals – mountain range in Russia, separating Europe from Asia

Ussuri – part of eastern Siberia bordering the Ussuri/Amur river border between China and Russia

Volga River – river in European Russia which flows into the Caspian Sea

Xizang Province – Autonomous Region of China, formerly Tibet

Xinjiang Province – Xinjiang Uygur Autonomous Region of China (Chinese Turkestan)

Yemen – country at south-western tip of the Arabian Peninsula

Yunnan Province – province in southern China bordering on Burma

Zagros Mountains – mountain range in south-western Iran

Appendix 4 – GLOSSARY

NOTE. For morphological terms, see also Text Figs 3–11 between pp. 21–32 of Introduction.

abdomen – the third or posterior part of the adult body, comprising ten segments (adj. **abdominal**).

aberration (ab.) – an example that differs from the common or typical form, usually in a striking manner and with a genetic basis.

acuminate – tapering to a long point.

-ad – towards; in the direction of.

aedeagus – the penis.

Afrotropical – zoographical area comprising sub-Saharan Africa, S.W. Saudi Arabia, Yemen, Madagascar and neighbouring islands.

allopatric – occupying different and disjunct geographical regions (cf. *sympatric*).

anal angle – point at which the inner (dorsal) and outer (terminal) wing margins meet (cf. *tornus*).

antenna (pl. **antennae**) – 'feelers'; paired segmented sensory organs situated on either side of the head (adj. **antennal**).

apex (pl. **apices**) – tip; point of wing where the costa and outer margin meet (adj. (**apical**).

apophysis – any tubercular or elongate process of the body-wall.

aposematic – having bright, contrasting colours and patterns to deter predators; warning coloration.

atrophied – no longer of any use; degenerated.

Australasian – Australia, New Zealand, the eastern Indonesian islands, New Guinea and Polynesia.

B.P. – before the present year.

basal – pertaining to the base; nearest the body.

bifurcate – forked.

bilateral – with two equal or symmetrical sides.

biosphere – the interacting community of plants and animals, and their environment.

biome – a geographical region having characteristic environmental, biological and climatic features which produce a distinct and stable climax community of organisms.

bivoltine – having two generations per year.

cariniform – keel-shaped.

cauda – the anal end of abdomen (adj. **caudal**).

caudad – towards the anal end of abdomen.

cell – in wings, an area enclosed by veins.

cephalad – pointing towards the head.

chemoreceptor – a sense organ with a group of cells presumed to be sensitive to chemical substances.

chitin – a horny substance which is the main constituent of an insect's exoskeleton (adj. **chitinous, chitinized**) (see also *sclerotin*).

chorion – the outer shell or covering of an insect's egg.

cilia – fine hairs; the fringes of a wing (adj. **ciliate**).

clavate – club-shaped; thickened at the tip.

cline – a graduation of differences of form within one species in contiguous (*q.v.*) populations (adj. **clinal**).

clypeus – that part of the forehead below the frons (*q.v.*), above the labrum (*q.v.*).

comb. nov. – indicates that a species or subspecies is being transferred to another genus for the first time.

community – a group of living organisms belonging to a number of different species that occur in the same habitat or area.

concolorous – of a uniform colour.

conspecific – belonging to the same species.

contiguous – in contact; touching.

cornutus (pl. **cornuti**) – a slender, heavily chitinized (*q.v.*) spine.

corpus bursae – in the female genitalia, a terminal pouch or sac at the end of the ductus bursae (*q.v.*).

cosmopolitan – distributed world-wide.

costa – the anterior margin of the wing (adj. **costal**).

cremaster – the apical point of the last segment of the abdomen of a pupa.

crepuscular – active at dusk or dawn.

cryptic coloration – protective, concealing coloration and/or pattern which resemble the background.

cuticle – the outer, non-cellular skin.

dentate – with indentations; toothed.

depressed – flattened.

dextro-laterad – towards the right side.

diapause – temporary cessation of development or activity in an insect to overcome unfavourable climatic conditions.

dimorphism – a difference in form or colour between individuals of the same species, characterizing two distinct types; may be seasonal, sexual or geographic (adj. **dimorphic**).

discal – pertaining to the central area of the wing.

discoidal – relating to the discal or central area of the wing.

distal – farthest from the body (cf. **proximal**).

diurnal – active during the day.

dorsolateral – situated between the dorsal and lateral regions (*qq.v.*).

dorsoventral – situated between the dorsal and ventral regions (*qq.v.*).

dorsum – the upper surface; back; the inner margin of the wing (adj. **dorsal**).

ductus bursae – the duct extending from the ostium to the corpus bursae (*qq.v.*) in the female genitalia.

ecdysis – the process of shedding or moulting the outer skin.

ecosystem – a community of organisms and their physical environment interacting as an ecological unit.

ectoderm – the outer layer of skin (adj. **ectodermal**).

ectoparasite – an external parasite (cf. **endoparasite**).

emarginate – notched; with an obtuse, rounded or quadrate section cut from the margin.

endemic – confined to a particular area.

endoparasite – an internal parasite (cf. **ectoparasite**).

epiphysis – a process on the inner side of the foretibia.

eye-spot – a pattern that resembles a vertebrate's eye.

F$_1$, F$_2$ – terms used to denote the first and second generations, etc., in cultures bred for genetic or other purposes.

falcate – hooked.

fasciculate – tufted; with tufts of hairs.

femur – the third segment of a leg; the thigh.

filamentous; filiform – thread-like.

flagellum – that part of the antenna beyond the pedicel (*q.v.*; see also *scape*).

form (f.) – a recognizable variant of a species that appears repeatedly; may be sexual, geographical or seasonal.

fusiform – spindle-shaped.

frass – larval excrement.

frenulum – the spine (simple in males, compound in females), arising from the base of the hindwing, projecting beneath the forewing, the function of which is to unite the wings during flight.

frons – the forehead; the front of the head, between and below the antennae.

galeae – the outer lobes of the maxillae (*q.v.*).

gena – the cheek (adj. **genal**).

genitalia – sexual appendages.

genus (pl. **genera**) – an assemblage of species agreeing in a given character or series of characters (adj. **generic**).

glaucous – pale bluish green.

gnathos – in the male genitalia, a pair of appendages arising from the caudal margin of the tegumen (*qq.v*).

granulose – covered with small, raised bumps.

gravid – pregnant.

gula – the throat sclerite (*q.v.*) forming the central part of the head underneath.

gustatory – relating to the sense of taste.

gynandromorph – an individual exhibiting both male and female characters.

haemolymph – lymph-like nutritive fluid in insects.

hair-pencil – a clump of specialized setae on the first abdominal segment of male sphingids which assists in the dispersal of pheromones.

Holarctic – the Palaearctic and Nearctic regions combined.

holotype – the actual or original specimen, designated by the author, on which the description of a new species or subspecies is established.

homonym – a name given to more than one taxon (cf. **synonym**).

hyaline – transparent or partly so.

imago (pl. **imagines**) – the fourth and final stage of the life-cycle in Lepidoptera; the adult insect.

in copula – referring to the male and female in the act of mating.

incrassate – thickened.

instar – any inter-moult stage of a larva.

juxta – a chitinized (*q.v.*) structure supporting and often surrounding the terminal part of the aedeagus (*q.v.*).

karyotype – the chromosome complement of a cell.

labium – the lower lip (adj. **labial**).

labrum – the upper lip.

lamella (pl. **lamellae**) – a thin plate or leaf-like process (adj. **lamellate**).

larva (pl. **larvae**) – the second, feeding, stage in the life-cycle of Lepidoptera; the caterpillar (adj. **larval**).

lateral – on the side.

lectotype – a specimen selected from a series of syntypes (*q.v.*) upon which a revised species is based, and having the same status as a holotype (*q.v.*).

linear – in the form of a line.

littoral – coastal.

lobate – lobed.

longitudinal – lengthwise.

luminescent – shiny.

lunule – a crescent-shaped mark, usually on a wing.

macular – spotted.

mandibles – the first pair of jaws.

maxillae (sing. **maxilla**) – the second pair of jaws (adj. **maxillary**).

maxillolabial – of or pertaining to both the maxilla and the labium (*qq.v.*)

meconium – waste products accumulated during the pupal stage and voided as a fluid by the adult during/after emergence.

median; mesal – in or at the middle; of or pertaining to the middle.

mesonotum – the dorsal part of the second thoracic segment.

mesothorax – second segment of the thorax.

metamorphosis – the sequence of changes through which an insect passes during its growth from the egg through the larva and pupa to the adult.

metathorax – third segment of the thorax.

micropyle – one of the minute openings in the egg through which spermatozoa enter during fertilization.

migrant – an individual that flies, often with others, from one place to another, usually for food or other purposes (cf. **vagrant**).

montane – mountainous.

morph – a form.

multivoltine – having several generations in a year.

musculature – a system of muscles.

Nearctic – zoographical area comprising Canada, the U.S.A. and temperate Mexico.

Neotropical – zoographical area comprising Central and South America.

nocturnal – active at night.

nominate subspecies – based on the holotype (*q.v.*) of a species, when the latter is divided into subspecies.

occipital – of or pertaining to the back part of the head.

ocellus (pl. **ocelli**) – (i) a simple eye in an adult insect; (ii) an eye-like spot (adj. **ocellate**).

olfactory – pertaining to the sense of smell.

ostium bursae – the female genital opening, often merely referred to as the ostium.

ovum (pl. **ova**) – the first stage in the life-cycle of an insect; the egg.

pabulum – food.

Palaearctic – zoographical area comprising Europe, North Africa, Asia Minor, the Himalaya and northern Asia.

palpi (sing. **palpus**) – paired, sensory organs either side of the proboscis which protect that structure; technically the labial palpi.

papillus (pl. **papilli**) – a minute soft projection.

parapodia – the false legs – i.e. the prolegs.

paratype – any specimen in a series from which the description of a species or subspecies has been made, other than the one specified as the type (*q.v.*) specimen or holotype (*q.v.*).

paronychium (pl. **paronychia**) – a bristle-like appendage at the base of the tarsal claws.

patagium (pl. **patagia**) – the lobe-like structure that covers the base of the wing.

pecten – a comb of scales or hairs; adj. **pectinate** – comb-like.

pedicel – the second segment of the antenna (see also *scape* and *flagellum*).

phenotype – refers to the visible characters of an individual, e.g. wing pattern.

pheromone – a chemical 'messenger' secreted by an organism and conveying information from one individual to another.

photoperiod – the light phase of a light-dark cycle.

pilifer – a small sclerite (*q.v.*) at each side of the clypeus (*q.v.*).

pilose – covered in short, fine hairs.

pleuron (pl. **pleurites**) – the lateral plate of the thorax (adj. **pleural**).

poikilothermic – cold-blooded.

polycentric – having more than one centre of distribution.

polymorphism – exhibiting more than one form.

polyphagous – feeding on many kinds of plants or food.

polypodous – having many feet.

porrect – extending forward horizontally.

postdiscal – beyond the discal area.

pretarsus – a terminal outgrowth of the tarsus (*q.v.*).

proboscis – the tubular feeding organ of an adult insect; the 'tongue' of Lepidoptera; haustellum.

process – the prolongation of a surface, of a margin, or of an appendage; any prominent part of the body not otherwise definable.

produced – extended; elongated.

proleg – the fleshy, unjointed abdominal legs of caterpillars.

prothorax – first segment of the thorax.

proximal – nearest to the body (cf. **distal**).

pulvillus (pl. **pulvilli**) – a pad, or lobe, between and beneath the tarsal claws.

punctate – with impressed points or punctures.

pupa (pl. **pupae**) – the third, non-feeding stage in the life-cycle of Lepidoptera; the chrysalis (adj. **pupal**).

race – often used as an alternative for a subspecies or geographical form.

range – the natural distribution of a species.

recurved – bent downwards, backwards or outwards.

refugium (pl. **refugia**) – an isolated habitat retaining the environmental conditions that were once widespread (adj. **refuge**, **refugial**).

relict – remnants of a formerly widespread fauna existing in certain isolated areas or habitats; the sole surviving group of an otherwise extinct species.

reticulate – covered with a fine net-like pattern.

retinaculum – a loop on the ventral surface of the forewing that engages the frenulum (*q.v.*).

retractile – capable of being produced and drawn back or retracted.

rostrate – having a rostrum, a long protraction bearing the mouth parts.

rugose – wrinkled; rough.

sacculus – in the male genitalia, the ventral part of the valva.

saccus – in the female genitalia, a median chitinized pocket formed by the invaginated or retracted sternal region of the ninth abdominal segment; in the male genitalia, a midventral cephalad (*q.v.*) projection of the vinculum.

sagittate – shaped like an arrowhead.

scape – the first, basal, often elongated, segment of the antenna (cf. **pedicel** and **flagellum**).

sclerite – any sclerotized piece of an insect's body-wall bounded by sutures.

sclerotin – a horny substance (adj. **sclerotized**).

sensilla – simple sense organs.

sensory – pertaining to sensation; having feeling.

sensu lato – in the broad sense.

sensu stricto – in the strict or narrow sense.

sequester – to obtain and store toxic substances from plants for defence purposes.

serrate – saw-like, with notched edges.

seta (pl. **setae**) – a hair or bristle.

setiform – thread-like.

sexual dimorphism – difference in appearance between the sexes of a species.

sinistrad – towards the left side.

sinuate – wavy.

spatulate – spoon-shaped.

species – a group of populations showing the same, or similar, characteristics and capable of interbreeding to produce fertile, viable offspring.

specific – associated with a particular species.

spermatophore – a covering around the spermatozoa.

spinneret – the larval apparatus by means of which silk is spun.

spinules – small spines.

spinose – spiny.

spiracle – a breathing pore (adj. **spiracular**).

spur – large, moveable spine, usually on the leg.

stat. nov. – signifies a change of taxonomic status.

stat. rev. – signifies that the taxonomic designation of a taxon is reverted to its previous status.

sternite; sternum – a ventral plate of a segment of the thorax or abdomen.

stria (pl. **striae**) – a fine line (adj. **striate**).

sub- – slightly less than, somewhat; below.

subquadrate – not quite a square.

subspecies – an aggregate of phenotypically similar populations of one species having recognizable, distinct characteristics which differentiate it from other population aggregates of that species.

sympatric – occurring in the same geographical area (cf. **allopatric**).

synonym – one of several names given to the same taxon (cf. **homonym**).

syn. nov. – denotes a new synonym or new synonymy.

syntype – any one of two or more specimens of the original type series (*q.v.*) when no holotype (*q.v.*) has been selected.

tactile – of or pertaining to the sense of touch.

tarsus (pl. **tarsi**) – the foot or jointed appendage attached to the tibia (*q.v.*) (adj. **tarsal**).

taxon (pl. **taxa**) – a biological category, e.g. genus, species, subspecies (*qq.v.*).

taxonomy – the theory and practice of the classification of taxa (adj. **taxonomic**).

tegula – a movable 'shoulder-pad' at the base of a wing.

tegumen – in the male genitalia, a structure dorsad of the anus to which the uncus (*q.v.*) is articulated.

termen – the outer margin of the wing.

terminal – situated at the tip.

tesselated – chequered.

tergite; tergum – the dorsal plate of a segment of the thorax or abdomen.

thorax – the second, middle region of an insect's body bearing the legs and wings (adj. **thoracic**).

tibia (pl. **tibiae**) – the fourth segment of a leg.

tornus – the anal angle of the wing (adj. **tornal**).

trimorphic – having three colour forms.

trivoltine – having three generations per year.

trochanter – a sclerite of the leg sometimes divided between the coxa and femur; sometimes fused with the femur.

truncate – squared off; ending abruptly.

tubercle – in larvae, a body structure forming a small, solid, often seta-bearing pimple.

type – the actual specimen on which the description of a new species is described (see also *holotype, lectotype, syntype, paratype*); the species on which a genus is established (cf. **type species**).

type locality – the place where the type or holotype (*qq.v.*) was collected.

type series – the original series of specimens on which the description of a new species or subspecies is established, when no holotype has been selected (cf. **syntype**).

type species – the species on which a genus is described and established.

uncus – in the male genitalia, the mid-dorsal structure extending caudad (*q.v.*) from the distal margin of tegumen (*q.v.*).

univoltine – having only one generation per year.

vagrant – an individual far outside its normal range; an occasional immigrant or 'accidental' (cf. **migrant**).

valva (pl. **valvae**) – in the male genitalia, one of a pair of lateral clasping organs.

vein – one of many rigid tubes supporting the wing membrane.

venation – the pattern of veins of a wing.

venter – the lower surface of the abdomen (adj. **ventral**).

ventrad – towards the underside.

vertex – the crown of the head.

vesica – a terminal, retractile (q.v.), membraneous part of the aedeagus.

vestigial – small or degenerate; in the process of disappearing.

vinculum – in the male genitalia, a V-shaped sclerite (*q.v.*) derived from the ninth abdominal sternite (*q.v.*).

wingspan – in this work, twice the distance from one wing-tip to the centre of the thorax.

References

Certain of the literature cited in the synonymies is given below in more detail, in particular where a change of status for a species or subspecies has been proposed. The list of references also includes literature not specifically cited in the text but which has been especially useful in the preparation of distribution maps for this work.

ABAFI-AIGNER, L., PÁVEL, J. & UHRYK, F., 1896. *A Magyar birodalom állatvilága (Fauna Regni Hungariae)*, III. Arthropoda (Insecta. Lepidoptera), 82pp., 1 map. Budapest.

ALBERTUS MAGNUS, 13th C. *De natura animalium*.

ALBIN, E., 1720. *A natural history of English insects*, [12] pp., 100 pls. London.

ALKEMEIER, F. & GATTER, W., 1990. Grössenordnung der herbstlichen Südwanderung des Windenschwärmers (*Agrius convolvuli*) in den Alpen. *Ent. Z., Frank. a. M.* **100**: 330–332.

ALPHÉRAKY, S., 1882. Lépidoptères du district de Kouldjà et des montagnes environnantes, II. Heterocera. *Horae Soc. ent. ross.* **17**: 15–103.

ANDUS, L., 1986. Hawk moths (Lep. Sphingidae) collection in Natural History Museum in Beograd. *Bull. nat. Hist. Mus. Belgrade* (B) Biol. (**41**): 89–107.

ASKEW, R. R., 1971. *Parasitic insects*, xvii, 316pp. 124 figs. London.

ATANASOV, A. Z., ĬONAĬTIS, V. P., KASPARYAN, D. R., KUSDITSKIĬ, V. S., RASNITSIN, A. P., SIĬTAN, U. V. & TODKONITS, V. D., 1981. Pereponchatokrȳlȳe. *In* Medvedev, G. S., *Opred. Nasek. europ. Chasti SSSR* 3 [Hymenoptera 3, Ichneumonidae (part)]: 687pp. Leningrad [in Russian].

AUSTAUT, J. L., 1905. Notice sur quelques lépidoptères nouveaux. *Ent. Z., Frankf. a. M.* **18**: 143.

BADR, M. A., OSHAIBAH, A. A., NAWABI, A. EL- & GAMAL, M. M. AL-, 1985. Classification of some species of the family Sphingidae – Lepidoptera in Egypt. *Ann. agric. Sci., Moshtohor* **23**: 885.

BAKER, G. T., 1891. Notes on the Lepidoptera collected by the late T. Vernon Wollaston. *Trans. ent. Soc. Lond.* **24**: 197–221, 1 pl.

BALDWIN, I. T., 1988. Short-term damage-induced increases in tobacco alkaloids protect plants. *Oecologia* **75**: 367–370.

BANG-HAAS, A., 1912. Neue oder wenig bekannte palaearktische Makrolepidopteren VI. *Dt. ent. Z. Iris* **26**: 229–230.

BANG-HAAS, O., 1927. *Horae Macrolepidopterologicae Regionis palaearcticae* 1: xxviii, 128pp., 11 pls. Dresden.

———, 1936. Neubeschreibungen und Berichtigungen der Palaearktischen Makrolepidopterenfauna, XXIII. *Ent. Z., Frankf. a. M.* **50**: 254–256.

———, 1937. Neubeschreibungen und Berichtigungen der Palaearktischen Makrolepidopterenfauna, XXVII. *Ent. Z., Frankf. a. M.* **50**: 562–564

———, 1938. Neubeschreibungen und Berichtigungen der Palaearktischen Makrolepidopterenfauna, XXXVI. *Ent. Z., Frankf. a. M.* **52**: 177–180.

———, 1939. Neubeschreibungen und Berichtigungen der Palaearktischen Macrolepidopterenfauna XXXVIII. *Dt. ent. Z. Iris* **53**: 49–60.

BÄNZIGER, H., 1988. Unsuspected tear drinking and anthropophily in thyatirid moths, with similar notes on sphingids. *Nat. Hist. Bull. Siam Soc.* **36**: 117–133.

BARBOSA, P., SAUNDERS, J. A., KEMPER, J., TRUMBULE, R., OLECHNO, J. & MARTINAT, P., 1986. Plant allelochemicals and insect parasitoids: effects of nicotine on *Cotesia congregata* (Say) (Hymenoptera: Braconidae) and *Hyposoter annulipes* (Cresson) (Hymenoptera: Ichneumonidae). *J. chem. Ecol.* **12**: 1319–1328.

BAROU, J., [1967]. *Contribution à la connaissance de la faune des Lépidoptères de l'Iran*, 18pp., 1 map. [Teheran].

BARRETT, C. G., 1893–95. *The Lepidoptera of the British Islands* (illus. edn) 2: [Sphingidae: pp. 1–75, pls 41–54 (1893)] 372pp., 46 pls. London.

BATRA, S. W. T., 1983. Establishment of *Hyles euphorbiae* (L.) (Lepidoptera: Sphingidae) in the United States for control of the weedy spurges *Euphorbia esula* L. and *E. cyparissias* L. *New York Ent. Soc.* **91**: 304–311.

BAUER, E. & TRAUB, B., 1980. Zur Macrolepidopterenfauna der Kapverdischen Inseln. Teil 1. Sphingidae und Arctiidae. *Ent. Z., Frankf. a. M.* **90**: 244–248.

BEARDSLEY, J. W., 1979. New immigrant insects in Hawaii: 1962 through 1976. *Proc. Hawaii. ent. Soc.* **23**: 35–44.

BECKAGE, N. E., TEMPLETON, T. J., NIELSEN, B. D., COOK, D. I. & STOLTZ, D. B., 1987. Parasitism-induced hemolymph polypeptides in *Manduca sexta* (L.) larvae parasitized by the braconid wasp *Cotesia congregata* (Say). *Insect Biochem.* **17**: 439–455.

BECKAGE, N. E., METCALF, J. S., NIELSEN, B. D. & NESBIT, D. J., 1988. Disruptive effects of azadirachtin on development of *Cotesia congregata* in host tobacco hornworm larvae. *Archs Insect Biochem. Physiol.* **9**: 47–65.

BELL, T. R. D. & SCOTT, F. B., 1937. *The Fauna of British India, including Ceylon and Burma. Moths.* 5 Sphingidae: xviii, 537pp., 15 pls, 1 map. London.

BELSHAW, R., (in press). Tachinid flies (Diptera: Tachinidae). *Handbk Ident. Br. Insects* **10**(4ai).

BERGMANN, A., 1953. *Die Großschmetterlinge Mitteldeutschlands* 3. Spinner und Schwärmer: xii, 552pp., 9 pls. Jena.

BERNAYS, E. A. & JANZEN, D. H., 1988. Saturniid and sphingid caterpillars: two ways to eat leaves. *Ecology* **69**: 1153–1160.

BESCHKOW, S. V., 1990. *Sphingonaepiopsis gorgoniades* (Hübner, [1819]) (Lepidoptera, Sphingidae) – a new genus and a new species for the Bulgarian fauna. *Acta zool. bulg.* **40**: 75–77.

BIENERT, T., 1870. *Lepidopterologische Ergebnisse einer Reise in Persien in den Jahren 1858 und 1859*. 56pp. Leipzig.

BIRCH, M. C., POPPY, G. M. & BAKER, T. C., 1990. Scents and eversible scent structures of male moths. *A. Rev. Ent.* **35**: 25–58.

BLEST, A. D., 1957. The evolution of protective displays in the Saturnioidea and Sphingidae (Lepidoptera). *Behaviour* **11**: 257–309, 4 pls.

BÖRNER, C., 1925. Lepidoptera, Schmetterlinge. *In* Brohmer, P. (Ed.), *Fauna aus Deutschland*: pp. 358–387.

[BOULENGER, G. A. *et al*], [1923]. *A survey of the fauna of Iraq, mammals, birds, reptiles, etc . . . made by members of the Mesopotamia Expeditionary Force 'D', 1915–1919.* xxi, 404pp., 20 pls. Bombay. [See also Watkins & Buxton, 1923.]

BRANDT, W., 1938. Beitrag zur Lepidopteren-Fauna von Iran. *Ent. Rdsch.* **55**: 698–699.

BRETHERTON, R. F. & WORMS, C. G. de, 1963. Butterflies of Corsica, 1962. *Entomologists' Rec. J. Var.* **75**: 93–104.

BROCK, J. P., 1990. Pupal protrusion in some bombycoid moths (Lepidoptera). *Entomologist's Gaz.* **41**: 91–97.

BUCKLER, W., 1887. *The larvae of the British butterflies and moths* 2: xi, 172 pp., pls 18–35. London.

BURESCH, I. & TULESCHKOW, K., 1931. *Rethera komarovi* Chr. (Lepidoptera) eine für die Fauna Europas neue Sphingide. *Izv. tsarsk. prirodonauch. Inst. Sof.* **4**: 121–138.

BUTLER, A. G., 1876. Revision of the Heterocerous Lepidoptera of the family Sphingidae. *Trans. zool. Soc. Lond.* **9**: 511–644, 5 pls.

CARCASSON, R. H., 1968. Revised catalogue of the African Sphingidae (Lepidoptera) with descriptions of the East African species. *Jl E. Afr. nat. Hist. Soc. natn. Mus.* **26**: 1–148, 17 pls. [Facsimile reprint, 1976. Faringdon.]

CHANG, F., 1982. Insects, poisons and medicine: the other one percent. *Proc. Hawaii. ent. Soc.* **24**: 69–74.

CHISTYAKOV, YU. A. & BELYAEV, E. A., 1984. Sphingidae of the genus *Hemaris* Dalm. (Lepidoptera) of the Russian Far East. *In* Ler, P. A. (ed.), The fauna and ecology of insects in the south of the Far East. Collected scientific papers. *Akad. Nauk SSSR, Vladivostok* **1984**: 50–59 [in Russian].

CHRISTOPH, H., 1894. *Deilephila peplidis* n. sp. *Ent. Nachr. Berlin* **20**: 333–334.

CHU, H. F. & WANG, L. Y., 1980a. New species and new subspecies of the family Sphingidae (Lepidoptera). *Acta zootaxon. sin.* **5**: 418–426 [in Chinese].

———, 1980b. Lepidoptera: Sphingidae. *In* Chu, H. F. & Fauna Edit. Cttee (Eds), *Economic insect fauna of China* **22**: x, 84pp., 26 pls. Beijing [in Chinese].

CLOSS, A., 1917. Einige neue Sphingiden (Lep.). *Ent. Mitt.* **6**: 33–34.

COCKERELL, T. D. A., 1923. The Lepidoptera of the Madeira Islands. *Entomologist* **56**: 243–247, 286.

COMMON, I. F. B., 1970. Lepidoptera. *In* Mackerras, I. M. (Ed.), *The Insects of Australia*, pp. 765–866.

COMSTOCK, J. H., 1893. *A contribution to the classification of the Lepidoptera*, pp. 37–113, 3 pls. Ithaca, N.Y.

——— & NEEDHAM, J. G., 1898. The wings of insects (the venation of the wings of Lepidoptera). *Am. Nat.* **32**: 253–257.

D'ABRERA, B., 1986. *Sphingidae Mundi. Hawk moths of the world*, [x], 226pp., 80 pls. Faringdon.

DANIEL, F., 1932, Zygaenidae – Cymatophoridae. *In* Osthelder, L. & Pfeiffer, E., Lepidopteren-Fauna von Marasch in türkisch Nordsyrien [Sphingidae: pp. 67–71]. *Mitt. münch. ent. Ges.* **22**: 52–82.

———, 1939. Zygaenidae – Hepialidae. *In* Osthelder, L. & Pfeiffer, E., Lepidopteren-Fauna von Marasch in türkisch Nordsyrien [Sphinghidae: pp. 94–95]. *Mitt. münch. ent. Ges.* **29**: 84–103.

———, 1960. Lepidoptera der Deutschen Nepal-Expedition 1955. Zygaenidae – Cossidae. *Veröff. zool. Staatssamml. Münch.* **6**: 151–162, 1 pl.

———, 1963. Ein Beitrag zur Spinner-und Schwärmerfauna des Iran und Afghanistans. *Z. Wien ent. Ges.* **48**: 145–155, pls 26, 27.

———, 1971. Österreichische Expeditionen nach Persien und Afghanistan. *Annln naturh. Mus. Wien* **75**: 651–660.

——— & WOLFSBERGER, J., 1955. Die Föhrenheidegebiete des Alpenraumes als Refugien wärmeliebender Insekten. *Z. wien. ent. Ges.* **40**: 49–71.

———, FORSTER, W. & OSTHELDER, L., 1951. Beiträge zur Lepidopterenfauna Mazedonien. *Veröff. zool. Staatssamml. Münch.* **2**: 1–78.

DANNEHL, F., 1929. Neue Formen und geographische Rassen aus meinen Ausbeuten und Erwerbungen der letzten Jahre. *Mitt. münch. ent. Ges.* **19**: 97–116.

DANNENBERG, D. 1942. Neue Schwärmerbastarde. *Z. wien Ent-Ver.* **27**: 225–230, 2 pls.

DEGEER, C., 1752. *Mémoires pour servir à l'histoire des insectes* **1**: xvi, 708pp., 37 pls. Stockholm.

———, 1771. *Mémoires pour servir à l'histoire des insectes* **2** (Pt. 1): xii, 616pp., 15 pls. Stockholm.

DEGTYAREVA, V. I. & SHCHETKIN, YU. L., 1982. First descriptions of the pre-imagal phases of two species of Sphingidae (Lepidoptera) from central Asia. *Izv. Akad. Nauk tadzhik. SSR*, (Otdel. Biol.) (2): 25–29 [in Russian].

DENSO, P., 1913a. *Celerio hippophaës*. *Dt. ent. Z. Iris* **27**: 22–45, 2 pls.

———, 1913b. Schwärmerhybriden. *In* Seitz, A., *Die Groß-Schmetterlinge der Erde* **2**: 260–270, pl. 43. Stuttgart.

DERZHAVETS, YU. A., 1977. Hawkmoths (Lepidoptera, Sphingidae) of Mongolia. *Nasekom. Mongol.* **5**: 642–648 [in Russian].

———, 1979a. Taxonomic status of *Hyles costata* Nordmann (Lepidoptera, Sphingidae). *Nasekom. Mongol.* **6**: 404–412 [in Russian].

———, 1979b. On the distribution and geographic variation of hawk-moths *Hyloicus pinastri* L. and *H. morio* Rotsch. et Jord., stat. n. (Lepidoptera, Sphingidae). *Ent. Obozr.* **58**: 112–115 [in Russian].

———, 1980. On the distribution area of the hawkmoths *Hyles euphorbiae* L. and *H. robertsi* Btlr. (Lepidoptera, Sphingidae) in Asia. *Ent. Obozr.* **59**: 346–349 [in Russian].

———, 1984. An account of the classification of the sphinx moths (Lepidoptera, Sphingidae) with a list of the species of the fauna of the USSR. *Ent. Obozr.* **63**: 604–620 [in Russian].

DJAKONOV, A. M., 1923. *O nekotorykh novykh ili malo izvestiykh. Lepidoptera Heterocera Palearkticheskoi Oblast.* De quibusdam Lepidopteris Heteroceris palaearcticis novis vel parum cognitis. *Ezheg. zool. Muz.* **24**: 104–123, 12 pls [in Russian].

DRURY, D., 1770–72. *Illustrations of natural history*, **1**: xxviii, 132pp., 50 pls; **2**: viii, 92pp., 50 pls; **3**: xxvi, 78pp., 50 pls. London.

———, 1837. *Illustrations of exotic entomology* (New edn. Westwood, J. O. (Ed.) **1**: xxii, 123pp., 50 pls; **2**: vi, 100pp., 50 pls; **3**: vi, 93, [i] pp., 50 pls. London.

DUBLITZKY, B., 1928. Eine neue Rasse *Celerio nicaea* Prun. v. *sheljuzkoi* Dub. aus der Umgebung der Stadt Alma-Ata (früher Wernyi) Djetissu Gouvernement (früher Semiretschje). *Ent. Z., Frankf. a. M.* **42**: 38–40.

DUTFIELD, J., 1748–1749. *A new and complete natural history of English moths and butterflies* (6 pts), 12pp., 12 pls. London.

EASTERBROOK, M., 1985. *Hawk-moths of the British Isles*, 24pp. Aylesbury.

EATON, J. L., 1988. *Lepidopteran anatomy*, xiii, 257pp. Chichester.

EBERT, G., 1969. Afghanische Bombyces und Sphinges. 3. Sphingidae (Lepidoptera). Ergebnisse der 2. Deutschen Afganistan-Expedition (1966) der Landessammlungen für Naturkunde in Karlsruhe. *Reichenbachia* **12**: 37–53.

———, 1974. Afghanische Bombyces und Sphinges, 12. Nachträge und Zusammenfassung. *Reichenbachia* **15**: 1–15.

———, 1976. Beiträge zur Kenntnis der Bombyces und Sphinges Irans. 1. Beitrag: Die Arten und Formen der Gattungen *Berutana* und *Deilephila* (Lep. Sphingidae). *J. ent. Soc. Iran* **3**: 85–99.

EICHLER, F. 1971. *Celerio galii tibetanica* ssp. n. sowie Bemerkungen zur Art (Lepidoptera, Sphingidae). *Ent. Abh. Mus. Tierk. Dresden* **38**: 315–324.

———, & FRIESE, G., 1965. Ergebnisse der Albanien-Expedition 1961 des Deutschen Entomologischen Institutes. 32.Beitrag. Lepidoptera: Sphingidae. *Beitr. Ent.* **15**: 633–640.

EISENSTEIN, I., 1984. *Bibleland hawkmoths*, 80pp., 80 pls. Tel Aviv [in Hebrew].

EITSCHBERGER, U., DANNER, F. & SURHOLT, B., 1989. Taxonomische Veränderungen bei den Sphingiden Europas und die Beschreibung einer neuen *Laothoe*-Unterart von der Iberischen Halbinsel (Lepidoptera, Sphingidae). *Atalanta, Würzburg* **20**: 261–271.

——, —— & ——, 1992. Zweiter Beitrag zu "Taxonomische Veränderungen bei den Sphingiden Europas" und die Beschreibung einer neuen *Sphinx*-Unterart aus der Türkei (Lepidoptera, Sphingidae). *Atalanta, Würzburg* **23**: 245–247.

ELLISON, R. E. & WILTSHIRE, E. P., 1939. The Lepidoptera of the Lebanon, with notes on their season and distribution. *Trans. R. ent. Soc. Lond.* **88**: 1–56, 1 pl.

ERNST, J. J., 1782. *Papillons d'Europe peints d'après nature* **3**: [iv], x, 132pp., frontispiece, pls 85–122. Paris.

EVANS, D. L. & SCHMIDT, J. O., 1990. *Insect defences. Adaptive mechanisms and strategies of prey and predators*, xv, 482pp. Albany, N.Y.

EVERSMANN, E., 1844. *Fauna Lepidopterologica Volgo-Uralensis*, xiv, 633pp. Moscow.

FABRICIUS, J. C., 1775. *Systema Entomologiae*, [xxviii], 832pp., Flensburg & Leipzig.

FELTWELL, J. & DUCROS, P., 1989. Sphingidae of the Cévennes. *Br. J. Ent. nat. Hist.* **2**: 69–76.

FIEDLER, A., 1927. Beitrag zu den Schmarotzern in Schmetterlingsraupen. *Int. ent. Z.* **21**: 186–187.

FLECK, E., 1901. *Die Macro-lepidopteren Rumäniens*, 200pp. Berlin.

FLEMING, R. C., 1968. Head musculature of Sphinx Moths (Lepidoptera: Sphingidae). *Contr. Am. ent. Inst.* **3**(3): 1–32.

FLETCHER, D. S. & NYE, I. W. B., 1982. *The generic names of the moths of the world* **4**: xiv, 192pp., 1 pl. London.

FORD, T. H. & SHAW, M. R., 1991. Host records of some west Palaearctic Tachinidae (Diptera). *Entomologist's Rec. J. Var.* **103**: 23–38.

FORSTER, W. & WOHLFAHRT, T. A., 1960. *Die Schmetterlinge Mitteleuropas* **3**: vii, 239pp., 28 pls. Stuttgart.

FRAENKEL, G., 1959. The Raison d'Être of Secondary Plant Substances. *Science* **129**: 1466–1470.

FREINA, J. J. de, 1979. Beitrag zur systematischen Erfassung der Bombyces- und Sphinges-Fauna Kleinasiens. *Atalanta, Würzburg* **10**: 175–224.

——, 1988. Bemerkungen über das fragliche Artrecht von *Hemaris dentata* (Staudinger, 1887) (Lepidoptera, Sphingidae). *Nota lepid.* **11**: 182–186.

——, 1991. Über Biologie und Morphologie der auf Madeira beheimateten *Hyles euphorbiae gecki* ssp. n. (Lepidoptera, Sphingidae). *NachrBl. bayer. Ent.* **40**: 65–72.

—— & WITT, T. J., 1984. Modifications taxinomiques dans les Sphingides et Syntomides d'Europe et de l'Afrique du Nord occidentale (Lepidoptera). *Alexanor* **13**: 283–288.

—— & WITT, T. J., 1987. *Die Bombyces und Sphinges der Westpalaearktis* **1**: 708pp., 46 pls. Munich.

FRIEDRICH, E., 1975. *Handbuch der Schmetterlingszucht: Europäische Arten*, 186pp. Stuttgart.

——, 1982. *L'élevage des papillons – espèces européenes* [French edn], 235pp., 18 pls. Paris.

——, 1986. *Breeding butterflies and moths – A practical handbook for British and European species* [revised English edn], 176pp. Colchester.

FREYER, C. F., 1836. *Neuere Beiträge zur Schmetterlingskunde mit Abbildungen nach der Natur* **2**: 164pp., 96 pls. Augsburg.

FRIONNET, M. C., 1910. *Les premiers états des Lépidoptères Français* **2** Sphingidae–Psychidae. Bombyces–Acronyctinae: 550, [2]pp. Saint-Dizier.

FROREICH, –. von, 1938. Eine neue Unterart von *Celerio zygophylli* O. *Ent. Rdsch.* **55**: 256–259.

GANEV, J., 1984. Catalogue of the Bulgarian Bombyces and Sphinges (Lepidoptera: . . .). *Entomofauna* **5**: 391–467.

GARDINER, B. O. C., 1978. The preparation and use of artificial diets for the rearing of insects. *Entomologist's Rec. J. Var.* **90**: 181–184, 267–270, 287–291.

——, 1986. Rearing with artificial or semi-synthetic diet. *In* Friedrich, E., *Breeding butterflies and moths – a practical handbook for British and European species*, pp. 36–38. Colchester.

GARDNER, A. E. & CLASSEY, E. W., 1960. Report on the insects collected by the E. W. Classey and A. E. Gardner expedition to Madeira in December 1957. *Proc. S. Lond. ent. nat. Hist. Soc.*, **1959**: 184–206.

GATTER, P. & GATTER, W., 1990. Das Migrationssystem des Windenschwärmers (*Agrius convolvuli*) zwischen Westafrika und dem Westen der Paläarktis – Ergebnisse und Hypothesen. *Ent. Z., Frankf. a. M.* **100**: 313–330.

GAULD, I. D. & BOLTON, B. (Eds), 1988. *The Hymenoptera*, xi, 332pp., 10 pls. Oxford.

GEHLEN, B., 1930. Neue Sphingiden. *Ent. Z., Frankf. a. M.* **44**: 174–176.

——, 1932a. Neue Sphingiden. *Ent. Rdsch.* **49**: 182–184.

——, 1932b. Familie: Sphingidae, Schwärmer; Schwärmerhybriden. *In* Seitz, A., *Die Groß-Schmetterlinge der Erde* **2** (Suppl.). pp. 137–156; 156–164; pls 12, 13. Stuttgart.

——, 1934a. Neue afrikanische und asiatische Sphingiden. *Ent. Z., Frankf. a. M.* **48**: 59–62.

——, 1934b. Liste der von S.K.H. dem Prinzen Leopold von Belgien im Jahre 1932 gesammelten Sphingidae. *Bull. Mus. r. Hist. nat. Belg.* **10**(3): 1–3.

——, 1941. Neue Sphingiden. *Ent. Z., Frankf. a. M.* **55**: 185–186.

GEOFFROY, E. L., 1799. *Histoire abrégée des insectes qui se trouvent aux environs de Paris . . .* **2** (Edn 2): 744pp., 11 pls. Paris.

GILCHRIST, W. L. R. E., 1979. Sphingidae, pp. 20–39, pls 1,2. *In* Heath, J. & Emmet, A. M. (Eds). *The Moths and Butterflies of Great Britain and Ireland*, **9**: 288pp., 16 pls. London & Colchester.

GNINENKO, YU. I., RASPOPOV, A. P., KOVALEVSKAYA, N. I., 1983. Features of the biology of the lime hawk-moth *Mimas tiliae* (Lepidoptera, Sphingidae) in the Transurals. *Zool. Zh.* **62**: 1009–1014 (in Russian).

GÓMEZ BUSTILLO, M. R. & FERNÁNDEZ-RUBIO, F., 1974. Los esfingidos de la Peninsula Ibérica. *Revta. Lepid.* **2** (Suppl.): 1–12.

—— & ——, 1976. *Mariposas de la Peninsula Ibérica* **3**, *Heteroceros I* [Sphingidae, pp. 96–107], 304pp. Madrid.

—— & MÉNDEZ GARNICA, J. M., 1981. Esos aerodinamicos esfingidos. *Vida Silv.* (37): 18–31.

GRAYSON, J. & EDMUNDS, M., 1989a. Growth, development, mortality and fecundity of the poplar hawkmoth (*Laothoe populi* (L.)) under laboratory conditions, with a note on the eyed hawkmoth (*Smerinthus ocellata* (L.)). *Entomologist's Gaz.* **40**: 211–219.

—— & ——, 1989b. The causes of colour and colour changes in caterpillars of the poplar and eyed hawkmoths (*Laothoe populi* and *Smerinthus ocellata*). *Biol. J. Linn. Soc.* **37**: 263–279, 3 figs.

GRAESER, L., 1892. Neue Lepidopteren aus Central-Asien. *Berl. ent. Z.* **37**: 299–318.

GRUM-GRSHIMAILO, G. E., 1890. Le Pamir et sa faune Lépidoptérologique. Heterocera, Sphinges, VIII Sphingidae B., pp.

510–514. *In* Romanoff, N. M., *Mémoires sur les Lépidoptères* **4**: xviii, 578pp., 21 pls, 1 map. St Petersburg.

GUNTHER, A., 1939. Die südlichste Rasse von *Celerio deserticola* Bart. *Ent. Z., Frankf. a. M.* **53**: 261–266.

GUYOT, H., 1990. Sur le présence en Corse de *Proserpinus proserpina* Pallas (Lepidoptera, Sphingidae). *Alexanor* **16**: 442–444.

HAGEN, R. H. & SCRIBER, J. M., 1989. Sex-linked diapause, color and allozyme loci in *Papilio glaucus*: linkage analysis and significance in a hybrid zone. *J. Hered.* **80**: 179–185.

——— & ———, 1991. Systematics of the *Papilio glaucus* and *Papilio troilus* Species Groups (Lepidoptera: Papilionidae): Inferences from Allozymes. *Ann. ent. Soc. Am.* **84**: 380–395.

———, LEDERHOUSE, R. C., BOSSART, J. L. & SCRIBER, J. M., 1991. *Papilio canadensis* and *P. glaucus* (Papilionidae) are distinct species. *J. Lep. Soc.* **45**: 245–258.

HAGMEIER, A., 1912. Beiträge zur Kenntnis der Mermithiden. 1. Biologische Notizen und systematische Beschreibung einiger alter und neuer Arten. *Zool. Jb. Abt. Syst.* **32**: 521–612, 4 pls.

HARBICH, H., 1976a. Isolationsmechanismen und Arterhaltung im Genus *Celerio* (Lep., Sphingidae). *Ent. Z., Frankf. a. M.* **86**: 33–42.

———, 1976b. Das photosensible Raupenstadium von *Celerio euphorbiae euphorbiae* (Lep.: Sphingidae). *Ent. Z., Frankf. a. M.* **86**: 233–237.

———, 1978a. Zur Biologie von *Acherontia atropos* (Lep.: Sphingidae) 1. Teil. *Ent. Z., Frankf. a. M.* **88**: 29–36.

———, 1978b. Zur Biologie von *Acherontia atropos* (Lep.: Sphingidae) 2. Teil. *Ent. Z., Frank. a. M.* **88**: 101–109.

———, 1980a. Ergebnisse von Hybridzucht zwischen *Hyles lineata* (Fabricius, 1775) und *Hyles livornica* (Esper, 1779) – 1. Teil. (Lep. Sphingidae). *Atalanta, Würzburg* **11**: 5–11.

———, 1980b. Weiteres zur Biologie von *Agrius convolvuli* (Linné, 1758) (Lep. Sphingidae). *Atalanta, Würzburg* **11**: 197–200.

———, 1980c. Zur Biologie von *Acherontia atropos* (Lep.: Sphingidae) 3. Teil. *Ent. Z., Frankf. a. M.* **90**: 11–13.

———, 1981. Zur Biologie von *Acherontia atropos* (Lep.: Sphingidae) 4. Teil. *Ent. Z., Frankf. a. M.* **91**: 57–62.

———, 1982. Ergebnisse von Hybridzuchten zwischen *Hyles lineata* (Fabricius, 1775) und *Hyles livornica* (Esper, 1779) – 2. Teil. *Atalanta, Würzburg* **13**: 294–301.

———, 1988. Der *Hyles euphorbiae*-Komplex – ein taxonomisches Problem? (Lepidoptera: Sphingidae), 1. Teil. *Ent. Z., Frankf. a. M.* **98**: 81–86.

———, 1989. Der *Hyles euphorbiae*-Komplex – ein taxonomisches Problem? (Lepidoptera: Sphingidae), 3. Teil. *Ent. Z., Frankf. a. M.* **99**: 241–248.

———, 1991. Der *Hyles euphorbiae*-Komplex – ein taxonomisches Problem? (Lepidoptera: Sphingidae), 4. Teil. *Ent. Z., Frankf. a. M.* **101**: 120–127.

———, 1992. Der *Hyles euphorbiae*-Komplex – ein taxonomisches Problem? (Lepidoptera: Sphingidae), 5. Teil. *Ent. Z., Frankf. a. M.* **102**: 53–60.

HARBORNE, J. B., 1988. *Introduction to ecological biochemistry* (Edn 3), xv, 356pp. London.

HARIRI, G. EL-, 1971. *A list of the recorded insect fauna of Syria* **2**: iii, 306pp. Aleppo, University of Aleppo.

HARRIS, M., 1766. *The Aurelian: or, natural history of English insects; namely moths and butterflies* . . ., 92pp., 44 col. pls. London.

———, 1775. *The English Lepidoptera: or, the Aurelian's pocket companion* . . ., xv, 66pp., 1 col. pl. London.

HEATH, J. & SKELTON, M. J. (Eds) 1973. *Provisional atlas of the insects of the British Isles.* **2**: Lepidoptera (Moths – pt. 1), 6pp., 102 maps [Sphingidae: maps 30–38]. Huntingdon.

HEINEMANN, H. von, 1859. *Die Schmetterlinge Deutschlands und der Schweiz* **1**: xxiii, 848 + 118pp. Braunschweig.

HEINIG, S., 1976. Nachzucht von *Daphnis nerii* (Lep., Sphingidae). *Ent. Z., Frankf. a. M.* **86**: 25–30.

———, 1981a. Ein Beitrag zur Biologie von *Macroglossum stellatarum* (Lep.: Sphingidae). *Ent. Z., Frankf. a. M.* **91**: 177–188.

———, 1981b. Einige Beobachtungen zur Biologie des Linienschwärmers (*Hyles livornica* Esper) (Lep.: Sphingidae). *Ent. Z., Frankf. a. M.* **91**: 241–247.

———, 1982. Untersuchungen zur Flügelgeometrie und Auftriebserzeugung an freifliegenden Sphingiden (Lepidoptera). *Zool. Jb. Abt. Anat.* **107**: 519–544.

———, 1984. Zum Nahrungsbedarf von *Macroglossum stellatarum* (Lep.: Sphingidae). *Ent. Z., Frankf. a. M.* **94**: 97–106.

———, 1987. *Macroglossum stellatarum* (Sphingidae) – Flugverhalten. *Publ. Wiss. Film., Sekt. Biol.* **19**(1) E 2850: 3–18.

HERBULOT, C., 1971. *Atlas des Lépidoptères de France, Belgique, Suisse, Italie du Nord. II Hétérocères*, 144pp., 16 pls. Paris.

HERTING, B., 1984. Catalogue of Palaearctic Tachinidae (Diptera). *Stuttg. Beitr. Naturk.* (A), **369**: 1–228.

HIGGINS, L. G. & RILEY, N. D., 1980. *A field guide to the butterflies of Britain and Europe* (Edn 4), 384pp., 63 col. pls. London.

HINTON, H. E., 1946. On the homology and nomenclature of the setae of lepidopterous larvae, with some notes on the phylogeny of the Lepidoptera. *Trans. R. ent. Soc. Lond.* **97**: 1–37.

HINZ, R., 1983. The biology of the European species of the genus *Ichneumon* and related species (Hymenoptera: Ichneumonidae). *Contr. Am. ent. Inst.* **20**: 151–152.

HODGES, R. W., 1971. Sphingoidea, Hawkmoths. *In* Dominick, R. B. *et al.* (Eds). *The moths of America north of Mexico (including Greenland)* **21**: xii, 158pp., 14 pls. London.

HOEFNAGEL, J., 1592. *Archetypa Studiaque Patris* (4 pts), 11pp., 11 pls. Amsterdam.

HOFFMANN, F. & KLOS, R., 1914. Die Schmetterlinge Steiermarks. 2. *Mitt. naturw. Ver. Steierm.* **51**: 249–441.

HOFFMEYER, S., 1960. *De Danske Spindere* (Edn 2), 270pp., 24 pls. Aarhus.

HOFMANN, E., 1893. *Die Raupen der Groß-Schmetterlinge Europas* [Sphingidae: pp. 27–31, pls 6, 7], xxiv, 318pp., 50 pls. Stuttgart.

———, 1894. *Die Groß-Schmetterlinge Europas* (Edn 2) [Sphingidae: pp. 28–31, pls 16–18], xxxxi, 240pp., 71 pls. Stuttgart.

HOLIK, O., 1948. *Catalog der in den Sammlungen Dr. O. Staudinger und Otto Bang-Haas enthaltenen Lepidopteren*, vi, 320, [27]pp. Dresden.

HOLLOWAY, J. D., BRADLEY, J. D. & CARTER, D. J., 1987. Lepidoptera. *In* Betts. C. R., (Ed.). *CIE Guides to insects of importance to man* **1**: [iv], 262pp. Wallingford.

HRUBÝ, K., 1964. *Prodromus Lepidopter Slovenska*, 962pp. Bratislava.

ILTSCHEV, D., 1919. Vŭrkhu biologirata na *Daphnis* nerii L. Ueber die Biologie von *Daphnis nerii* L. (Sphingidae – Lepidoptera). *Spis. bulg. Akad. Nauk* **17**: 135–174, 2 pls (in Bulgarian & German).

INOUE, H., SUGI, S., KUROKO, H., MORIUTI, S. & KAWABE, A., 1982. *Moths of Japan* **2**. Plates and synonymic catalogue: 552pp., 392 pls. Tokyo [in Japanese and English].

INTERNATIONAL COMMISSION ON ZOOLOGICAL NOMENCLATURE, 1961. *International Code of Zoological Nomenclature*, xvi, 176pp. London.

————, 1985. *International Code on Zoological Nomenclature* (Edn 3), xx, 338pp. London, Berkeley & Los Angeles.

JORDAN, K., 1912. Familie: Sphingidae, Schwärmer. *In* Seitz, A., *Die Groß-Schmetterlinge der Erde* (Abt. 1) **2**: 229–260, pls 36–42. Stuttgart.

————, 1931. On the geographical variation of the pine hawk-moth, *Hyloicus pinastri*. *Novit. zool.* **36**: 243–249.

KALALI, GH.-H., 1976. A list of Lepidoptera from province of Khorasan (Iran). *J. ent. Soc. Iran* **3**: 131–135.

KARSCH, F., 1898. Giebt es ein System der recenten Lepidopteren auf phytogenetischer Basis? *Ent. Nachr. Berlin* **24**: 296–303.

KAZLAUSKAS, R., 1984. *Lietuvos drugiai*, 190pp., 70 pls. Vilnius [in Lithuanian].

KERNBACH, K., 1958. Über einige paläarktische Sphingidenarten und -unterarten (Lep. Sphingidae). *Dt. ent. Z.* (N.F.) **5**: 376–381.

————, 1959. Die Sphingidengattungen *Dolbina* Stgr. und *Rethera* R. & J. *Ent. Z., Frankf. a. M.* **69**: 253–260.

KIRBY, W. F., 1903. *The butterflies and moths of Europe*, lxxii, 432pp., 55 pls. London.

KISHIDA, Y., 1987. *Sphinx constricta* Butler, a distinct species to be separated from Eurasian *S. ligustri* Linnaeus (Lepidoptera: Sphingidae). *Tinea* **12** (Suppl.): 296–302 [in Japanese with English summary].

KOCH, J. & HEINIG, S., 1976. *Daphnis nerii* – ein Labortier? (Lep., Sphingidae). *Ent. Z., Frankf. a. M.* **87**: 57–62.

KOMÁREK, O. & TYKAČ, J., 1952. *Atlas Mótylů*, 113pp., 48 pls. Prague.

KOUTSAFTIKIS, A., 1970. *Vergleichend zoogeographische Untersuchung über die Lepidopterenfaunen der Nordägäischen Inseln Thassos, Samothraki und Limnos*, i, 134pp. Thesis, Universität des Saarlandes, Germany.

————, 1973. Ökologische und zoogeographische Untersuchungen der Sphingidae Griechenlands (Lepidoptera). *Ent. Z., Frankf. a. M.* **83**: 93–95.

————, 1974. Die Sphingiden-Arten der nordägaischen Insel Limnos. *Annls Mus. Goulandris* **2**: 89–91.

KOVÁCS, L., 1953. A magyarországi nagylepkék és elterjedésük. Die Groß-Schmetterlinge Ungarns und ihre Verbreitung. *Folia ent. hung.* (N.S.) **6**: [77]–164 [in Hungarian and German].

KUMAKOV, A. P., 1977. Heterocera and Rhopalocera (Macrolepidoptera) of the chalk hills in the vicinity of Saratov. *Ent. Obozr.* **56**: 765–775 [in Russian].

KUZNETSOV, V. I. & STEKOLNIKOV, A. A., 1985. Comparative and functional morphology of the male genitalia of bombycoid moths (Lepidoptera, Papilionomorpha: Lasiocampoidea, Sphingoidea, Bombycoidea) and their systematic position. *Trudy. zool. Inst. Leningr.* [Proc. zool. Inst. USSR Acad. Sci.) **134**: 3–48 [In Russian with English summary].

KUZNETSOVA, N. Ya., 1906. A review of the family Sphingidae of the Palearctic and Chinese-Himalayan Faunas. *Horae Soc. ent. ross.* **37**: 293–346 [in Russian]. [English edn, 1972, 43pp. New Delhi].

LACASA, A., GARRIDO, A., RIVERO, J. M. *et al.*, 1979. Algunos Lepidópteros de los arrozales de Sevilla y Valencia. *Revta Lepid.* **7**: 41–46.

L'ADMIRAL, J., 1740. *Naauwkeurige waarneemingen van veele gestaltverwisselende gekorvene Diertjes*, iv + 16pp., 33 pls. Amsterdam.

————, 1744. *Naauwkeurige waarneemingen omtrent de veranderingen van veele Insekten*, ii, 34pp., 33 pls. Amsterdam.

LATREILLE, P. A., [1802]. Familles naturelles des genres, Sphingides. *In* Buffon, Comte G. L. L. de, *Histoire naturelle genérale et particulière des crustacés et des insectes* (Ed. C. S. Sonnini). (Suites) **3**: 400–402. Paris.

LAVERY, T. A., 1991. The hawkmoths of Ireland. *Bull. amat. Ent. Soc.* **50**: 73–80.

LEDERER, G., 1944. Das Auftreten des Wanderschwärmers *Deilephila nerii* L. in der Mainebene sowie Freilandbeobachtungen über die Lebensweise dieser Art. *Z. wien. ent. Ges.* **29**: 293–299.

LEMPKE, B. J., 1937. Catalogus der Nederlandsche Macrolepidoptera. *Tijdschr. Ent.* **80**: 244–303.

LHONORÉ, J., 1988. Notes sur *Proserpinus proserpina* Pallas, 1772 (Lepidoptera, Sphingidae). *Alexanor* **15**: 322.

LINNAEUS, C., 1746. *Fauna Suecica . . .*, xxviii, 412pp., 2 pls. Stockholm.

————, 1758. *Systema naturae* (Edn 10) **1**: 824pp. Stockholm.

LITVINCHUK, L. N., 1986. Ethological observations on the moths and caterpillars of the larch sphinx *Hyloicus morio* Rothschild & Jordan (Lepidoptera, Sphingidae) in pine forests near the Ob'. *Izv. sib. Otdel. Akad. Nauk SSSR*, (Biol. Nauk) (6): 129–134 [in Russian].

LUCAS, W. J., 1895. *The book of British hawk-moths*, x, 157pp., frontispiece + 14 pls. London.

MABBERLEY, D. J., 1987. *The plant-book, a portable dictionary of the higher plants*, xii, 706pp. Cambridge.

MANN, J., 1855. Die Lepidopteren gesammelt auf einer entomologischen Reise in Corsika. *Verh. zool.-bot. Ver. Wien* **5**: 529–572.

MARIANI, M., 1939. Fauna Lepidopterorum Siciliae. *Mem. Soc. ent. ital.* **17**: 129–187.

MATHUR, R. N., 1959. *Mermis* sp. (Mermithidae, Ascaroidea, Nematoda) and its insect hosts. *Curr. Sci.* **28**: 255–256.

MATSUMURA, S., 1929. New moths from Kuriles. *Insecta matsum.* **3**: 165–168.

MEERMAN, J. C., 1987. De Nederlandse Pijlstaartvlinders (Lepidoptera: Sphingidae). *Wetenschappelijke Mededeling van de Koninklijke Nederlandse Natuurhistorische Vereniging* **180**: 60pp. Hoogwoud.

————, 1988a. Overlevingsstrategieën bij pijlstaartvlinders en hun rupsen. *Natura* **85**(5): 139–144.

————, 1988b. The subspecies of *Hyles tithymali* with a description of a new subspecies (Lepidoptera: Sphingidae). *Ent. Ber., Amst.* **48**: 61–67.

———— & SMID, G., 1988. Der *Hyles euphorbiae*-Komplex; die Wolfsmilchschwärmer von Kreta (Lepidoptera: Sphingidae). *Ent. Z., Frankf. a. M.* **98**: 161–170.

————, 1991. *Hyles euphorbiae himyarensis* from Yemen (Lepidoptera: Sphingidae). *Trop. Lepid.* **2**: 107–109.

MENTZER, E. von, 1974. *Sphingonaepiopsis pfeifferi* Zerny bona species und *S. pfeifferi* ssp. nova *chloroptera* aus Jugoslawien (Lep., Sphingidae). *Acta ent. Jugosl.* **10**: 147–153.

MERIAN, M. S., 1679. *Der Raupen wunderbare Verwandelung und sonderbare Blumen-nahrung . . .* [7], 102pp., 50 pls. Nuremberg.

————, 1683. *Der Raupen wunderbare Verwandelung und sobderbare Blumen-nahrung . . . Anderer Theil*, 100pp, 50 pls. Frankfurt-am-Main.

————, 1705. *Metamorphosis insectorum Surinamensium*, [iv], 60pp., 60 pls. Amsterdam.

MERZHEEVSKAYA, I., LITVINOVA, A. N. & MOLCHANOVA, R. V., 1976. *Cheshnekrylye (Lepidoptera) Belorussii, Katalog*, 129pp. Minsk [in Russian].

MEYER, J. H., 1953. Die Bluttransfusion als Mittel zur Überwindung letaler Keimkombination bei Lepidopteren-Bastarden. *Z. wien. ent. Ges.* **38**: 44–62, 3 pls.

MEYER, M., 1991. Les Lépidoptères de la région macaronésienne. II. Liste des Macro-Hétérocères observés en juillet-août 1990 aux Açores (Lepidoptera: Geometridae, Sphingidae, Noctuidae). *Linn. Belg.* **13**: 117–134.

MILYANOVSKIĬ, E. S., 1959. The adaptive coloration of *Celerio vespertilio* Esp. (Lepidoptera: Sphingidae) moths and caterpillars. *Ent. Obozr.* **38**: 223–224 [in Russian].

MINA-PALUMBO, F. & FAILLA-TEDALDI, L., 1889. *Materiali per la fauna Lepidopterologica della Sicilia.* 144pp. Palermo. [Reprinted from *Naturalista Sicil.*: (1887–89) **6–8**].

MINET, J., 1986. Ébauche d'une classification moderne de l'orde des Lépidoptères. *Alexanor* **14**: 291–313.

MOUFFET, T. 1634. *Insectorum sive minimorum animalium theatrum* [20], 326, [4]pp. London.

MYARTZEVA, S. N. & TOKEAEW, R. T., 1972. *Nasekomÿe Yuzhnoĭ Turkmenii.* Ashkhabad [in Russian].

NEWMAN, L. H., 1965. *Hawk-moths of Great Britain and Europe*, [xx], 148pp., 24 pls. London.

NORDSTRÖM, F., 1955. De Fennoskandiska Dagfjärilarnas Utbredning. Lepidoptera diurna (Rhopalocera & Hesperioidea). *Acta Univ. Lund.* (N.F.) Avd. 2, **51**(1): 1–176 [incl. 181 distribution maps], 1 map.

——, OPHEIM, M. & SOTAVALTA, O., 1961. De Fennoskandiska Svärmarnas och Spinnarnas Utbredning (Sphinges, Bombycimorpha, etc.). *Acta Univ. Lund.* N.F. Avd. 2, **57**(4): 1–92, 182 maps.

PACKARD, A. S., 1895. On a new classification of the Lepidoptera. *Am. Nat.* **29**: 636–647, 788–809.

PELZER, A., 1982. Zur Kenntnis der frühen Stände von *Hyles centralasiae siehei* (Püngeler) (Sphingidae). *Nota lepid.* **5**: 134–140.

——, 1989. Die Präimaginalstadien von *Laothoe amurensis* – ein Vergleich mit *L. populi* (Lepidoptera: Sphingidae). *Nota lepid.* **11**: 274–278.

——, 1991. Illustrierter Bestimmungsschlüssel für die Präimaginalstadien der Schwärmer Europas und Nordafrikas (Lepidoptera: Sphingidae). Teil 1: Erwachsene Raupen. *Nota lepid.* **14**: 220–233.

PÉREZ LOPEZ, F.J., 1989. Citas interesantes de Heteróceros en la provincia de Granada (Lepidoptera: Heterocera). *Revta Lepid.* **17**: 159–164.

PETERSEN, W., 1924. *Lepidopteren-Fauna von Estland (Eesti)* Teil I (Edn 2): 316pp. Tallinn – Reval.

PIERCE, F. N. & BEIRNE, B. P., 1941. *The genitalia of the British Rhopalocera and the larger moths*, 66pp., 21 pls. Peterborough. [Facsimile reprint, 1975. Faringdon.]

PIERRON, M., 1990. Contribution à la connaissance de la biologie de *Papilio machaon saharae* Obth. Différences avec *Papilio machaon machaon* L. et hybridations expérimentales (Lep. Papilionidae). *Alexanor* **16**: 331–340.

PIHLAJAMÄKI, J., VÄÄNÄNEN, V.-M., KOSKINEN, P. & NUORTEVA, P., 1989. Metal levels in *Laothoe populi* and *Sphinx pinastri* (Lepidoptera, Sphingidae) in Finland. *Ann. Entomol. Fennici* (*Suom. hyönt. Aikak.*). **55**: 17–21.

PITTAWAY, A. R., 1976. *Hyles (Celerio) tithymali. Bull. amat. Ent. Soc.* **35**: 54–57.

——, 1979a. On *Rethera komarovi manifica* (Brandt) (Lepidoptera: Sphingidae). *Entomologist's Gaz.* **30**: 3–6, 1 pl.

——, 1979b. The butterflies and hawk-moths of eastern Saudi Arabia. *Proc. Trans. Br. ent. nat. Hist. Soc.* **12**: 90–101, 2 pls.

——, 1981. Further notes on the butterflies and hawk-moths

(Lepidoptera) of eastern Saudi Arabia. *Entomologist's Gaz.* **32**: 27–35.

——, 1982a. *Hyles hippophaes hippophaes* (Esper) (Lepidoptera: Sphingidae). *Entomologist's Gaz.* **33**: 97–98, 1 pl.

——, 1982b. Notes on the subspecies and biology of *Clarina kotschyi* (Kollar, (1849)) (Lepidoptera: Sphingidae). *Entomologist's Gaz.* **33**: 173–174, 1 pl.

——, 1983a. Of foodplants and hawkmoths. *Bull. amat. Ent. Soc.* **42**: 64–69.

——, 1983b. An annotated checklist of the western Palaearctic Sphingidae (Lepidoptera). *Entomologist's Gaz.* **34**: 67–85.

——, 1987. The 'Arabian Sphinx'. *Ahlan Wasahlan* **11**: 42–45.

PLANAS, A. M., DE-GREGORIO, J. P. & CASTELLS, L. S., 1979. Primera cita d'*Hyles dahlii* (Lep. Sphingidae) al continent Europeu. *Treb. Soc. catal. Lepidòpterol.* **2**: 11–17.

PLUGARU, S. G., 1971. A hawkmoth new to the USSR, *Dolbina elegans* (Lepidoptera, Sphingidae). *Izv. Akad. Nauk moldav.* (1): 88–89 [in Russian].

PODA, N., 1761. *Insecta musei graecensis*, [vi], 127, [12]pp., 2 pls. Graz. [Facsimile reprint, 1915. Berlin.]

POLUNIN, O., 1969. *Flowers of Europe*, 662pp., 192 pls. London.

—— & WALTERS, M., 1985. *A guide to the vegetation of Britain and Europe*, x, 238pp., 170 pls. Oxford.

POPESCU-GORJ, A., 1971. *Dolbina elegans* Bang-Haas (Lep.: Sphingidae) en Roumanie. *Trav. Mus. Hist. nat. "Gr. Antipa"*, **11**: 219–225.

PRATT, C., 1985. *Proserpinus proserpina* Pall. (Lep.: Sphingidae) new to Britain. *Entomologist's Rec. J. Var.* **97**: 147.

PÜNGELER, R., 1903. *Deilephila siehei* n. sp. *Berl. ent. Z.* **47** (1902): 235–238, 1 pl.

RANGNOW, R., 1935. Neue Lepidopteren aus Lappland. *Ent. Rdsch.* **52**: 188–191.

RAY, J., 1710. *Historia Insectorum*, xv, 400pp. London.

RÉAUMUR, R. A. F. de, 1736. *Mémoires pour servir à l'histoire des insectes* **2**: xlvi, 514pp., 40 pls. Paris.

REAVEY, D. & GASTON, K., 1991. Caterpillars in Wonderland. *New Sci.* **129** (1757): 52–55.

REBEL, H., 1926. Lepidopteren von den Balearen. *Dt. ent. Z. Iris* **40**: 135–146.

——, 1933. Neue Lepidopteren von Ankara. *Z. öst. EntVer.* **18**: 23–24.

——, 1934. Lepidopteren von den Balearen und Pityusen. *Dt. ent. Z. Iris* **48**: 122–138.

—— & ZERNY, H., 1934. Wissenschaftliche Ergebnisse der im Jahre 1918 entsendeten Expedition nach Nordalbanien. Die Lepidopterenfauna Albaniens (mit Berücksichtigung der Nachbargebiete). *Denkschr. Akad. Wiss. Wien* **103**: 37–161, 1 pl., 1 map.

REMOROV, V. V., 1984. The role of chemoreceptors of the mouthparts and antennae in reactions of caterpillars to food-attractants and repellents. *Zool. Zh.* **63**: 62–68 [in Russian].

——, 1985. Growth and development of larvae of the vine [elephant] hawkmoth (*Deilephila elpenor* L.) and the oak silkmoth (*Antheraea pernyi* Guer.) on hostplants other than their own. *Vest. leningr. gos. Univ., Biol.* (3): 16–22 [in Russian].

RILEY, D. & YOUNG, A., 1968. *World vegetation*, 96pp. Cambridge.

ROBINSON, G. S. & KIRKE, C. M. St G., 1990. Lepidoptera of Ascension Island – a review. *J. nat. Hist.* **24**: 119–135.

ROEDER, K. D., 1975. Acoustic interneuron responses compared in certain hawk moths. *J. Insect Physiol.* **21**: 1625–1631.

RONDOU, J.-P., 1903. *Catalogue Raisonné des Lépidoptères des Pyrénées*, 179pp. Paris.

RÖSEL VON ROSENHOF, A. J., [1746]. *Insecten Belustigung* 1: [xlvi], 64pp., pls 1–10, [viii], 60pp., pls 1–10, [viii], 64pp., pls 1–8, [viii], 312pp., pls 1–63, [viii], 48pp., pls 1–13, 48pp., pls 1–17, xxiv pp. Nuremberg.

———, 1755. *Insecten Belustigung* 3: [viii], 624, [viii]pp., 101 pls. Nuremberg.

———, 1761. *Insecten Belustigung* 4: [x], 264, [iv]pp., 40 pls. Nuremberg.

ROTHSCHILD, W., 1917. Supplemental notes to Mr. Charles Oberthür's *Faune des Lépidoptères de la Barbarie* with lists of the specimens contained in the Tring Museum. *Novit. zool.* 24: 61–120, 325–373, 393–409.

——— & JORDAN, K., 1903. A revision of the lepidopterous family Sphingidae. *Novit. zool.* 9 (Suppl.): i–cxxxv, 1–972, 67 pls.

ROTHSCHILD, M., 1985. British Aposematic Lepidoptera. *In* Heath, J. & Emmet, A. M. (Eds), *The moths and butterflies of Great Britain and Ireland* 2: 9–62, 2 pls. Colchester.

———, 1989. Moths and Memory. *Endeavour* (N.S.) 13: 15–19.

ROÜAST, G., 1883. *Catalogue des chenilles Européennes connues*, 196pp. Lyons.

ROUGEOT, P.-C., 1972. Capture de *Macroglossum stellatarum* L. au Congo-Brazzaville (Sphingidae). *Alexanor* 7: 335.

——— & VIETTE, P., 1978. *Guide des Papillons nocturnes d'Europe et d'Afrique du Nord. Héterocères (Partim)*, 228pp., 40 pls. Paris.

RUNGS, C. E., 1977. Notes de Lépidoptérologie Corse. *Alexanor* 10: 178–188.

———, 1981. *Catalogue raisonné des Lépidoptères du Maroc 2*: 365pp. Rabat. [Reprinted from *Trav. Inst. scient. chérif., Série Zool.* (40): 223–583.]

———, 1988. Liste-Inventaire systématique et synonymique des Lépidoptères de Corse. *Alexanor* 15 (Suppl.): 1–86.

SBORDONI, V. & FORESTIERO, S., 1985. *The world of butterflies*, 312pp. Poole.

SCHROEDER, H., 1990. *Menyanthes trifoliata*, eine weitere Nahrungspflanze von *Deilephila elpenor*? (Lepidoptera: Sphingidae). *Ent. Z., Frankf. a. M.* 100: 31–32.

SCHROEDER, L. A., 1986. Changes in tree leaf quality and growth performance of lepidopteran larvae. *Ecology, USA* 67: 1628–1636.

SCHURIAN, K. G. & GRANDISCH, H., 1991. Anmerkung zur Biologie von *Hyles euphorbiae tithymali* (Lepidoptera: Sphingidae). *Nachr. ent. Ver. Apollo* 11: 253–256.

SCHWEIZERISCHER BUND FÜR NATURSCHUTZ, 1987. *Tagfalter und ihre Lebensräume. Arten, Gefährdung, Schutz*, xi, 516pp., 25 pls. Basle.

SCLATER, P. L., 1858. On the general Geographical Distribution of the Members of the class Aves. *J. Proc. Linn. Soc., Zoology* 2: 130–145.

SCOBLE, M. J., 1991. Classification of the Lepidoptera. *In* Emmet, A. M. & Heath, J. (Eds), *The Moths and Butterflies of Great Britain and Ireland*, 7(2): 11–45. Colchester.

———, 1992. *The Lepidoptera – form, function and diversity*, xii, 404pp., 321 figs. Oxford.

SCOPOLI, J. A., 1763. *Entomologia Carniolica* . . ., xxxvi, 422pp., 43 pls. Vienna.

SCORTICHINI, M., 1986. Frutti minori. Il corbezzolo. *Riv. Fruttic. Ortofloric.* 48: 43–49.

SCRIBER, J. M., LINTEREUR, G. L. & EVANS, M. H., 1982. Foodplant suitabilities and a new oviposition record for *Papilio glaucus canadensis* (Lepidoptera: Papilionidae) in northern Wisconsin and Michigan. *Gr. Lakes Ent.* 15: 39–46.

——— & LEDERHOUSE, R. C., 1983. Temperature as a factor in the development and feeding ecology of tiger swallowtail caterpillars, *Papilio glaucus* (Lepidoptera). *Oikos* 40: 95–102.

SEITZ, A., 1906–13. *Die Groß-Schmetterlinge der Erde* 2: viii, 479pp., 56 pls. Stuttgart.

———, 1930–33. *Die Groß-Schmetterlinge der Erde* 2 (Suppl.): vii, 315pp. 16 pls. Stuttgart.

SEPP, C. & SEPP, J. C., 1762–1811. *De Nederlandsche insekten* 1–4 [many col. pls]. Amsterdam.

SEPPÄNEN, E. J., 1954. Die suurperhostoukkien ravintokasvit. Die Futterpflanzen der Großschmetterlingsraupen Finnlands. *Suom. Eläim.* 8: xvi, 414pp. Helsinki.

SHAW, M. R. & ASKEW, R. R., 1976. Parasites. *In* Heath, J. (Ed.), *The Moths and Butterflies of Great Britain and Ireland* 1: 24–56. London & Colchester.

———, 1982. Parasitic control. *In* Feltwell, J., Large White Butterfly: the biology, biochemistry and physiology of *Pieris brassicae* (Linnaeus). *Series Ent.* 18: 401–407.

———, 1990. Parasitoids of European butterflies and their study. 12.3 Host Associations. *In* Kudrna, O. (Ed.), *Butterflies of Europe* 2: 452–455. Wiesbaden.

SHCHETKIN, Yu. L., 1949. A hawkmoth new to our fauna. *Soobshch. Akad. Nauk tadzhik. SSR* (19): 24–27 [in Russian].

———, 1956. On the Sphingidae fauna of Tadzhikistan (Lepidoptera). *Izv. Akad. Nauk tadzhik. SSR*, (Otdel. Estestvennỹkh Nauk) (16): 143–156 [in Russian].

———, 1981. New subspecies of *Eumorpha suellus* Stgr. from Pamir-Alaya (Lepidoptera: Sphingidae). *Izv. Akad. Nauk tadzhik. SSR*, (Otdel. Biol.) (4): 90–92 [in Russian].

———, DEGTYAREVA, V. I. & SHCHETKIN, Yu. Yu., 1988. *Smerinthus kindermanni* in Tajhikistan (Lepidoptera, Sphingidae). *Izv. Akad. Nauk tadzhik. SSR*, (Otdel. Biol. Nauk.) (4): 67–70 [in Russian].

SHELJUZHKO, L., 1933. Zwei neue Rassen von *Celerio hippophaes* Esp. *Mitt. münch. ent. Ges.* 23: 43–45.

SIEBOLD, C. T. E. von, 1842. Ueber die Fadenwürmer der Insekten. *Stett. Ent. Ztg.* 3: 146–161.

SKINNER, B., 1984. *Colour identification guide to moths of the British Isles*, x, 267pp., 42 pls. Harmondsworth.

SKVORTSOV, V. & THOMSON, E., 1974. Über das Vorkommen und die Biologie von *Laothoe amurensis* in Estland (Lep., Sphingidae). *Ent. Z., Frankf. a. M.* 84: 59–63.

SOFFNER, J., 1959. *Dolbina elegans* O. Bang-Haas in Europa (Lep. Sphingidae). *Ent. Z., Frankf. a. M.* 69: 269–270.

SOKOLOFF, P., 1984. *Breeding the British and European Hawkmoths*, 56pp. Feltham.

SOUTH, R., 1907. *The moths of the British Isles* 1: vi, 343pp., 159 pls. London.

———, 1961. *The moths of the British Isles* (Edn. 4, ed. and rev. by Edelstein, H. M. & Fletcher, D. S.) 1: 427pp., 148 pls London & New York.

SPEIDEL, W. & KALTENBACH, T., 1981. Eine neue Unterart des Abendpfauenauges aus Sardinien (Lepidoptera, Sphingidae). *Atalanta, Würzburg* 12: 112–116.

——— & HASSLER, M., 1989. Die Schmetterlingsfauna der südlichen algerischen Sahara und ihrer Hochgebirge Hoggar und Tassili n'Ajjer (Lepidoptera). *Nachr. ent. Ver. Apollo* 8 (Suppl.): 1–156.

SPENCER, K. C., 1988. *Chemical mediation of coevolution*, xv, 609pp., maps. San Diego & London.

SPERLING, F. A. H., 1990. Natural hybrids of *Papilio* (Insecta: Lepidoptera): poor taxonomy or interesting evolutionary problem? *Can. J. Zool.* 68: 1790–1799.

SPULER, A., 1908. *Die Schmetterlinge Europas* 1: cxxviii, 385 pls. Stuttgart.

STAUDER, H., 1921a. *Celerio lineata livornica* Esp. subsp. nova *saharae* Stdr. *Dt. ent. Z. Iris* **35**: 179–181.

——, 1921b. Neues aus Unteritalien. *Dt. ent. Z. Iris* **35**: 26–31.

——, 1928. Neue Lepidopterenformen aus Sizilien. *Lepid. Rdsch.* **2**: 113–116.

——, 1930. *Celerio euphorbiae* L. *dolomiticola* Stauder, nova subspecies (Lep.). *Ent. Z., Frankf. a. M.* **43**: 268–270.

STAUDINGER, O., 1874. Einige neue Lepidopteren des europäischen Faunengebiets. *Stettin. ent. Ztg* **35**: 87–98.

——, [1879]–81. Lepidopteren-Fauna Kleinasien's. *Horae Soc. ent. ross.* **14** (1878): 176–482; **15** (1879): 159–435; **16** (1881): 65–135 [Sphingidae: pp. 98–105].

—— & REBEL, H., 1901. *Catalog der Lepidopteren des palaearctischen Faunengebietes* 1. Theil: Famil. Papilionidae – Hepialidae, xxxii, 368pp. Berlin.

STAVRAKIS, G. N., 1976. Outbreak of *Acherontia atropos* L. on olive trees. *Pl. Prot. Bull F.A.O.* **24**: 28.

SUTTON, S. L., 1963. South Caspian Insect Fauna 1961. I. Systematic List of Lepidoptera with notes. *Ann. Mag. nat. Hist.* **(13) 6**: 353–374.

SWAMMERDAM, J., 1669. *Historia generalis insectorum* [Flemish edn], [xxviii], 168 + 48pp, 13 pls. Utrecht.

——, 1758. *The book of nature; or the history of insects* [English edn], (trans. by Flloyd, T., rev. and ed. by Hill, J.) Pt. 1: xxvi, 236pp.; Pt. 2: 153, lxiii, 12pp., 53 pls. London.

TASHLIEV, A. O. (Ed.), 1973. *Ekologiya Nasekomykh Turkmenii*, 188pp. Ashkhabad.

TATCHELL, L., [1926]. *Observations & notes on the Sphingidae (Hawk moths) of the Swanage District*, 12pp. Swanage.

THEUNERT, R., 1990. *Zur Ökologie der Raupen des Mittleren Weinschwärmers* (Deilephila elpenor) *(Lepidoptera: Sphingidae)*. Thesis, Technischen Universität Braunschweig, Germany.

THOMPSON, J. N., 1988. Variation in preference and specificity in monophagous and oligophagous swallowtail butterflies. *Evolution* **42**: 118–128.

THOMSON, E., 1967. *Die Großschmetterlinge Estlands*, 203pp., 1 map. Stollhamm (Oldb.).

THORPE, K. W. & BARBOSA, P., 1986. Effects of consumption of high and low nicotine tobacco by *Manduca sexta* (Lepidoptera: Sphingidae) on survival of gregarious endoparasitoid *Cotesia congregata* (Hymenoptera: Braconidae). *J. chem. Ecol.* **12**: 1329–1337.

TILLYARD, R. J., 1918. The Panorpoid complex. Part 1 – The wing coupling apparatus, with special reference to the Lepidoptera. *Proc. Linn. Soc. N.S.W.* **43**: 285–319, 2 pls.

——, 1919. The Panorpoid complex. Part 3 – The wing venation. *Proc. Linn. Soc. N.S.W.* **44**: 533–718, 5 pls.

TINBERGEN, N., IMPEKOVEN, M. & FRANCK, D., 1967. An experiment on spacing-out as a defence against predation. *Behaviour* **28**: 307–321, 1 pl.

TUTT, J. W., 1902. *A natural history of the British Lepidoptera; a text-book for students and collectors* **3**: xi, 558pp. London & Berlin.

——, 1904. *A natural history of the British Lepidoptera; a text-book for students and collectors* **4**: xvii, 535pp. London & Berlin.

USINGER, R. L., 1964. The role of Linnaeus in the advancement of Entomology. *Ann. Rev. Ent.* **9**: 1–16.

VALETTA, A, 1973. *The moths of the Maltese islands.* 118pp. Malta.

VAN DER HEYDEN, T. 1990. Ergebnisse von Lichtfalleneinsätzen auf Gran Canaria/Spanien im Hinblick auf Arctiidae, Ly-

mantriidae, Notodontidae und Sphingidae (Lepidoptera). *Ent. Z., Frankf. a. M.* **100**: 153–160.

——, 1991. Freilandfunde von Sphingiden und deren Präimaginalstadien auf Gran Canaria/Spanien (Lepidoptera). *Nachr. ent. Ver. Apollo* **11**: 247–250.

VAN DER SLOOT, M., 1967. La classification des Smerinthes. *Lambillionea* **66**: 96–99.

VARIS, V., 1976. *Histriosphinx* Varis, 1976, a junior objective synonym of *Daphnis* Hübner, 1819 (Lepidoptera, Sphingidae). *Notulae ent.* **57**: 58.

VERITY, R., 1911. Alcuni Lepidotteri inediti o non ancora figurati. *Boll. Soc. ent. ital.* (1910) **42**: 266–281, 1 pl.

VIEDMA, M. G. de & GÓMEZ BUSTILLO, M. R., 1981. *Hippotion osiris* (Dalman, 1823) y otras nuevas citas del Coto Nacional de las Sierras de Cazorla y Segura (Lepidoptera-Heterocera). *Revta Lepid.* **9**: 125–126.

VIIDALEPP, J., 1979. Contribution to the lepidopterous fauna of the Tuvinian ASSR, II. Lepidoptera, Heterocera (Families Zygaenidae – Cossidae). *Uchen. Zap. tartu gos. Univ.* **12**: 17–39 [in Russian].

VIVES MORENO, A., 1981. *Hippotion osiris* (Dalman, 1823), en España. Historia bibliográfica (Lep. Sphingidae). *Revta Lepid.* **9**: 223–229.

VORBRODT, K. & MÜLLER-RUTZ, J., 1911. *Die Schmetterlinge der Schweiz* **1**: lv, 489pp., 2 maps. Berne.

WAHLBERG, J. A. & WALLENGREN, H. D. J., 1857. Kafferlandets Dag-fjärilar, insamlade åren 1838–1845. Lepidoptera Rhopalocera, in Terra Caffrorum. Annis 1838–1845. *K. svenska VetenskAkad. Handl.* **2**(4): 5–55.

—— & ——, 1864. Heterocer-Fjärilar, samlade i Kafferlandet. *K. svenska VetenskAkad. Handl.* **5**(4): 1–83.

WARING, P., 1992. Progress on Atlas of Red Data Book and Nationally Scarce Macro-moths. *Antenna* **16**: 136.

WATKINS, H. T. G. & BUXTON, P. A., 1923. Moths of Mesopotamia and N.W. Persia, part II. Sphinges & Bombyces. *J. Bomb. nat. Hist. Soc.* **28**: 184–186.

WEI, H. & ZHENG, L., 1987. Observations on the biological characteristics of *Microplitis ocellatae* Bouché. *Natural Enemies of Insects* **9**: 10–12 (in Chinese).

WHITCOMBE, R. P. & ERZINÇLIOGLU, Y. Z., 1989. The convolvulus hawk-moth, *Agrius convolvuli* (Lep.: Sphingidae) and its parasite *Zygobothria ciliata* (Dipt.: Tachinidae) in Oman. *J. Oman Stud.* **10**: 77–84.

WILKES, B., 1742. *Twelve new designs of English butterflies*, frontispiece, 12 pls. London.

——, [1749]. *The English moths and butterflies . . .*, 8, [22], 64, [4] pp., 120 pls. London.

WILSON, O. S., 1880. *The larvae of the British Lepidoptera and their food plants*, xxiv, 367pp., 40 pls. London.

WILTSHIRE, E. P., 1957. *The Lepidoptera of Iraq*, 162pp., 17 pls. London.

——, 1975a. Lepidoptera: Part I. Families Cossidae, Pyralidae, Geometridae, Sphingidae, Arctiidae, Lymantriidae and Noctuidae. *In* Scientific results of the Oman flora and fauna survey, 1975. *J. Oman Stud.*, Spec. Rep. No. 1: 155–160.

——, 1975b. Lepidoptera: Part II. A list of further Lepidoptera – Heterocera from Oman, with remarks on their economic importance and descriptions of one new genus, four new species, and two new subspecies. *In* Scientific results of the Oman flora and fauna survey, 1975, *J. Oman Stud.*, Spec. Rep. No. 1: 161–176.

——, 1980. Insects of Saudi Arabia. Lepidoptera: Fam. Cossidae, Limacodidae, Sesiidae, Lasiocampidae, Sphingidae, Notodontidae, Geometridae, Lymantriidae, Nolidae,

Arctiidae, Agaristidae, Noctuidae, Ctenuchidae. *Fauna of Saudi Arabia* **2**: 179–240.

———, 1986. Lepidoptera of Saudi Arabia: Fam. Cossidae, Sesiidae, Metarbelidae, Lasiocampidae, Sphingidae, Geometridae, Lymantriidae, Arctiidae, Nolidae, Noctuidae (Heterocera); Fam. Satyridae (Rhopalocera) (Part 5). *Fauna of Saudi Arabia* **8**: 262–323.

———, 1990. An Illustrated, Annotated Catalogue of the Macro-Heterocera of Saudi Arabia. *Fauna of Saudi Arabia* **11**: 91–250.

WOLFF, N. L., 1971. Lepidoptera. *The Zoology of Iceland* **3**(45): 193pp. Copenhagen & Reykjavik.

WORMS, C. G. M. de, 1964. Madeira in the spring, April 1964. *Entomologist's Rec. J. Var.* **76**: 252–254.

ZERNY, H., 1933. Lepidopteren aus dem nördlichen Libanon. *Dt. ent. Z. Iris* **47**: 60–109.

———, 1936. Die Lepidopterenfauna des Grossen Atlas in Marokko und seiner Randgebiete. *Mém. Soc. Sci. nat. Phys. Maroc.* (42) (1935): 1–157pp., 2 pls.

Notes and Addenda

Notes and Addenda

.

Notes and Addenda

Notes and Addenda

Notes and Addenda

Notes and Addenda

Notes and Addenda

Notes and Addenda

COLOUR PLATES

Plate 1 – Sphinginae (Larvae)

Figs 1, 2, 5, 8, 9, natural size; Figs 3, 4, × 1.25; Figs 6, 7, × 1.1

1 *Mimas tiliae* (Linnaeus), Germany
(*Page* 96)

2 *Acherontia atropos* (Linnaeus), yellow form, Canary Islands
(*Page* 82)

3 *Marumba quercus* ([Denis & Schiffermüller]), Spain
(*Page* 94)

4 *Sphinx ligustri ligustri* Linnaeus, Austria
(*Page* 85)

5 *Smerinthus ocellatus ocellatus* (Linnaeus), green form, Austria
(*Page* 100)

6 *Sphinx pinastri pinastri* Linnaeus, England
(*Page* 87)

7 *Smerinthus kindermanni kindermanni* Lederer, green form, Iran
(*Page* 98)

8 *Acherontia styx styx* (Westwood), green form, Saudi Arabia
(*Page* 84)

9 *Agrius convolvuli convolvuli* (Linnaeus), green form, Saudi Arabia
(*Page* 81)

Plate 1

Plate 2 – Sphinginae; Macroglossinae (Larvae)

Figs 1, 3–6, 8–9, 11, 13, 14, natural size; Figs 2, 7, 12, × 1.1; Fig. 10, × 1.5

1 *Daphnis nerii* (Linnaeus), green form, Saudi Arabia
   (*Page* 115)
2 *Hippotion osiris* (Dalman), brown form, West Africa
   (*Page* 161)
3 *Laothoe populi populi* (Linnaeus), green form with red spots, Germany
   (*Page* 102)
4 *Hyles tithymali deserticola* (Staudinger), Morocco
   (*Page* 140)
5 *Clarina kotschyi kotschyi* (Kollar), Iran
   (*Page* 118)
6 *Sphingonaepiopsis gorgoniades chloroptera* Mentzer, Macedonia
   (*Page* 128)
7 *Hemaris tityus tityus* (Linnaeus), Austria
   (*Page* 113)
8 *Macroglossum stellatarum* ((Linnaeus), blue-green form, Iran
   (*Page* 134)
9 *Hyles euphorbiae conspicua* (Rothschild & Jordan), Izmir, Turkey
   (*Page* 138)
10 *Hyles euphorbiae euphorbiae* (Linnaeus), half-grown final instar, N.E. Austria
   (*Page* 137)
11 *Proserpinus proserpina proserpina* (Pallas), Alto Adige/Südtirol, Italy
   (*Page* 131)
12 *Hyles zygophylli zygophylli* (Ochsenheimer), pale form, Turkey
   (*Page* 149)
13 *Hemaris fuciformis fuciformis* (Linnaeus), England
   (*Page* 111)
14 *Rethera komarovi manifica* Brandt, Iran
   (*Page* 125)

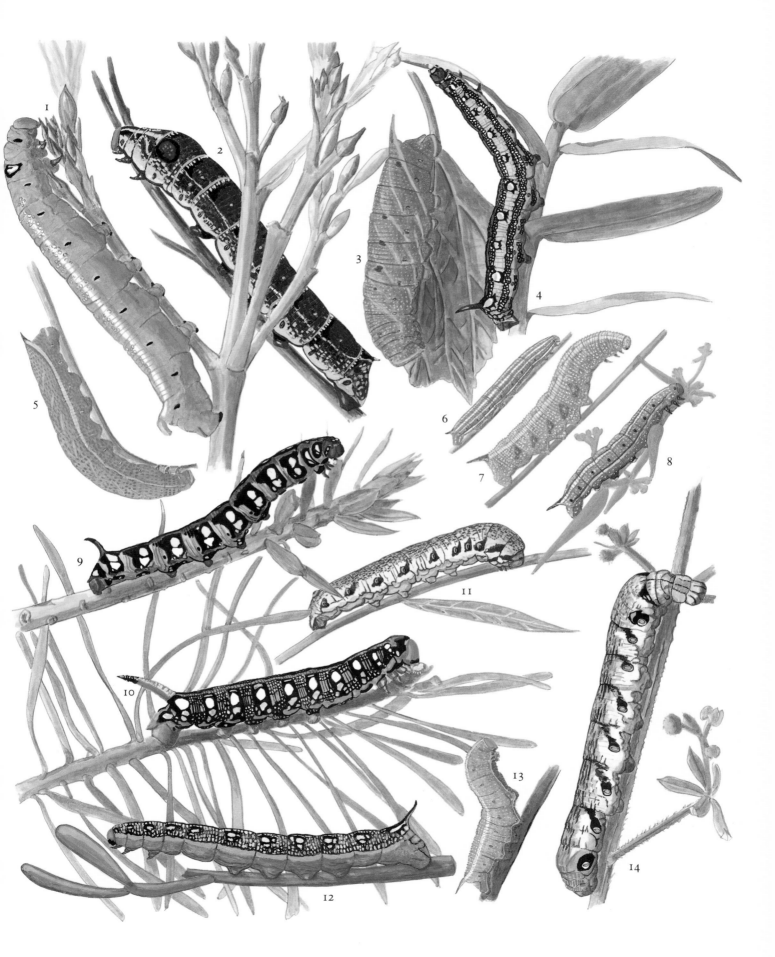

Plate 2

Plate 3 – Macroglossinae (Larvae)

Figs 1–3, 5–9, 11, natural size; Figs 4, 10, 12, × 1.1

1 *Hippotion celerio* (Linnaeus), green form, Saudi Arabia
(*Page* 162)

2 *Hemaris croatica* (Esper), blue-green form, Trieste, Italy
(*Page* 112)

3 *Hyles vespertilio* (Esper), Austria
(*Page* 150)

4 *Hyles nicaea nicaea* (de Prunner), pale form with yellow eye-spots, France
(*Page* 147)

5 *Deilephila elpenor elpenor* (Linnaeus), brown form, England
(*Page* 156)

6 *Deilephila porcellus porcellus* (Linnaeus), brown form, Austria
(*Page* 159)

7 *Theretra alecto* (Linnaeus), Cyprus
(*Page* 165)

8 *Hyles livornica livornica* (Esper), black and yellow form, Saudi Arabia
(*Page* 155)

9 *Hyles dahlii* (Geyer), Sardinia
(*Page* 142)

10 *Hyles hippophaes hippophaes* (Esper), Ayvalik, W. Turkey
(*Page* 151)

11 *Hyles gallii gallii* (Rottemburg), brown form, Austria
(*Page* 146)

12 *Hyles centralasiae centralasiae* (Staudinger), Turkey
(*Page* 144)

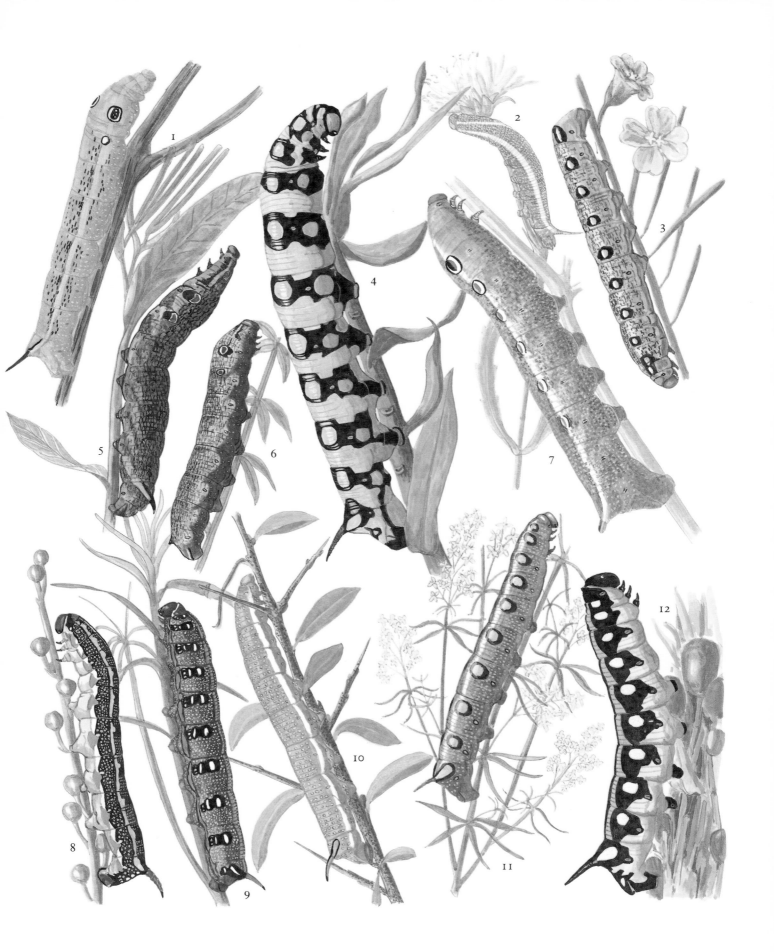

Plate 3

Plate 4 – Sphinginae; Macroglossinae (Larvae)

Figs 1–10, natural size

1 *Hyles euphorbiae robertsi* (Butler), Shiraz, Iran
(*Page* 139)

2 *Laothoe populi populi* (Linnaeus), green form, England
(*Page* 102)

3 *Acosmeryx naga hissarica* Shchetkin, Tajikistan
(*Page* 120)

4 *Hyles zygophylli zygophylli* (Ochsenheimer), dark form, Turkey
(*Page* 149)

5 *Laothoe amurensis* (Staudinger), blue-grey form, Estonia
(*Page* 106)

6 *Hyles nervosa* (Rothschild & Jordan), Kashmir
(*Page* 143)

7 *Hyles chamyla* (Denso), heavily spotted, green form, Tajikistan
(*Page* 154)

8 *Smerinthus caecus* Ménétriés, blue-green form, Hokkaido, Japan
(*Page* 98)

9 *Laothoe austauti* (Staudinger) green form, Morocco
(*Page* 104)

10 *Hyles euphorbiae conspicua* (Rothschild & Jordan), Abha, S.W. Saudi
Arabia (Asir Mountain race)
(*Page* 138)

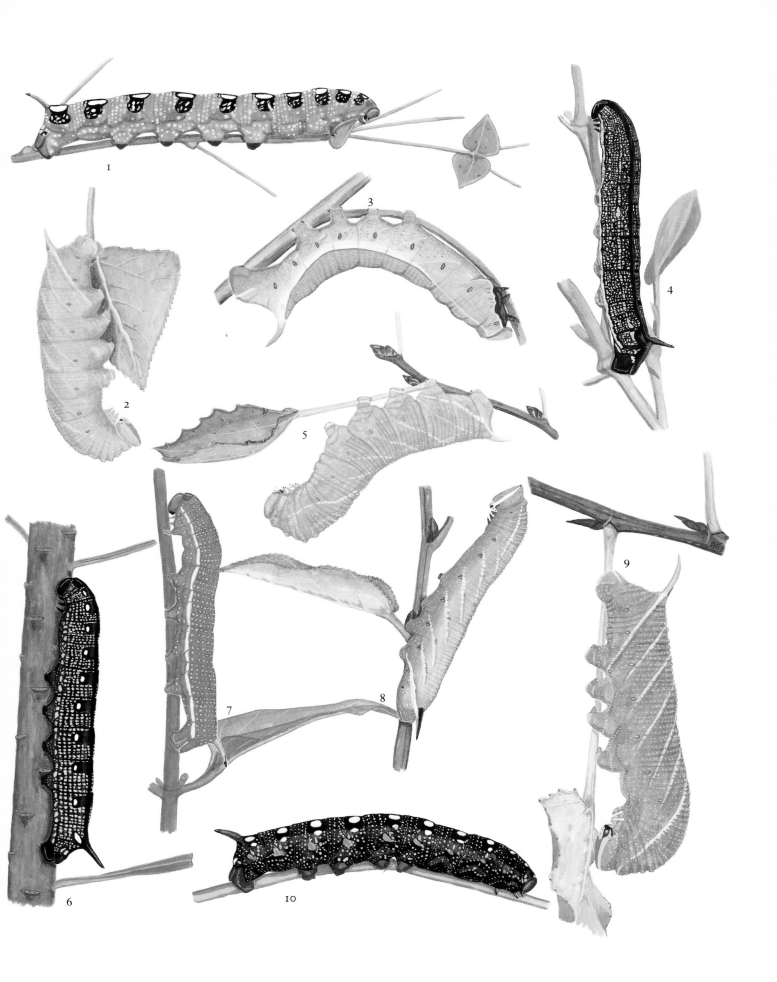

Plate 4

Plate 5 – Sphinginae: Sphingini

Figs 1–9, natural size

1 *Agrius cingulatus* (Fabricius) ♂
Jamaica, Caribbean, coll. BM(NH)
(*Page 79*)

2 *Agrius convolvuli convolvuli* (Linnaeus) ♂
?Algeria, leg. Capt. Holl, coll. BM(NH)
(*Page 80*)

3 *Agrius convolvuli convolvuli* (Linnaeus) ♀
Austria, Steiermark, Mureck, 5.viii.1970, leg. et coll. A. R. Pittaway
(*Page 80*)

4 *Acherontia atropos* (Linnaeus) ♂
Italy, Sicily, leg. E. Ragusa, coll. BM(NH)
(*Page 82*)

5 *Acherontia styx styx* (Westwood) ♀
Saudi Arabia, Ras Tanura, x.1979 (ex larva, *Clerodendrum inerme*), leg.
et coll. A. R. Pittaway
(*Page 83*)

6 *Sphinx ligustri ligustri* Linnaeus ♂
Algeria, Hammam-Meskoutine, 22.iv.1914, leg. Dr Nissen, coll. BM(NH)
(**HOLOTYPE** of *Sphinx ligustri nisseni*)
(*Page 85*)

7 *Sphinx ligustri ligustri* Linnaeus ♀
Austria, Steiermark, Mureck, 2.viii.1970, leg. et coll. A. R. Pittaway
(*Page 85*)

8 *Sphinx pinastri pinastri* Linnaeus ♀
Germany, Duben [near Lübben], 3.vii.1929 (ex larva), leg. W. Ude,
coll. BM(NH)
(*Page 86*)

9 *Sphinx pinastri maurorum* (Jordan) ♂
Algeria, Les Pias, Province of Oran, 28.viii.1918, leg. M. Rotrou,
coll. BM(NH)
(*Page 88*)

Plate 5

Plate 6 – Sphinginae: Sphingini; Smerinthini

Figs 1–14, natural size

1 *Dolbinopsis grisea* (Hampson) ♂
Afghanistan, Salang Pass (north side), Khinjan, 2100m, 5–11.vii.1966,
leg. G. Ebert, coll. BM(NH)
(*Page* 90)

2 *Kentrochrysalis elegans* (A. Bang-Haas) ♂
Syria, Akbés [Maydan Ikbis], leg. Böttcher, coll. BM(NH)
(*Page* 91)

3 *Kentrochrysalis elegans* (A. Bang-Haas) ♂
Bulgaria, Nessebar [Nesebur], 17–31.vii.1963, leg. J. Soffner,
coll. J.-M. Cadiou
(*Page* 91)

4 *Akbesia davidi* (Oberthür) ♂
Syria, Akbés [Maydan Ikbis], coll. BM(NH)
(*Page* 92)

5 *Marumba quercus* ([Denis & Schiffermüller]) ♂
Greece, coll. BM(NH)
(*Page* 93)

6 *Mimas tiliae* (Linnaeus) ♂
Germany, Leipzig, leg. F. Albrecht, coll. BM(NH)
(*Page* 95)

7 *Mimas tiliae* (Linnaeus) ♀
Germany, Saxony, Zwickau, leg. F. Albrecht, coll. BM(NH)
(*Page* 95)

8 *Smerinthus kindermanni kindermanni* Lederer ♀
Syria, Akbés [Maydan Ikbis], coll. BM(NH)
(*Page* 97)

9 *Smerinthus caecus* Ménétriés ♂
Russia, Vladivostok [Primorsk Kray], coll. BM(NH)
(*Page* 98)

10 *Smerinthus caecus* Ménétriés ♀
Russia, Chabar [Khabarovsk], (Amur River), coll. BM(NH)
(*Page* 98)

11 *Smerinthus ocellatus ocellatus* (Linnaeus) ♀
Germany, Berlin, coll. BM(NH)
(*Page* 99)

12 *Smerinthus ocellatus atlanticus* Austaut ♂ **LECTOTYPE**
Algeria, Meridje [Meridja, near Kenadsa], 15.viii.1890, leg. Austaut,
coll. BM(NH)
(*Page* 100)

13 *Smerinthus ocellatus atlanticus* Austaut ♂
Corsica, Ponte Leccia, 180m, 16.vii.1980, leg. W. J. Tennent,
coll. A. R. Pittaway
(*Page* 100)

14 *Smerinthus ocellatus atlanticus* Austaut ♀
Sardinia, Province Nuoro, Siniscola, 3.viii.1979, leg. W. Speidel, coll. BM(NH)
(**HOLOTYPE** of *Smerinthus ocellatus protai*)
(*Page* 100)

Plate 6

Plate 7 – Sphinginae: Smerinthini

Figs 1–9, natural size

1 *Laothoe populi populi* (Linnaeus) ♂
England, London, 14.viii.1980, leg. et coll. A. R. Pittaway
(*Page* 102)

2 *Laothoe populi populeti* (Bienert) f. *populetorum* Staudinger ♂
Iran (south Caspian region), Shahi, 9.ix.1961, leg. S. L. Sutton, coll.
BM(NH)
(*Page* 103)

3 *Laothoe populi populeti* (Bienert) ♂
[Azerbaijan, Nakhichevan], Ordubad, 900m, vi.—, coll. BM(NH)
(*Page* 103)

4 *Laothoe austauti* (Staudinger) ♀
Algeria, Meridje [Meridja, near Kenadsa], coll. BM(NH)
(*Page* 104)

5 *Laothoe philerema* (Djakonov) ♂
Tajikistan, Tigrovaja Balka Reserve, 37°15'N/68°42'E, 22.iv.1988, leg.
O. Gorbunov, coll. J.-M. Cadiou
(*Page* 105)

6 *Laothoe philerema* (Djakonov) ♂
Tajikistan, Tigrovaja Balka Reserve, 37°15'N/68°42'E, 22.iv.1988, leg.
O. Gorbunov, coll. J.-M. Cadiou
(*Page* 105)

7 *Laothoe amurensis* (Staudinger) ♂
Russia, Amurland, coll. BM(NH)
(*Page* 106)

8 *Laothoe amurensis* (Staudinger) f. *baltica* Viidalepp ♂
Finland, Lempäälä, vi.1962, leg. O. Sotavalta, coll. BM(NH)
(*Page* 106)

9 *Smerinthus ocellatus ocellatus* (Linnaeus) × *Laothoe populi populi* (Linnaeus) ♂
Germany, Saxony, Plauen, coll. BM(NH)
(*Page* 64)

Plate 7

Plate 8 – Macroglossinae: Dilophonotini; Macroglossini

Figs 1–14, natural size

1 *Hemaris rubra* Hampson ♀
[Pakistan/India], Kashmir, Scind [Sind] Valley, 2100m, vi.1887, leg.
J. H. Leech, coll. BM(NH)
(*Page* 108)

2 *Hemaris dentata* (Staudinger) ♂
Syria, leg. J. H. Leech, coll. BM(NH)
(*Page* 109)

3 *Hemaris ducalis* (Staudinger) ♂
[China, Tian Shan, Xinjiang Province], Juldus, Kuldsha [near Yining/
Gulja], coll. BM(NH)
(*Page* 109)

4 *Hemaris fuciformis fuciformis* (Linnaeus) ♀
Germany, Graitschen, Thüring, 21.v.1926, leg. W. Ude, coll. BM(NH)
(*Page* 110)

5 *Hemaris fuciformis jordani* (Clark) ♂
Morocco, Middle Atlas Mountains, Azrou, v.1981 (ex larva, *Lonicera*), leg.
et coll. A. R. Pittaway
(*Page* 111)

6 *Hemaris croatica* (Esper) ♀
Yugoslavia [Croatia], Dalmatian region, coll. BM(NH)
(*Page* 112)

7 *Hemaris tityus tityus* (Linnaeus) ♂
Switzerland, Canton Vaud, Bex, 24.iv.1918, leg. K. Jordan &
N. C. Rothschild, coll. BM(NH)
(*Page* 113)

8 *Hemaris tityus aksana* (Le Cerf) ♀
Morocco, Middle Atlas Mountains, near Ifrane, 11.iv.1935, leg.
D. Aubertin, coll. BM(NH)
(*Page* 114)

9 *Daphnis nerii* (Linnaeus) ♀
Germany, Berlin, coll. BM(NH)
(*Page* 115)

10 *Daphnis hypothous hypothous* (Cramer) ♂
Malaya [Malaysia], Frasers Hill [near Kuala Lumpur], 900m, iv.1954, leg.
J. L. C. Banks, coll. BM(NH)
(*Page* 117)

11 *Clarina kotschyi kotschyi* (Kollar) ♀
Iran, Shiraz, 9.vii.1978 (ex larva, *Vitis*), leg. et coll. A. R. Pittaway
(*Page* 118)

12 *Clarina kotschyi syriaca* (Lederer) ♂
Lebanon, Beirut, (ex larva), coll. BM(NH)
(*Page* 119)

13 *Acosmeryx naga hissarica* Shchetkin ♂
Tajikistan, Tigrovaja Balka Reserve, 37°15'N/68°42'E, 22.iv.1988, leg.
O. Gorbunov, coll. J.-M. Cadiou
(*Page* 120)

14 *Acosmeryx naga hissarica* Shchetkin ♂
Afghanistan, Kabul, 18.v.1973, leg. Dr. Liedgens, coll. J.-M. Cadiou
(*Page* 120)

Plate 8

Plate 9 – Macroglossinae: Macroglossini

Figs 1–17, natural size

1 *Rethera afghanistana* Daniel ♂
Afghanistan, Khurd-Kabul (S.E. of Kabul), 1900m, 20.v.1965, leg. Kasy
& Vartian, coll. ZSM
(*Page 122*)

2 *Rethera amseli* Daniel ♀ **PARATYPE**
Afghanistan, Herat, 970m, 15.iv.1956, leg. H. G. Amsel, coll. ZSM
(*Page 123*)

3 *Rethera komarovi drilon* (Rebel & Zerny) ♂
Turkey, Amasya, -.1892, leg. Korb, coll. BM(NH)
(*Page 124*)

4 *Rethera komarovi rjabovi* O. Bang-Haas ♂ **SYNTYPE**
Azerbaijan, [Nakhichevan], Ordubad, 900m, vi.—, leg. Bang-Haas, coll.
BM(NH)
(*Page 124*)

5 *Rethera komarovi manifica* Brandt ♀
Iran, Shiraz, 1500m, 14.v.1950, leg. E. P. Wiltshire, coll. BM(NH)
(*Page 125*)

6 *Rethera brandti brandti* O. Bang-Haas ♂
Iran, Tehran, 11.v.1939, leg. E. P. Wiltshire, coll. BM(NH)
(*Page 126*)

7 *Rethera brandti euteles* Jordan ♂ **HOLOTYPE**
Iran, C'urum (Shiraz to Bushehr road), 20.iii–10.v.1937, leg.
F. H. Brandt, coll. BM(NH)
(*Page 126*)

8 *Sphingonaepiopsis gorgoniades gorgoniades* (Hübner) ♂
Turkmenistan, Kisyl Art [Kyzyl-Arvat], coll. BM(NH)
(*Page 127*)

9 *Sphingonaepiopsis gorgoniades pfeifferi* Zerny ♂
Lebanon, Cedar Mountain, 16.viii.1931, leg. R. E. Ellison, coll. BM(NH)
(*Page 128*)

10 *Sphingonaepiopsis gorgoniades chloroptera* Mentzer ♂
Yugoslavia [Croatia, near Senj], Zengg Admin. area, Dobiasch, 25.v.1916,
coll. BM(NH)
(*Page 128*)

11 *Sphingonaepiopsis kuldjaensis* (Graeser) ♂
[China, Tian Shan, Xinjiang Province], Kuldja [Yining/Gulja], early vii.—,
coll. BM(NH)
(*Page 129*)

12 *Sphingonaepiopsis nana* (Walker) ♂
Saudi Arabia, Djidda (North Creek) [near Jeddah], iv.1979, leg.
V. S. Nielsen, coll. Zool. Mus. Copenhagen
(*Page 129*)

13 *Proserpinus proserpina proserpina* (Pallas) ♂
Austria, Wien [Vienna], coll. BM(NH)
(*Page 131*)

14 *Proserpinus proserpina proserpina* (Pallas) ♀
Sicily, Siracusa [Syracuse], 6.vi.1844, coll. BM(NH)
(*Page 131*)

15 *Proserpinus proserpina gigas* Oberthür ♀ **SYNTYPE**
Morocco, Middle Atlas Mountains, vi.1921 (ex pupa), coll. BM(NH)
(*Page 132*)

16 *Proserpinus proserpina japetus* (Grum-Grshimailo) ♀ **HOLOTYPE**
[Tajikistan], Kabadian [Shaartuz], 12.v.1885, leg. Grum-Grshimailo, coll.
BM(NH)
(*Page 132*)

17 *Macroglossum stellatarum* (Linnaeus) ♂
Austria, Niederösterreich, Oberweiden, coll. BM(NH)
(*Page 133*)

Plate 9

Plate 10 – Macroglossinae: Macroglossini

Figs 1–12, natural size

1 *Hyles euphorbiae euphorbiae* (Linnaeus) f. *lucida* Derzhavets ♂
[China, Tian Shan, Xinjiang Province], Kuldscha [Yining/Gulja],
19.v.1924, coll. BM(NH)
(*Page* 136)

2 *Hyles euphorbiae euphorbiae* (Linnaeus) ♀
Austria, Burgenland, Lake Neusiedl, 29.vi.1973 (ex larva, *Euphorbia*), leg.
et coll. A. R. Pittaway
(*Page* 136)

3 *Hyles euphorbiae conspicua* (Rothschild & Jordan) ♂
Syria, Aleppo, coll. BM(NH)
(*Page* 138)

4 *Hyles euphorbiae conspicua* (Rothschild & Jordan) ♂
Saudi Arabia, Asir Province, 150km N. of Abha, 2500m, 19.viii.1983, leg.
et coll. A. R. Pittaway
(*Page* 138)

5 *Hyles euphorbiae robertsi* (Butler) ♂
Turkmenistan, Merw [Mary], coll. BM(NH)
(*Page* 138)

6 *Hyles euphorbiae robertsi* (Butler) ♀
Turkmenistan, Dortkuju [Dzhu-Dzhu-Klu, W. of Mary], v.1903, coll.
BM(NH)
(*Page* 138)

7 *Hyles tithymali tithymali* (Boisduval) ♂
Canary Islands, Gran Canaria, Las Palmas, -.1975 (ex ovo), leg. et coll.
A. R. Pittaway
(*Page* 139)

8 *Hyles tithymali tithymali* (Boisduval) ♀
Canary Islands, Gran Canaria, Las Palmas, -.1975 (ex ovo), leg. et coll.
A. R. Pittaway
(*Page* 139)

9 *Hyles tithymali mauretanica* (Staudinger) ♂
Algeria, ?Casba Alaer, 6.vi.1906, leg. Capt. Holl, coll. BM(NH)
(*Page* 139)

10 *Hyles tithymali deserticola* (Staudinger) ♂
Algeria, Bissa [?Mt. Bissa, near Ténès], leg. Capt. Holl, coll. BM(NH)
(*Page* 140)

11 *Hyles tithymali gecki* de Freina ♀
Madeira, -.1990, (ex larva, *Euphorbia*), leg. M. Geck, coll. J.-M. Cadiou
(*Page* 141)

12 *Hyles tithymali himyarensis* Meerman ♀
Yemen, Dhufar, 2500m, -.1988 (ex larva, *Euphorbia*), leg. J. Meerman,
coll. A. R. Pittaway
(*Page* 142)

Plate 10

Plate 11 – Macroglossinae: Macroglossini

Figs 1–12, natural size

1 *Hyles dahlii* (Geyer) ♂
Sardinia, Gonnosfanadiga, coll. BM(NH)
(*Page* 142)

2 *Hyles nervosa* (Rothschild & Jordan) ♂
Afghanistan, Safed Koh Mountains (south side), Shahidan, 2700m,
21.vi.1966, leg. G. Ebert, coll. BM(NH)
(*Page* 143)

3 *Hyles salangensis* (Ebert) ♂
Afghanistan, Salang Pass (north side), 69°E 35°40'N, 2400m,
11–12.vii.1971, leg. Vartian, coll. ZSM
(*Page* 144)

4 *Hyles centralasiae centralasiae* (Staudinger) ♂
[Turkmenistan/Uzbekistan], Achal-Teke region, coll. BM(NH)
(*Page* 144)

5 *Hyles centralasiae centralasiae* (Staudinger) ♀
[Turkmenistan/Uzbekistan], Achal-Teke region, coll. BM(NH)
(*Page* 144)

6 *Hyles centralasiae siehei* (Püngeler) ♀
Turkey, Toros Mountains, Bulghar Dag [Bolkar Dağları], 25.vi.1902,
coll. BM(NH)
(*Page* 145)

7 *Hyles gallii gallii* (Rottemburg) ♀
Germany, Saxony, Grimma, 11.vii.1928, leg. W. Ude, coll. BM(NH)
(*Page* 146)

8 *Hyles nicaea nicaea* (de Prunner) ♂
France, Cannes, viii.1905, coll. BM(NH)
(*Page* 147)

9 *Hyles nicaea castissima* (Austaut) ♀
Algeria, Puits Baba (18km S.E. of Guelt es Stel), -.1913, leg. V. Faroult,
coll. BM(NH)
(*Page* 148)

10 *Hyles nicaea libanotica* (Gehlen) ♂
Lebanon, Zahlé, coll. BM(NH)
(*Page* 148)

11 *Hyles nicaea libanotica* (Gehlen) ♂
Turkestan [Kazakhstan], Wernyi [Almaty/Alma-Ata], 28.ix.1926, leg.
B. N. Dublitzky, coll. BM(NH)
(**SYNTYPE** of *Hyles nicaea sheljuzkoi*)
(*Page* 148)

12 *Hyles nicaea crimaea* (A. Bang-Haas) ♂
[Ukraine], Crimea, Sevastopol, 24.vi.1902, (?ex pupa), coll. BM(NH)
(*Page* 149)

Plate 11

Plate 12 – Macroglossinae: Macroglossini

Figs 1–12, natural size

1 *Hyles zygophylli zygophylli* (Ochsenheimer) ♂
[Kazakhstan], Baigacum, Syr. Daria, [Baygakum (Syrdar'ya River)], coll.
BM(NH)
(*Page* 149)

2 *Hyles vespertilio* (Esper) ♂
Austria, Kärnten, Villach, vii.1978, (ex larva, *Epilobium*), leg. et coll.
A. R. Pittaway
(*Page* 150)

3 *Hyles hippophaes hippophaes* (Esper) f. *kiortsii* Koutsaftikis ♂
Turkey, Ayvalik, -.1980 (ex larva, *Elaeagnus angustifolia*), leg. et coll.
A. R. Pittaway
(*Page* 151)

4 *Hyles hippophaes hippophaes* (Esper) ♀
Switzerland, Canton Valais/Wallis, coll. BM(NH)
(*Page* 151)

5 *Hyles hippophaes bienerti* (Staudinger) ♀
[Kazakhstan], Baigakum, Syr. Daria, [Baygakum (Syrdar'ya River)],
25.vii.1912, coll. BM(NH)
(*Page* 152)

6 *Hyles hippophaes bienerti* (Staudinger) f. ♂
[Azerbaijan, Nakhichevan], Ordubad, 900m, VI.—, coll. BM(NH)
(*Page* 152)

7 *Hyles hippophaes caucasica* (Denso) ♂
[Azerbaijan], Caucasus Mountains, Elisabethpol [Kirovabad], coll.
BM(NH)
(*Page* 153)

8 *Hyles chamyla* (Denso) ♂ **SYNTYPE**
China, Xinjiang Province, Chamyl [Hami/Kumul], leg. Denso, coll.
BM(NH)
(*Page* 153)

9 *Hyles chamyla* (Denso) ♂ **SYNTYPE**
China, Xinjiang Province, Chamyl [Hami/Kumul], leg. Denso, coll.
BM(NH)
(*Page* 153)

10 *Hyles livornica livornica* (Esper) ♂
Algeria, Hammam-R'irha [near Miliana], 29.v.1913, leg. W. R. & E. H.,
coll. BM(NH)
(*Page* 154)

11 *Hyles livornica livornica* (Esper) f. *saharae* Stauder ♀
Egypt, leg. C. C. Dudgeon, coll. BM(NH)
(*Page* 154)

12 *Hyles euphorbiae euphorbiae* (Linnaeus) × *Hyles vespertilio* (Esper) ♀
Austria, Niederösterreich (ex ovo), coll. BM(NH)
(*Page* 65)

Plate 12

Plate 13 – Macroglossinae: Macroglossini

Figs 1–11, natural size

1 *Deilephila elpenor elpenor* (Linnaeus) ♂
England, Tring, 9.vii.1906, leg. A. T. Goodson, coll. BM(NH)
(*Page* 156)

2 *Deilephila rivularis* (Boisduval) ♀
[India, Uttar Pradesh], Mussoorie, [near Dehra Dun], 19.iii.1927, leg.
F. B. Scott, coll. BM(NH)
(*Page* 157)

3 *Deilephila porcellus porcellus* (Linnaeus) ♂
Austria, Wien [Vienna], coll. BM(NH)
(*Page* 158)

4 *Deilephila porcellus porcellus* (Linnaeus) ♂
(temperature breeding experiment)
England, leg. E. Cornell, coll. BM(NH)
(*Page* 158)

5 *Deilephila porcellus colossus* (A. Bang-Haas)
Algeria, El-Maouna (S. of Guelma), 18.vi.1919, leg. V. Faroult, coll.
BM(NH)
(*Page* 159)

6 *Deilephila porcellus suellus* Staudinger ♂
[Azerbaijan], Caucasus Mountains, Elisabethpol [Kirovabad], 3.v.1911,
coll. BM(NH)
(*Page* 160)

7 *Deilephila porcellus suellus* Staudinger f. *gissarodarvasica* Shchetkin ♀
[China, Tian Shan, Xinjiang Province], Juldas, Kuldscha [near Yining/
Gulja], coll. BM(NH)
(*Page* 160)

8 *Hippotion osiris* (Dalman) ♂
Seychelles [Indian Ocean], Mahé, 30.iii.1906, leg. H. Thomasset, coll.
BM(NH)
(*Page* 161)

9 *Hippotion celerio* (Linnaeus) ♂
Saudi Arabia, Al-Hassa Province, Hofuf Oasis, 19.iii.1978, leg. et coll.
A. R. Pittaway
(*Page* 162)

10 *Theretra boisduvalii* (Bugnion) ♀
Malaya [Malaysia], Maxwells Hill [near Kuala Lumpur], 1200m,
6.ii.1954, leg. J. L. C. Banks, coll. BM(NH)
(*Page* 164)

11 *Theretra alecto* (Linnaeus) ♂
Lebanon, Beirut, coll. BM(NH)
(*Page* 164)

Plate 13

Systematic Index

Principal entries are given in bold type followed by reference to colour plates, as **100** (Pls C:4; 6:12–14). Reference to figures and tables in the text are given as 161 (text fig.50) and 52 (table 4). Page numbers in square brackets, as [67] or [68], are for references to species on Habitat Pls A or B. Distribution maps are to be found with each species principal entry; map numbers correspond with species check list numbers on pp. 75–77.

Additionally, an Index of Plants will be found on pp. 237–240 and a short Subject Index on p. 240.

Index of Plants

See also Subject Index on p. 240

Subject Index

The entries below are drawn principally from the Introductory chapters (I–VIII) and Appendixes 1 and 2.